The Future of the Brain Sciences

The Future of the Brain Sciences

Proceedings of a Conference held at the New York
Academy of Medicine, May 2-4, 1968

Edited by Samuel Bogoch

Foundation for Research on the Nervous System
Boston, Massachusetts

Ⅽⱷ Springer Science+Business Media, LLC 1969

ISBN 978-1-4899-6166-2   ISBN 978-1-4899-6323-9 (eBook)
DOI 10.1007/978-1-4899-6323-9

Library of Congress Catolog Card Number 75-79571

© Springer Science+Business Media New York 1969
Originally published by Plenum Press in 1969.
Softcover reprint of the hardcover 1st edition 1969

PREFACE

This book attempts to report the flavor and content of a particular interdisciplinary conference on the brain and its disorders. But more importantly, through the reports and discussions herein contained, it is hoped that some indication will be given of the extraordinary breadth, vitality and progress which presently characterize the field of the brain sciences.

The exploration of the "inner space" of man, with the dual hope of understanding the basis of its order and correcting its disorders, is one of the most exciting pursuits, and perhaps the most pressing task before us. Indeed, with the prospect of the rapidly increasing ability to control both genetic and environmental influences, both brain-mind illness and dimensions of health are becoming accessible.

So broad and active is the area of the brain sciences that a single volume can only offer some selected examples from some of the fields which are encompassed. Through the choice of participants in the advancing fronts of the computer sciences, anatomical fine structure, neurophysiology, biochemistry, genetics and clinical disorders, it is hoped that the sample achieved is somewhat representative.

The spirit which characterized the conversations between the participants during the Conference is not adequately captured in this volume. All of the participants recognized our still profound fundamental ignorance, and the complexity of the task of integrating knowledge from diverse specialties.

The gracious atmosphere provided by our hosts of the two sponsoring organizations did much to facilitate communication both in and out of the meeting room. The active participation of Student Delegates nominated by their graduate or medical schools across the country, and brought to the Conference by the sponsors, was a further stimulus to top performance.

Prediction of the future is of course a most unsatisfactory preoccupation. However, when the growth of the brain sciences is plotted against time, it is clear that in just the last few years we have entered a phase which holds very great promise for the future. The critical mass of basic information needed for conceptual leaps is being gathered at an ever-increasing rate by a rapidly expanding number of scientists. When that information becomes digested and integrated, when the information of genetics reveals the genetics of information, when the knowledge of non-biological computer nets is correctly applied to the biology of nerve cell nets, and when the chemistry of recognition at the unicellular level is correctly translated into the chemistry and psychology of recognition in the brain, it appears that we shall have the potential to experience changes in our lives without parallel in the history of man.

Samuel Bogoch
Conference Chairman

ACKNOWLEDGMENTS

The generous financial and personal support of the Manfred Sakel Institute, Inc. and the Foundation for Research on the Nervous System in making possible the conference, The Future of the Brain Sciences, upon which this book is based, is gratefully acknowledged.

The cooperation of the scientific contributors to this volume, and of Miss Joan Henry, Mr. Karr Wolfe and Mr. Kenneth Demac, and of the Plenum Press facilitated the timely publication of the book. The permission of the artist, Mr. Ernest Trova, and of the Pace and the Harcus-Krakow Galleries for the reproduction upon the jacket of this book of what had become the conference insignia, is also gratefully acknowledged.

Samuel Bogoch

CONTENTS

INTRODUCTION

ON THE BASIS OF ORDER IN THE NERVOUS SYSTEM

A. BRAIN CIRCUITRY AND ITS STRUCTURAL BASIS

B. NUCLEIC ACIDS IN MEMORY AND LEARNING

ON THE BASIS OF DISORDER IN THE NERVOUS SYSTEM

A. INBORN ERRORS OF NERVOUS SYSTEM METABOLISM

B. THE GENETICS AND BIOCHEMISTRY OF SCHIZOPHRENIA

CONTRIBUTORS

Bernard Agranoff, Ph.D.
Department of Psychiatry
Mental Health Research Institute
The University of Michigan
Ann Arbor, Michigan

Joseph Altman, Ph.D.
Department of Biological Sciences
Purdue University
Lafayette, Indiana

James H. Austin, M.D.
Division of Neurology
University of Colorado Medical Center
Denver, Colorado

Samuel H. Barondes, M.D.
Departments of Psychiatry
  and Molecular Biology
Albert Einstein College of Medicine
Bronx, New York

George W. Bensch
The University of North Carolina
  at Chapel Hill
Chapel Hills, North Carolina

John R. Bergen, Ph.D.
The Worcester Foundation for
  Experimental Biology
Shrewsbury, Massachusetts

Samuel Bogoch, M.D., Ph.D.
Foundation for Research on the Nervous System,
and Departments of Biochemistry and Psychiatry
Boston University School of Medicine
Boston, Massachusetts

Robert J. Campbell, M.D.
Department of Psychiatry
St. Vincent's Hospital and
   Medical Center of New York
New York, New York

Carlo L. Cazzullo, M.D.
Institute of Psychiatry
University of Milan Medical School
Milan, Italy

Stephan L. Chorover, Ph.D.
Department of Psychology
Massachusetts Institute of Technology
Cambridge, Massachusetts

Robert A. Cleghorn, M.D., D.Sc., F.R.C.P.C.
Department of Psychiatry
McGill University
Montreal, Canada

Alec J. Coppen, Ph.D.
Medical Research Council Laboratories
Carshalton, Surrey
England

Allen C. Crocker, M.D.
The Children's Hospital Medical Center
and Harvard Medical School
Boston, Massachusetts

Peter F. Davison, Ph.D.
Department of Biology
Massachusetts Institute of Technology
Cambridge, Massachusetts

John C. Eccles, M.B., Ph.D., F.R.S., Nobel Laureate
Laboratory of Neurobiology
State University of New York at Buffalo
Buffalo, New York

Austin Fitzjarrell
Department of Psychiatry
Tulane University School of Medicine
New Orleans, Louisiana

Frank Fremont-Smith, M.D.
Interdisciplinary Communications Program
The New York Academy of Sciences
New York, New York

Simon Freed , Ph.D.
Department of Biochemistry
New York Medical College
New York, N.Y.

Arnold J. Friedhoff, M.D.
Center for Study of Psychotic Disorders
Department of Psychiatry and Neurology
New York University School of Medicine
New York, New York

Robert Galambos, Ph.D.
Department of Neurosciences
University of California
La Jolla, California

Gerald E. Gaull, M.D.
New York State Institute for Basic Research
   in Mental Retardation
Staten Island, New York

Ralph W. Gerard, M.D., Ph.D.
Graduate Division
University of California
Irvine, California

Mrs. Louis S. Gimbel, Jr.
The Manfred Sakel Institute, Inc.
New York, New York
          and
The Foundation for Research
   on the Nervous System
Boston, Massachusetts

Edward Glassman, Ph.D.
Department of Biochemistry
The University of North Carolina
Chapel Hill, North Carolina

Robert G. Heath, M.D., D.M.Sc.
Department of Psychiatry
Tulane University School of Medicine
New Orleans, Louisiana

Harold E. Himwich, M.D.
Thudichum Psychiatric Research Laboratory
Galesburg State Research Hospital
Galesburg, Illinois

Williamina A. Himwich, Ph.D.
Thudichum Psychiatric Research Laboratory
Galesburg State Research Hospital
Galesburg, Illinois

Hudson Hoagland, Ph.D., Sc.D.
The Worcester Foundation for
  Experimental Biology
Shrewsbury, Massachusetts

David Yi-Yung Hsia, M.D.
Northwestern University School of Medicine
Chicago, Illinois

Holger Hydén, Ph.D.
Institute of Neurobiology
Faculty of Medicine
University of Göteborg
Göteborg, Sweden

E. Roy John, Ph.D.
Brain Research Laboratories
New York Medical College
Flower and Fifth Avenue Hospitals
New York, New York

Heinrich Klüver, Ph.D.
University of Chicago
Chicago, Illinois

Jerzy Konorski, Ph.D.
The Nencki Institute of Experimental Biology
Warsaw, Poland

Iris M. Krupp, Ph.D.
Department of Psychiatry
Tulane University School of Medicine
New Orleans, Louisiana

Abel Lajtha, Ph.D.
New York State Research Institute
   for Neurochemistry and Drug Addiction
Ward's Island, New York, N.Y.

H. Peter Laqueur, M.D.
Creedmore State Hospital,
and the Mount Sinai School of Medicine
New York, New York

David J. McClure, M.D., M.Sc., F.R.C.P.I.
Department of Psychiatry
McGill University
Montreal, Canada

James L. McGaugh, Ph.D.
Department of Psychobiology
School of Biological Sciences
University of California
Irvine, California

Paul Mandel, M.D.
Centre de Neurochimie du C.N.R.S.,
Institut de Chimie Biologique
   de la Faculté de Médecine
Strasbourg, France

R. M. Marchbanks, Ph.D.
Department of Biochemistry
University of Cambridge, U.K.

B. David Polis, Ph.D.
Biochemistry Division
Aerospace Medical Research Department
U.S. Naval Air Development Center
Johnsville, Warminster, Pennsylvania

Karl H. Pribram, M.D.
Stanford University School of Medicine
Palo Alto, California

Maurice Rapport, Ph.D.
Albert Einstein College of Medicine
Yeshiva University
Bronx, New York

Mark R. Rosenzweig, Ph.D.
Department of Psychology
University of California
Berkeley, California

William Sacks, Ph.D.
Research Center
Rockland State Hospital
Orangeburg, New York

M.R.J. Salton
Department of Microbiology
New York University
  School of Medicine
New York, N.Y.

Samuel Seifter, Ph.D.
Albert Einstein College of Medicine
Yeshiva University
Bronx, New York

Elvin V. Semrad, M.D.
Massachusetts Mental Health Center
and Department of Psychiatry,
Harvard Medical School,
Boston, Massachusetts

George M. Simpson, M.B.
Research Center
Rockland State Hospital
Orangeburg, New York

F. Marott Sinex, Ph.D.
Department of Biochemistry
Boston University School of Medicine
Boston, Massachusetts

Fritiof S. Sjöstrand, Ph.D.
Department of Zoology
University of California
Los Angeles, California

James R. Stabenau, M.D.
Section on Twin and Sibling Studies,
   Adult Psychiatry Branch, Clinical Investigations
National Institute of Mental Health
Bethesda, Maryland

Georges Ungar, M.D., D.Sc.
Department of Pharmacology
Baylor University College of Medicine
Houston, Texas

Heinz Von Foerster, Ph.D.
Department of Electrical Engineering
University of Illinois
Urbana, Illinois

Annemarie S. Welch, Ph.D.
Memorial Research Center
The University of Tennessee
Knoxville, Tennessee

V. P. Whittaker, Ph.D.
Department of Biochemistry
University of Cambridge, U.K.

Introduction

INTRODUCTION

It is not often that lay people can be of help in furthering both the search for scientific knowledge and furthering the application of this knowledge for the good of mankind. This conference on "THE FUTURE OF THE BRAIN SCIENCES" is such an occasion.

As Chairman of the Boards of the two sponsoring organizations: The Manfred Sakel Institute, Inc., and The Foundation for Research on the Nervous System, I express great hope and great joy that this conference will bring forth many of the tangible results the world needs and is waiting for. The suffering in the world of nervous and mental disorders will be enormously ·helped because of this conference and when this book of its proceedings is published and distributed. One of the objectives of this, The Third International Conference, is the bringing together of basic research scientists and clinical medical doctors. It is also vital to have the scientists who are working in the various disciplines which pertain to the Brain Sciences meet and exchange their knowledge and views. Through this effort there can be cross pollenization, inspiration and concrete medical benefits for all of mankind.

In behalf of The Manfred Sakel Institute and The Foundation for Research on the Nervous System,

and the goals they seek, I extend to the scientists, the doctors and to the lay audience deepest
thanks for their splendid participation in making
this Scientific Conference a memorable one.

    Mrs. Louis S. Gimbel, Jr.

    Chairman of the Board of
    The Manfred Sakel Insitute, Inc.
       and
    The Foundation for Research on
    the Nervous System

ON THE INTERNATIONAL AND
INTERDISCIPLINARY CONFERENCE

FRANK FREMONT-SMITH

The problem of an international and inter-disciplinary conference has two aspects to it.

First the interdisciplinary aspect. May I put it this way. As Dr. Bogoch has said, the different disciplines may reside happily within a given brain, but they don't reside happily within any or almost any institution. Indeed there is very little communication between the disciplines in our universities and scientific institutions. Moreover, most of our scientific conferences are organized to make it almost impossible to have genuine cross-discipline communication because they consist of a series of "speeches at", rather than "conversations with", with little time allotted for and protected for discussion.

I am very glad that in this meeting the discussants are as important as the speakers. I would like to say they are far more important than the speakers because they can really get into the level of communication.

Each of the disciplines develop their own special language, their own acronyms, their own patois and these are very good for rapid communication among those at the front line of a particular branch of science. But these short cuts are very poor for communication with any other group which uses a different patois.

Bear this in mind and please try to speak. Maybe this is unfair at an international conference, but try to avoid the acronyms and special phrases which are meaningful only to your own discipline or subdiscipline.

One other aspect often neglected is that the special languages of individual disciplines often contain hidden assumptions. The use of plain English may reveal such assumptions and bring into the open their limitations.

Accordingly, please make your assumptions clear and bring them into the open.

Point Number Two. International conferences are conferences in which many people come and hear what is said not in their native tongue. They must listen to a language with which they are not familiar and, therefore this is a very fatiguing process especially when it continues on for several days.

So, I am asked to get up here and asked to be a good example. First, a good example would be to get through on time and sit down immediately, but I am trying to speak slowly so that those of you who come from other lands may listen with less fatigue.

Lastly, I would like to say that humility rather than arrogance is really a basic aspect of science and that we should be more and more aware of the fact that every observation is partly in error. Most of the things we were sure were true ten years ago are no longer quite so sure and the things we are quite certain about today, we will not be so certain about in even five years because of the progressively accelerated rate of change.

I like to tell my friends that the half-life of facts is getting shorter and shorter.

Now, in order to really set you an example, I will end with an anecdote, what I consider to be a most important scientific statement made by a lady who is not thought of primarily as a scientist--Gertrude Stein. On her deathbed, so the story goes, with Alice Toklas at her side. Gertrude kept turning to Alice demanding, "Alice, what is the answer, what is the answer", and Alice didn't know how to reply. She tried to quiet her but Gertrude wouldn't be quieted. "What is the answer what _is_ the answer?" Finally, Alice thought there is only one thing she could do and that is what she had always done with her friend, to tell her the truth as best she knew it and when Gertrude once again said, "Alice, what is the answer?" Alice said, "Gertrude, there is no answer." Quick as a flash, Gertrude said: "If no answer, then what is the question."

THE FUTURE OF THE BRAIN SCIENCES

John C. Eccles

Brain research is the ultimate problem confronting man. Ever since the realization of his existence, man has been trying to understand what he is and the meaning of his life. This is a much bigger problem than cosmology. But for man's brain no problem would exist. The whole drama of the cosmos would be played out before empty stalls. Nothing would be known to exist because there would be no knowing. A better understanding of the brain is certain to lead man to a richer comprehension both of himself, of his fellow man, and of society, and in fact of the whole world with its problems. It is the greatest adventure undertaken by man. However, there is the ultimate misgiving--how far can brain be effective in fully understanding brain?

In my own lifetime I have lived through very great developments in neurosciences. Forty years ago when I was beginning research we had the great examples of Ramón y Cajal in neuroanatomy and Sherrington in neurophysiology. Many of the presently important neurosciences virtually did not exist, and there were of course no technical developments in the various micro-methods of investigation that play such an important role today. However, my theme tonight is not the history of the past decades but rather to attempt some imaginative appraisal of

developments in the next few decades. Of course,
in this attempt one inevitably is guided by some
extrapolation from the very recent developments.
I will organize my ideas about the future developments under various headings.

1. Neurogenesis. The building of the nervous
system provides the most wonderful example of organic growth. Nothing in the natural world matches
the brain for its organized complexity. Yet it
arises by a process of progressive neurogenesis--
nerve cell multiplication, migration and fiber out-
growths to give specific patterns of neuronal connectivity. We suppose that the growth of nerve
fibers and all the specific synaptic connectivity
that they develop arises by chemical instructions
that ultimately stem from the genetic code of DNA,
acting through a multitude of secondary chemical
instructions. This detailed connectivity is built
before there is any physiological performance of
the brain. Evidently here we have the most chal-
lenging problem. Work on tissue culture of brain
explants is now providing a means of investigating
these problems of neurogenesis, and there is an
urgent need to discover and exploit the most appropriate preparations for this work.

2. Neuro-anatomy. Already there has been an
immense and laborious study of the pathways of the
central nervous system. However, much has yet to
be known in this most complex field, particularly
in the brain stem region. Also within each nucleus
there is an important task of elucidating the principal operative machinery with the corresponding
principle of operation. The most important anatomical investigations are now being carried out with
electronmicroscopy, which gives a refinement of
operation matching the individual nerve cells and
their components, in particular the synaptic connections with their detailed microstructures of
vesicles and membranes.

3. Neurophysiology. Investigations on the
functional performance of the nervous system have
been carried out now for many decades but only in

the last 10 or 15 years have the microelectrode techniques been effectively utilized. These techniques are now the most important in all investigations concerning the operation of the neuronal machinery, both in localized regions of interaction and also in the long and complex pathways of operation. There are now several regions of the brain which are fairly well understood functionally, all the neuronal constituents being recognized and their synaptic connections established. However, much more investigation is required on the organization of the neuronal systems operating more widely and in the more complex integrations. For this reason much more effort must be made on topographic exploration. Another very important development for the future concerns firing frequency analysis of the discharges of the individual units. One must regard the nervous system as being designed essentially for conveying information that is coded in these patterns of impulse discharge.

4. <u>Neurocommunications</u>. Neurocommunication is essentially concerned with the transfer and integration of information. The patterns of impulse frequency referred to under item 3 provide the raw material for theoretical developments of neurocommunications theory. However, it must be recognized that the nervous system does not operate simply by single lines of communication but by multitudes of lines in parallel and superimposed upon this there is an amazing complexity of dynamic loop operation with all manner of feedback controls and cross-linkages.

5. <u>Neurochemistry</u>. This relatively new subject is now developing in an amazing manner by the utilization of the many micro-chemical procedures and also by the utilization of radio-tracer techniques. Some aspects of neurochemistry are very closely related to neurophysiology and neuropharmacology because they are concerned with the manufacture both of the specific synaptic transmitters and of the specific receptor sites for these transmitters. Another aspect of neurochemistry will ultimately relate to the problems of the trophic

influences between the individual units of the nervous system and also to the surface sensing which
is postulated as forming the basis of neurogenesis.
However, there are many other aspects of neurochemistry with its detailed studies of enzyme systems
and metabolic cycles. The brain has a very high
metabolism which is undoubtedly to a large extent
concerned with the operation of ionic pumps that
are so importantly involved in maintaining the
ionic concentrations of nerve cells and the electrical potentials across the surfaces of the membranes. Despite the high metabolism of the nervous system (10 watts) the energy per unit of the
nervous system is extremely low, about 3×10^{-10}
watts.

6. Neuropharmacology. This subject is closely
related both to neurophysiology and neurochemistry,
and is especially concerned with the identification
and mode of action of the various synaptic transmitters. In this study the recent developments of
micropharmacology with the electrophoretic applica-
cation of specific substances to the exterior of
cell membranes has become of great importance.
Several central nervous system transmitters are now
known with reasonable assurance: acetylcholine,
gamma-aminobutyric acid, and glycine. Many more
have yet to be discovered or more convincingly displayed: noradrenaline and glutamic acid for
example. There is a most important field of investigation in studying the homologues of these various
transmitters and of the substances which control
their manufacture, emission and removal.

7. Memory and behaviour. The study of conditioned reflexes in all their varieties is one way
of investigating the plasticity of the nervous system. Undoubtedly much more refined methods will be
developed in relationship to behavioural studies.
An example of such methods is provided by Evarts
in his investigation of discharges of single cortical pyramidal cells during skilled and learned
movements. Investigations on memory have tended
to develop in two quite separate ways. One line
of investigation is physiological and has a counter-

part in neuro-anatomy. It is concerned with the
effect of use and disuse on synaptic efficacy and
an attempt to correlate such changes with the
microgrowths and microatrophies that are postulated
to occur in the synapses concerned. This work is
in the manner of the classical concepts of the
neuronal mechanisms of memory. The other line of
investigation is particularly concerned with the
postulated chemical mechanisms of memory and seeks
to discover specific chemical molecules that have
a unique relationship each with a specific memory.
This work can be criticized because it is not effectively correlated with neuro-anatomy and neurophysiology and provides no basis for recall, which
is the essence of memory. Nevertheless the chemical investigations are of great interest because
with any explanation of memory there must be some
change in neural structure and association,
and this must depend upon a chemical and metabolic
basis.

8. Perception. The physiological operation
of receptor organs and pathways properly belongs
as a special section of neurophysiology. Perception in the psychological sense of conscious aware-
ness gives rise to quite separate and immensely
complex problems that have most important philosophical implications. Conscious experiences by observers have long been used as the basis of investigations of sense organs as, for example, in psy-
chophysics. Furthermore, cortical stimulations by
Penfield and more recently by Libet have given most
interesting information about the complexities of
neuronal operation that are required before a conscious experience actually arises. It can be predicted that such is the complexity of the spatio-
temporal patterns of neuronal activation that many
decades will elapse before these postulated patterns
have been experimentally demonstrated. Nevertheless
most significant investigations are now being done
at the simpler levels of neuronal responses evoked
by various sensory inputs. At the level of investigations on consciousness, very special emphasis
must be given to the work of Sperry and his colleagues on the conscious experiences of patients having

split brains.

9. Neurological disorders. These various
experimental attempts to understand the structure
and functioning of the brain have been described
as basic scientific investigations. However, it
must be realized that they all are essential to
the development of logical methods of treating
neurological and psychiatric disorders. It will
be appreciated that the adequate treatment of a
malfunctioning machine such as the brain must depend upon an understanding of its mode of operation.
At present our understanding is very imperfect and
even may be described as no more than primitive.
Hence it can be expected that a great many of the
disorders cannot be satisfactorily treated. In
fact it is remarkable that treatments often empiri-
cally developed have been so successful. Necessarily the treatment of disorders has to be carried
on despite the inadequacy of our understanding.
All levels of investigation must be carried out in
parallel. It can be predicted that the great successes in all the basic scientific investigations
will be paralleled by successes in the understanding
and treatment of neurological disorders. However,
it must be appreciated that the brain has a different level of complexity from anything else in the
natural world and hence our present efforts both
to understand its mode of operation and to treat
it effectively when malfunctioning can only be in
part successful. Much more effort is required.

At the present time we can think of the total
problem of brain operation as a pile of disconnected
fragments of a jigsaw puzzle with only a relatively
few fragments pieced together to give a meaningful
understanding; but every year many more intelligible
pictures emerge as pieces of the puzzle are put into position and special successes occur when various
small organized regions are seen to fit together
giving a still larger understanding. At the present
it can be stated that all of the various fields of
investigation outlined above are being very effectively pursued in many laboratories throughout the
world, but much more effort could be fruitfully used.

It is important to recognize that brain research
is not a restricted field of scientific investigation but involves a study of the most complex
structure in nature, and the understanding of this
structure will require the highest intellectual
efforts, not only of experts in the various biological disciplines specified here but also of
theoreticians from all the physical sciences, in
particular of mathematics, chemistry, biophysics
and bio-engineering.

ON THE BASIS OF BEHAVIOR IN SINGLE CELLS

Cell Surface Components in Relation to Bacterial Behavior

Milton R.J. Salton

While the investigations on the nature of bacterial surfaces may not have any direct bearing on problems related to 'brain research', I think it is worthwhile emphasizing that these studies have provided the biochemists, microbial chemists and geneticists with extremely useful model systems for correlating structure and function of cell surfaces at the biochemical, chemical and genetic levels. It is also fairly true to say that at the present time, we know as much about the detailed chemistry and architecture of some bacterial cell surfaces as we do of any other type of cell. The knowledge we have gained from the intensive work in this field could provide a background for the exploration of cell surface receptors, cell to cell contact and general properties of the structure and function of the membrane systems of mammalian cells.

There is now a great body of information on the chemistry and biochemistry of bacterial and microbial cell walls, envelopes and membranes. No attempt will be made to review this aspect of the field in any detail and the reader is referred to a number of contributions covering various facets of wall and membrane biochemistry (Gelman, Lukoyanova and Ostrovskii, 1967; Nikaido, 1967; Rogers and Perkins, 1968; Salton, 1964, 1967, 1968; Strominger and Ghuysen, 1967; Strominger and Tipper, 1965).

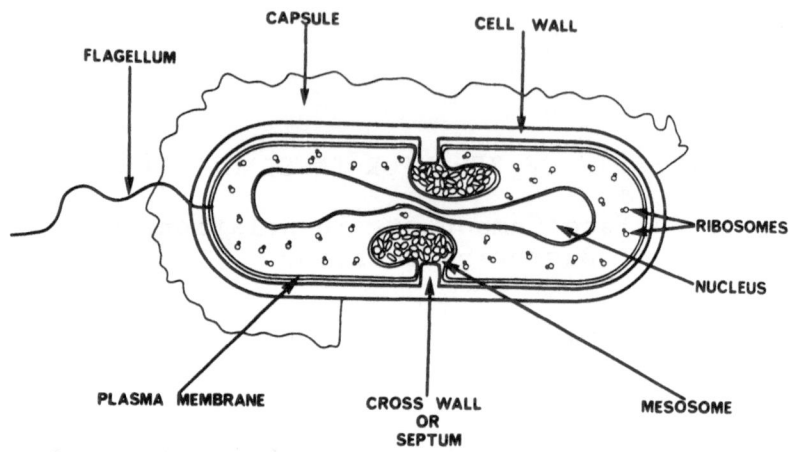

Fig. 1. A diagrammatic representation of the anatomy of a Gram-positive bacterium, indicating surface appendages, surface layers (capsules, walls and plasma membranes) and internal structures. The discontinuity of the capsule indicates that not all species possess this surface component.

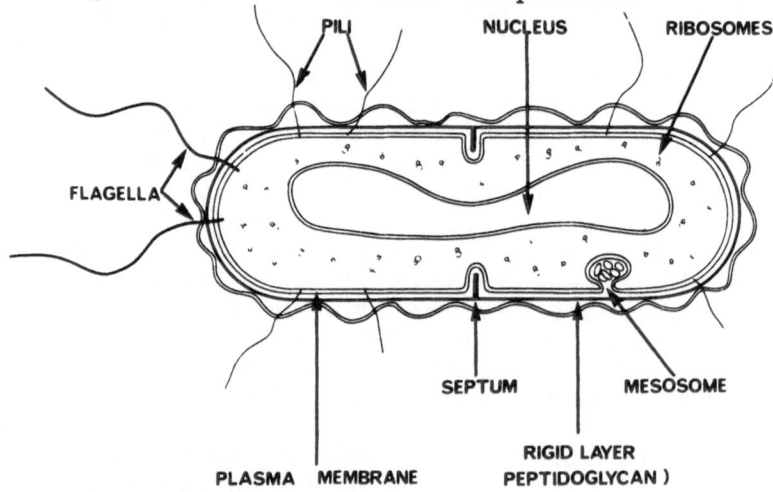

Fig. 2. A diagrammatic representation of the anatomy of a Gram-negative bacterium. Note the two types of surface appendages, flagella and pili. The outer envelope component (lipopolysaccharide, protein, lipid complex) is illustrated as a convoluted double-track structure touching the peptidoglycan layer. Capsules have not been shown although they occur in some species.

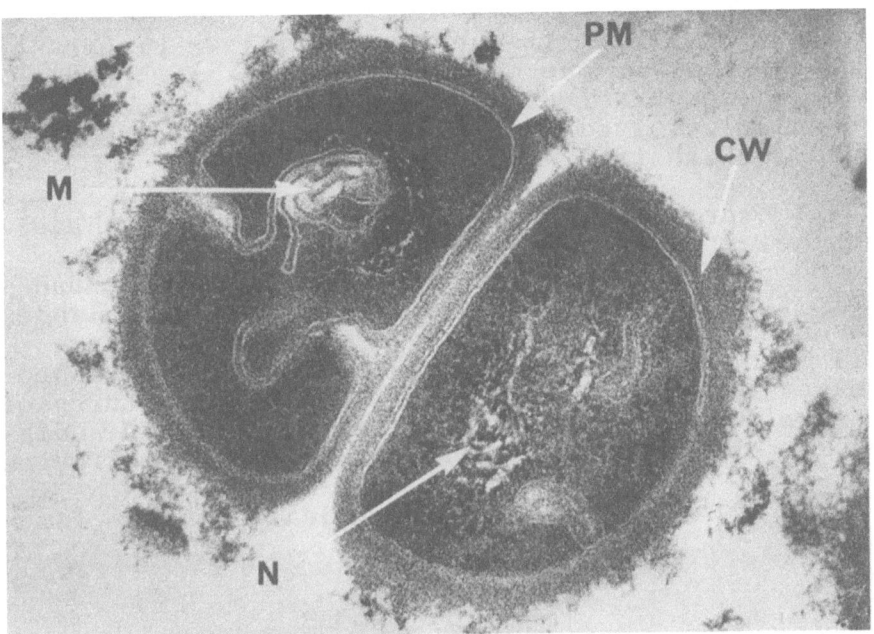

Fig. 3. Electron micrograph of a thin section of
Micrococcus lysodeikticus illustrating the thick cell
wall (CW), the plasma membrane (PM) and its invagina-
tion to form the mesosome (M) and the nuclear region
(N). x 137,000.(reduced 23% for reproduction).

In this contribution all that can be done is
to give you a brief glimpse of some of the types of
structures and proposed functions of the bacterial
cell surface components. Discussions of the structure and anatomy of bacterial cells invariably require an understanding of the distinctions between
the two broad groups separated by the Gram strain
reaction into the Gram-positive and Gram-negative
bacteria. The essential structural differences between these two groups are presented diagrammatic-
ally in Figures 1 and 2 where the bacterial cells
are depicted as rod-shaped organisms. The anatomy
revealed by the examination of thin sections of
fixed bacterial cells in the electron microscope,
is illustrated for a typical Gram-positive coccus
(Micrococcus lysodeikticus) in Figure 3 and a Gram-

negative bacterium (Escherichia coli) in Figure 4.
Although most of the major structures are shown in
such sections, not all of the cellular components
are fully resolved by a single preparative technique for electron microscopy.

The principal "surface structures" that have
been detected in various types of bacteria, some
of the proposed functions of these structures and
their chemical composition are summarized in Table 1.
Many additional structures are present as intracellular components (e.g. ribosomes, nuclei, photo-
synthetic membranes and other complex structures of
unknown functions) of the bacterial cell but a discussion of these is beyond the scope of this review.

Bacteria have provided an ideal 'tissue' for

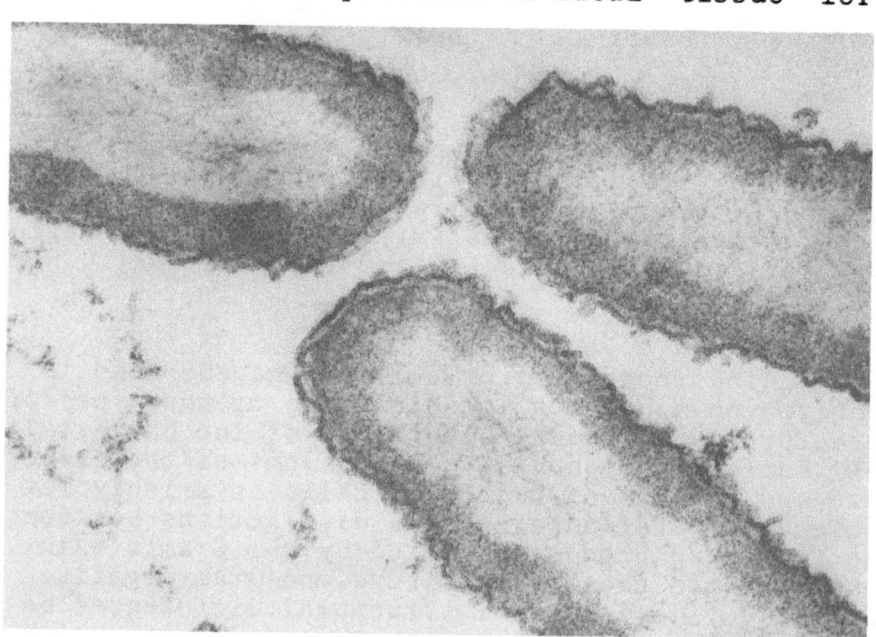

Fig. 4.   Electron micrograph of a thin section of
Escherichia coli showing the multiple-layered outer
envelope structure and the internal organization of
the cell.   The rigid layer is not resolved by the
preparative procedure used.   x 105,000 (reduced 30%
for reproduction).

the isolation and characterization of their major
structural components, since they can be grown as
uniform cultures of strains possessing specific
genetic properties. From the list of the surface
structures presented in Table 1, it is evident
that the major components of the bacterial cell
have been isolated as separate morphological entities and characterized chemically and in some instances their functional properties defined. All
of this has been achieved in the past two decades.

SURFACE APPENDAGES

Bacteria possess several filamentous appendages which can be recognized as either the flagella, the hairlike structures responsible for the
motility or locomotion of bacterial cells, or as
the pili. The latter type of filament was identified after the flagella had been well characterized
and the pili could be readily distinguished from
the flagella as they are generally thinner and
more numerous (Brinton, 1965). Flagella were isolated in homogeneous form by Weibull (1948) and the
protein nature of these structures firmly established. Their role as the organs of locomotion in
bacteria is indisputable (Newton and Kerridge,
1965).

On the other hand the precise function of the
different types of pili seen on the bacterial surface is not always known. There is, however,
abundant evidence that one type of pili is involved
in the mating of donor and recipient cell strains
of Escherichia coli. The male pilus believed to
be responsible for the conduction of DNA from one
cell to the other can be recognized by the ability
of the male-specific bacteriophage to adsorb to
the surface of the pili (Brinton, 1965). In this
way Brinton (1965) was able to deduce that there
were approximately 1-5 such pili on the cell surface. Other types of pili are probably responsible for the adhesion of bacterial cells to various
kinds of cell surfaces and this can be demonstrated
by mixing bacteria and erythrocytes (Duguid, et al,
1955). Brinton's (1965) extensive investigations

TABLE 1

FUNCTIONS OF SURFACE STRUCTURES

OF BACTERIAL CELLS

STRUCTURE	FUNCTION	CHEMICAL COMPOSITION
FLAGELLA	LOCOMOTION	PROTEIN
PILI	CONJUGATION TUBE CELL ADHESION	PROTEIN
CAPSULES and EXTRACELLULAR MATERIAL	PROTECTIVE (?) PHAGE RECEPTOR CELL ADHESION	POLYSACCHARIDES, POLYPEPTIDE POLYSACCHARIDE
CELL WALL (GRAM + ive)	MECHANICAL PROTECTION PHAGE RECEPTORS	PEPTIDOGLYCAN, TEICHOIC ACIDS, POLYSACCHARIDES
"WALL", OUTER ENVELOPE (GRAM-ive)	MECHANICAL PROTECTION PERMEABILITY PHAGE RECEPTORS	PEPTIDOGLYCAN LIPOPOLYSACCHARIDE LIPID, PROTEIN
PLASMA MEMBRANES + MESOSOMES	PERMEABILITY, BIOSYNTHETIC, ELECTRON TRANSPORT, CHROMOSOME ANCHORING & PARTITION	LIPID PROTEIN

on these structures have shown that pili are composed of protein.

In contrast to flagella that are distributed
throughout various genera of Gram-positive bacteria,
the pili are restricted to Gram-negative species.
The latter may possess both flagella and pili.
Mutant strains lacking these structures (flagella
and/or pili) have been isolated from most species.

BACTERIAL CELL SURFACE STRUCTURES

As part of the cell surface, the capsules are
perhaps the most conspicuous structures one can see
on a bacterial cell. They are often very thick,
mucilagenous structures and may be anything up to
10-20 times the overall dimensions of the bacterial
cell itself. Capsules are not invariably present

(see diagram, Figure 1) and with some strains of
bacteria it has been possible to isolate a range
of mutants producing capsular structures of the
very thick mucilagenous type all the way down to
a very thin slime layer (Wilkinson, 1958). As with
flagella, capsules are distributed throughout various groups of Gram-positive and Gram-negative
bacteria. They are dispensable components for the
cell since many organisms are nonencapsulated,
apparently lacking the ability to produce any capsular material at all.

Capsules are generally devoid of any fine
structure detectable in the electron microscope,
but several instances of complexity within the
capsular matrix have been reported (see Salton,
1964). Polysaccharides constitute by far, the
most important class of chemical substance in the
capsules. A large variety of polysaccharide structures have been found in the pneumococcal capsular
polysaccharides and other groups of bacteria (Salton, 1960; Wilkinson, 1958). The capsules of some
of the Bacillus group, including Bacillus authracis
and Bacillus subtilis are composed of the unusual
polypeptide of the γ-D-glutamyl type.

The precise functions of the capsules are not
clearly understood. There is some suggestion that
the presence of a capsule may protect the organism
against desiccation and abnormal or harsh environmental conditions. With pathogenic bacteria in-
vading the tissues of the body, the encapsulated
cells can be protected from phagocytosis, so in
this case again a "protective function" for survival is indicated. One other function suggested
for capsular substances is that they may act as a
"reserve" source of carbon which may be utilized
after exhaustion of nutrients in the medium or
natural environment. The evidence for this capsular function is somewhat meager and more extensive
studies are needed before the reutilization of
capsular material is unequivocally established.
The final property of capsules worth mentioning
is their role as bacteriophage receptors and in at
least one instance, a role in covering the phage

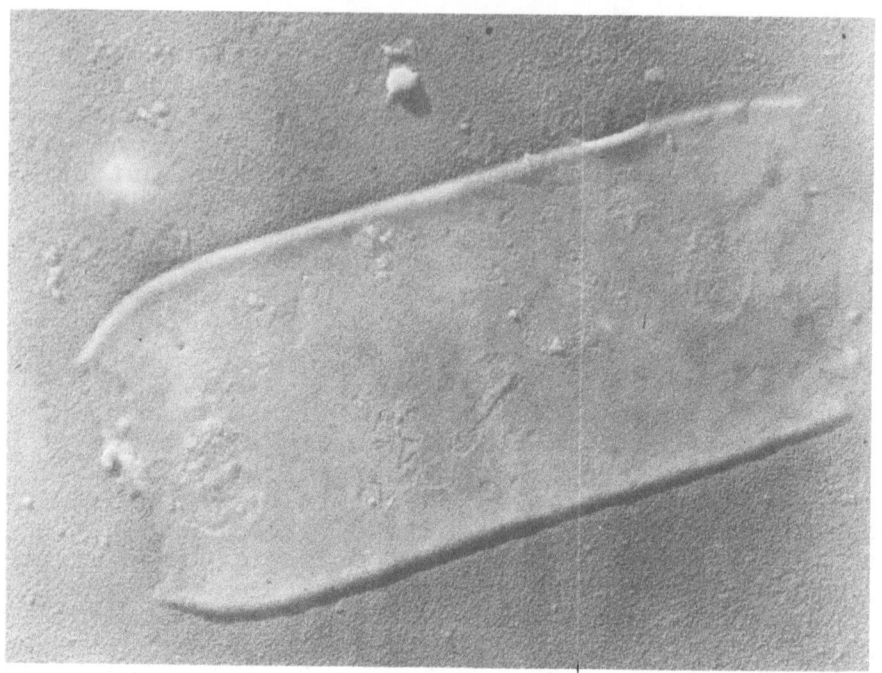

Fig. 5.   Electron micrograph of a "shadowed" preparation of the isolated cell wall of <u>Bacillus cereus</u>
showing the characteristic rod-shape. (Reduced 40%
for reproduction.)

receptors so that the organism is rendered less
sensitive to the bacterial virus (see review,
Salton, 1960).

The more "rigid" parts of the bacterial cell
surface constitute the cell-wall structures of the
Gram-positive bacteria or the multilayered complex
forming the envelopes of Gram-negative bacteria.
Walls of Gram-positive bacteria can be readily
isolated by mechanical disruption of the cells,
although it is likely that in some instances the
'non-rigid' components may be sheared off the surface.   Nermut and Murray (1967) have observed a
fine-structured component on the outer wall of
<u>Bacillus polymyxa</u> and it is possible that such com-
ponents are more sensitive to fragmentation during
cell disruption.

For Gram-positive bacteria, there is abundant
evidence that the function of the rigid wall is
largely mechanical. The isolated cell walls retain
the shapes of the original organisms as shown by the
rod-shape of <u>Bacillus cereus</u> walls in Figure 5 and
the coccal or spherical form of the walls for <u>Staph-
ylococcus aureus</u> in Figure 6. The mechanical and
form-conferring functions of the wall can be further
demonstrated by the removal of the wall with a selective wall-degrading enzyme such as lysozyme.
Weibull (1953) showed that when the rod-shaped organism, <u>Bacillus megaterium</u>, was treated with lysozyme in the presence of an osmotic stabilizer
(e.g. sucrose or polyethylene glycol) wall removal
was accompanied by a transformation to spherical
protoplasts. The protoplasts undergo osmotic lysis
on dilution of the external stabilizer thus indica-

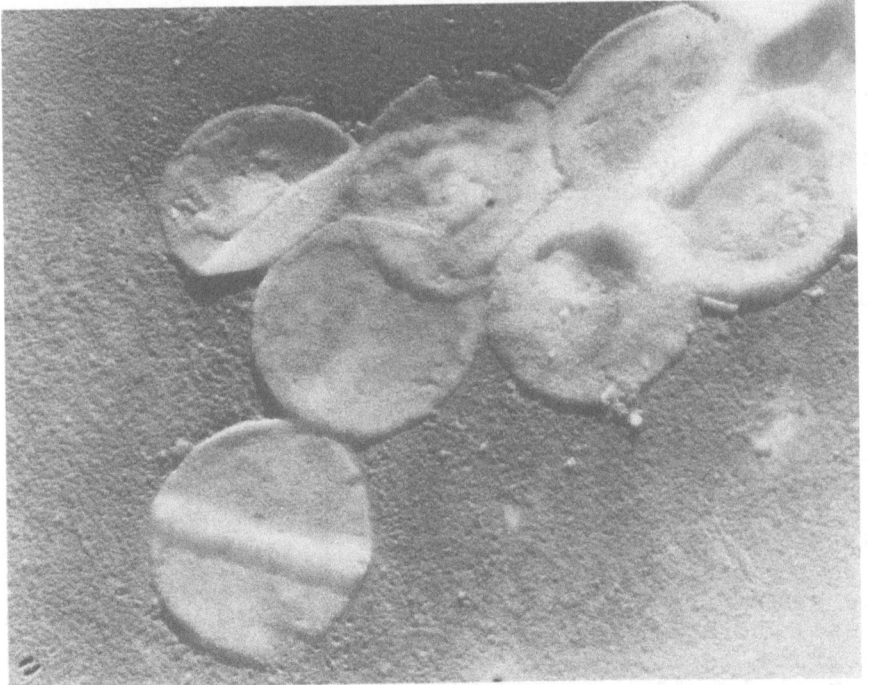

Fig. 6. Electron micrograph of a "shadowed" preparation of the isolated cell walls of <u>Staphylococcus
aureus</u> showing the typical coccal form characteristic
of this organism. (Reduced 40% for reproduction.)

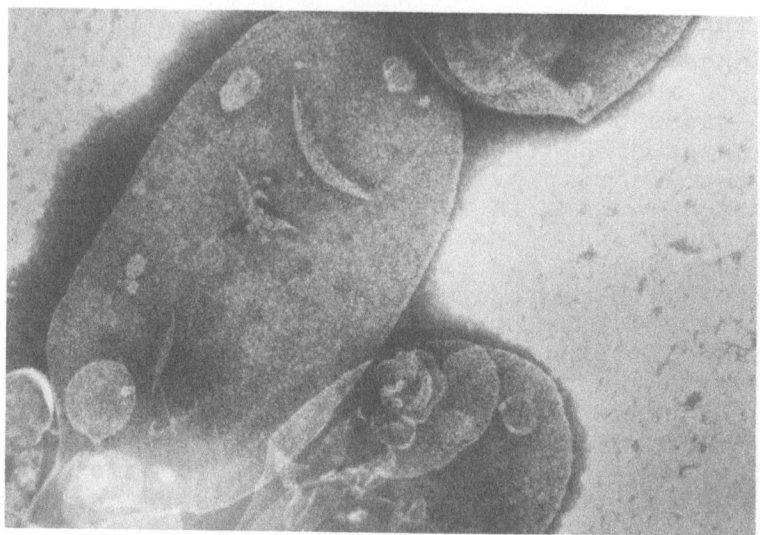

Fig. 7. Electron micrograph of isolated envelopes
of <u>Escherichia</u> <u>coli</u> negatively stained with phospho-
tungstate, illustrating the complexity of the envelope
fraction. Many of the circular discs may arise from
fragmentation of the plasma membrane. Note the pit-
ted appearance. x 106,000 (reduced 40% for reproduction).

ting the protective function of the wall in preventing osmotic explosion of the more fragile protoplast structures.

The rigid component of the bacterial cell is
classified chemically as a peptidoglycan and its
occurrence in Nature is confined to the bacteria
and several related groups such as the blue-green
algae and Rickettsias (Salton, 1968). The possession of this type of structure forms the basis for
the selective action of penicillin against bacteria
thus indicating an interesting biochemical and
chemical correlation for this component of the
bacterial cell. Associated with the peptidoglycan
structure are other types of polymers including
polysaccharides and teichoic acids. Most of the
wall polymers of Gram-positive bacteria are negatively charged so that the cell-wall structure may

play a role in ion exchange across the bacterial
surface. It is clear however, that the wall is
fairly porous and is not the site of the permeability barrier in these organisms (Salton, 1964;
Rogers and Perkins, 1968).

The peptidoglycans may account for about 50-
95% of the cell wall structure in Gram-positive
bacteria, depending upon the particular species,
growth conditions, etc. (Salton, 1964; Rogers and
Perkins, 1968). However, in Gram-negative bacteria,
the peptidoglycan accounts for a smaller fraction
of the substance of the cell envelope. In strains
of Escherichia coli it constitutes about 20% of
the envelope, but it is, nonetheless, most important for cell shape and mechanical protection.
"Spheroplasts" formed by treating Gram-negative
bacteria with ethylenediamine tetraacetic acid
(EDTA) and lysozyme (Repaske, 1956) are more osmo-
tically fragile after the dissolution of the peptidoglycan, despite the presence of the other outer
envelope components such as the lipopolysaccharides,
proteins and lipids.

The multilayered nature of the envelope of a
Gram-negative bacterium such as Escherichia coli,
is illustrated in Figure 7. Envelopes of this
kind, prepared by mechanical disruption of the
cells are rod-shaped and possess the rigid peptidoglycan layer as well as the outer envelope components of the lipopolysaccharide-protein-lipid complex and fragments or sheets of membrane believed
to originate from partial disintegration of the
plasma membranes. The complex organization of the
whole envelope structure of Gram-negative bacteria
has been resolved in profiles of thin sections examined by Murray, et al, (1965) and by de Petris
(1965, 1967) and in negatively-stained preparations
by Bayer and Anderson (1965). Both thin sections
and negatively-stained specimens of Gram-negative
bacteria have shown that the surface envelope components may present a highly convoluted appearance
in some species, while in others, complex-fine-
structured layers with arrays of hexagonally packed structures are seen.

The outermost layers of the envelopes of Gram-negative organisms may also confer some "permeability" properties on the bacterial cell. The outer lipopolysaccharide layers may prevent the escape of certain enzymes that are "trapped" in the periplasmic region between the plasma membrane and the external surface of the cell. The release of certain enzymes on treating the cells with EDTA to disrupt the external barrier without damaging the plasma membrane was first reported by Lamaly and Horecker (1961) and by Neu and Heppel (1964). Leive (1965) has shown that under carefully controlled conditions it is possible to disrupt the lipopolysaccharide layer without killing the cell and at the same time induce their sensitivity to actinomycin D which does not otherwise attack the 'intact' cell. Thus, this layer of the Gram-negative cell can prevent certain substances from passing into the bacterial cell and it is also responsible for the retention of enzymes and possibly other cellular components in the periplasmic regions. Although this structure of the Gram-negative envelope exhibits these 'permeability' properties it appears less likely that it is involved in any of the transport functions of the cell.

The lipopolysaccharides which form the dominant surface antigens of Gram-negative bacteria are also important as bacteriophage receptors. The nature of these components has been investigated in some detail at the chemical and genetic levels (Nikaido, 1967). The direct interaction between specific lipopolysaccharides isolated from the bacterial surface and the bacterial viruses, has been demonstrated in a number of studies (Weidel, 1958). The bacterial cell surface obviously has a large number of specific phage receptors as shown in the elegant sections of E. coli phage attaching to and penetrating the cell surface (Simon and Anderson, 1967). In this study it was evident that there were sufficient receptors to accommodate a closely packed layer of the bacteriophages around the cell surface. Mutant strains lacking the receptor or possessing biochemically modified receptor molecules, are resistant to infection with the phages.

Bacteria obviously fulfill all of the other essential functions found in other types of cells, although the biochemical machinery may be organized somewhat differently in the bacterial cell. Despite the absence of organized mitochondrial structures, nuclear membranes and spindle structures, and a membranous endoplasmic reticulum, these cellular functions reside in specialized regions of the bacterial membrane systems. From Weibull's investigations (1953) it became apparent that the plasma or protoplast membrane was the site of the permeability barrier. The electron transport system is localized in the bacterial membrane (Gelman, et al, 1967) and these components can be isolated as lipid-depleted, membranous sheets from the rest of the bacterial membrane (Salton, et al, 1968). The bacterial membrane obviously has biosynthetic properties since some of the enzymes involved in the assembly and synthesis of the peptidoglycan cell wall, have been found in the membranes or in particles derived from the bacterial membranes (Strominger and Tipper, 1965; Salton, 1967; Rogers and Perkins, 1968).

Thus, functionally, the bacterial membrane is a very important structure and in addition to being involved in active transport and selective permeability, it has to fulfill the role of anchoring the DNA of the chromosomes and ensuring nuclear separation during cell division (Ryter, 1968).

In addition to the plasma membrane forming a 'layer' next to the inner face of the thick cell wall or the rigid, peptidoglycan layer of the Gram-negative bacteria, it invaginates to give rise to the internal membrane system known as the "mesosomes" (Fitz-James, 1960). The mesosomes look like 'sacs' of lamellae or vesicles and although they have a superficial anatomical resemblance to mitochondria there is still some uncertainty that they are the mitochondrial structures of the bacterial cell. Indeed, it is possible that mesosomes having similar profiles may have different functions in different bacterial cells. A role in secretion of proteins has recently been suggested by Ghosh et al. (1968).

The plasma and mesosome membranes are essentially lipid-protein structures. The whole membrane systems from a variety of Gram-positive bacteria have been isolated and chemically characterized (Salton, 1967; Rogers and Perkins, 1968) and in overall chemical composition and physico-chemical properties they are very similar to other types of cell membranes. Unlike the peptidoglycan cell-wall structure of covalently bonded peptide and glycan, the bacterial membranes are held together by relatively weak forces involving hydrophobic interactions between phospho-lipids and proteins. The membranes are thus susceptible to dissociation with surface-active agents, a property which has led to a better understanding of the variety of proteins in the membranes and enabled the isolation of some of the functional proteins.

Finally, in discussing the nature and behavior of bacterial surface components, brief mention of cell adhesion and contact seems appropriate. Bacterial cells usually exist as single, separated cells or in the form of long chains or small clumps and packets. There is only one instance where a bacterial cell produces anything analogous to a "tissue" and this is in the unusual organism, Lampropedia hyalina. This organism produces sheets or tablets of cells that are held together by a cementing layer and surrounded by a specialized, structured envelope or layer keeping the tablet together. This organism forms a flat sheet of cells, apparently one cell in thickness (Chapman, et al. 1963) and could provide an interesting system in which to investigate cell-to-cell contact, cell adhesion and the function of the outer 'corset' surrounding the cell aggregate.

The brief descriptions of some of the properties of the surface structures of the bacterial cell from investigations carried out in a number of laboratories and from our own studies, serve to emphasize that they have provided model systems in which to investigate structure and function in relation to cellular behavior. The approaches to the study of these problems in elucidating the nature of the bacterial surfaces may provide a background for the exploration of more complex phenomena in animal cell surface structures.

ACKNOWLEDGMENTS

The author's work was supported by a National Science Grant (GB4603). I wish to thank Dr. John H. Freer for electron micrographs of some of the unpublished work on bacterial wall and envelope preparations, Miss Roula Christie for the diagrams and Mr. Charles Harman for all the photographic work.

REFERENCES

1. Bayer, M.E. and Anderson, T.F., Proc. Natl. Acad. Sci., U.S., 54, 1592, 1965.
2. Brinton, C.C., Trans. N.Y. Acad. Sci., 27, 1003, 1965.
3. Chapman, J.A., Murray, R.G.E.and Salton, M.R.J., Proc. Roy. Soc., B., 158, 498, 1963.
4. Duguid, J.P., Smith, I.W., Dempster, G. and Edmunds, P.N., J. Pathol. Bacteriol., 70, 335, 1955.
5. Fitz-James, P.C., J.Biophys.Biochem.Cytol., 8, 507, 1960.
6. Gel'man, N.S., Lukoyanova, M.A. and Ostrovskii, D.N., Respiration and Phosphorylation of Bacteria, Plenum Press, New York, 1967.
7. Ghosh, B.K., Sargent, M.G. and Lampen, J.O., J. Bacteriol., 96, 1314, 1968.
8. Leive, L., Proc. Natl. Acad. Sci. U.S., 53, 745, 1965.
9. Leive, L., Biochem. Biophys. Res. Commun., 21, 290, 1965.
10. Malamy, M. and Horecker, B.L., Biochem. Biophys Res. Commun., 5, 104, 1961.
11. Murray, R.G.E., Steed, P. and Nelson, H.E., Can. J. Microbiol., 11, 547, 1965.
12. Nermut, M.V. and Murray, R.G.E., J. Bacteriol., 93, 1949, 1967.
13. Neu, H.C. and Heppel, L.A., Proc. Natl. Acad. Sci. U.S., 51, 1267, 1964.
14. Newton, B.A. and Kerridge, D., in: Soc. Gen. Microbiol. Symp. No. 15, 220, 1965.

15. Nikaido, H., In: The Specificity of Cell Surfaces, B.D. Davis and L. Warren (Eds.), Prentice-Hall, Englewood Cliffs, New Jersey, 3, 1967.
16. de Petris, S., J. Ultrastruct. Res., 12, 247, 1965.
17. de Petris, S., J. Ultrastruct. Res., 19, 45, 1967.
18. Repaske, R., Biochim. Biophys. Acta, 22, 189, 1956.
20. Rogers, H.J. and Perkins, H.R., Cell Walls and Membranes, Spon, London, 1968.
21. Ryter, A., Bacteriol. Rev., 32, 39, 1968.
22. Salton, M.R.J., In: The Bacteria, Academic Press, New York, 1960.
23. Salton, M.R.J., The Bacterial Cell Wall, Elsevier, Amsterdam, 1964
24. Salton, M.R.J., Ann. Rev. Microbiol., 21, 417, 1967.
25. Salton, M.R.J., In: Comprehensive Biochemistry, 23, 127, Elsevier, Amsterdam, 1968.
26. Salton, M.R.J., Freer, J.H. and Ellar, D.J., Biochem. Biophys. Res. Commun., 33, 909, 1968.
27. Simon, L.D. and Anderson, T.F., Virology, 32, 279, 1967.
28. Strominger, J.L. and Ghuysen, J.M., Science, 156, 213, 1967.
29. Strominger, J.L. and Tipper, D.J., Amer. J. Med., 39, 708, 1965.
30. Weibull, C., Biochim. Biophys. Acta, 2, 351, 1948.
31. Weibull, C., J. Bacteriol., 66, 688, 1953.
32. Wilkinson, J.F., Bacteriol. Ref., 22, 46, 1958.
33. Weidel, W., Ann. Rev. Microbiol., 12, 27, 1958.

On the Basis of
Order in the Nervous System

A. BRAIN CIRCUITRY AND ITS STRUCTURAL BASIS

WHAT IS MEMORY THAT IT MAY HAVE

HINDSIGHT AND FORESIGHT AS WELL?

Heinz Von Foerster

"What is Time?" According to Legend, Augustine's reply to this question was: "If no one asks me, I know: but if I wish to explain it to one that asketh, I know not." Memory has a similar quality, for if not asked, we all know what memory is, but when asked, we have to call for an International Conference on the Future of Brain Sciences. However, with a minimal change of the question, we could have made it much easier for Augustine. If asked "What's the time?" he may have observed the position of the sun and replied: "Since it grazes the horizon in the west, it is about the sixth hour after noon."

A theory of memory that is worth its name must not only be able to account for Augustine's or anybody else's intelligent conduct in response to these questions, moreover, it also must be able to account for the recognition of the subtle but fundamental difference in meaning of the two questions regarding time or memory of before, a distinction that is achieved by merely inserting a syntactic "operator"--the definite article "the"-- at a strategic point in the otherwise unchanged string of symbols. At first glance it seems that to aim at a theory of memory which accounts for such subtle distinctions is overambitious and preposterous. On second thought, however, we shall

see that models of mentation that ignore such
aims and merely account for a hypothetical map-
ping of sensations into indelible representations
on higher levels within the neural fabric of the
brain or--slightly less naively--account for habituation, adaptation and conditioning by replac-
ing "indelibility" by "plasticity", do not only
fall pitiably short of explaining anything that
may go on at the semantic level or, to put it
differently, that is associated with "information"
in the dictionary sense, i.e., "knowledge acquired in any manner" (1), but also appear to inhibit
the development of notions that will eventually
account for these so-called "higher functions" of
cerebral activity.

Since an approach that attempts to integrate
the enigmatic faculty of memory into the even more
enigmatic processes of cognition veers off under a
considerable angle from well established modes of
thinking about this problem, it may be profitable
to develop the argument carefully step by step,
first exposing and circumventing some of the semantic traps that have become visible in the course
of this study, and then showing that even at the
possible risk of losing track of some operational
details a conceptual frame work is gained which,
hopefully, allows the various bits and pieces to
fall smoothly into place.

At this moment it appears to me that this
objective may be best achieved by delivering the
argument in four short "chapters". I shall open
the discussion with an attempt to clarify some of
the most frequently used terms in discussing memory and related mental functions. In the second
chapter I shall state my thesis which is central
to the whole argument, and I shall develop this
thesis in details that are commensurate with the
scope of this paper in Chapter III. Finally, I
shall venture to present a conjecture regarding
the possibility of computing recursive function
on the molecular level.

Throughout this paper I shall be using examples and metaphors as explanatory tools, rather than the frightful machinery of mathematical and logical calculi. I am aware of the dangers of misrepresentation and misunderstanding that are inherent in these explanatory devices, and I shall try to be as unambiguous as my descriptive powers permit me to be. For those who wish to become acquainted with a more rigorous treatment of this subject matter I must refer to the widely scattered technical literature as much--or as little--as there exists such literature today (2, 3, 4, 5, 6, 7, 8, 9, 10).

I. Clarification of Terminology

There are two pairs of terms that occur and re-occur with considerable frequency in discussions of memory and related topics. They are (i) "storage and retrieval" and (ii) "recognition and recall". Unfortunately--in my opinion-- they are used freely and interchangeably as if they were to refer to the same processes. Permit me, therefore to restore their distinctive features:

(i) Storage and Retrieval

I wish to associate with these terms a certain invariance of quality of that which is stored at one time and then retrieved at a later time.

Example: Consider Mrs. X who wishes to store her mink coat during the hot months in summer, takes this coat to her furrier for storage in his vault in spring and returns in the fall for retrieving it in time for the opening night at the opera.

Please note that Mrs. X is counting on getting precisely her mink coat back and not any other coat, not to speak of a token of this coat. It is up to everybody's imagination to predict what would happen if in the fall her furrier would tell her "Here is your mink coat" by handing over to her a slip on which is printed "HERE IS YOUR MINK COAT".

At this point I don't believe that anybody may disagree with my insistence on invariance of quality of entities when stored and retrieved and with my choice of example illustrating this invariance. Consequently, one may be tempted to use this concept for somewhat more esoteric entities than mink coats as, for instance, "information". Indeed, it may be argued that there exist reasonable well functioning and huge information storage and retrieval systems in the form of some advanced library search and retrieval systems, the nationwide Educational Resources Information Center (ERIC), etc., etc., which may well serve as appropriate models or analogies for the functional organization of physiological memory.

Unfortunately, there is one crucial flaw in this analogy inasmuch as these systems store books, tapes, micro-fiches or other forms of documents, because, of course, they can't store "information". And it is again these books, tapes, micro-fiches or other documents that are retrieved which only when looked upon by a human mind, may yield the desired "information". By confusing vehicles for potential information with information, one puts the problem of cognition nicely into one's blind spot of intellectual vision, and the problem conveniently disappears. If indeed the brain were seriously compared with one of these document storage and retrieval systems, distinct from these only by its quantity of storage rather than by quality of process, such theory would require a little demon, bestowed with cognitive powers, who zooms through this huge storage system in order to extract the necessary information for the owner of this brain to be a viable organism.

It is the aim of this paper to explore the brain of this demon in terms of the little I know of neurophysiology so that we may ultimately dismiss the demon and put his brain right there where ours is.

If there should be any doubt left as to the distinction between vehicles of potential information and information proper, I suggest experimenting with existent so-called "information storage and retrieval systems" by actually requesting answers to some queries. He who did not yet have the chance to work with these systems may be amused or shocked--depending on his view of such systems--by the sheer amount of pounds or tens of pounds, of documents that may arrive in response to a harmless query, some of which--if he is lucky --may indeed carry the information he requested in the first place.

I shall now turn to the second pair of terms I promised to discuss, namely "Recognition and Recall".

(ii) Recognition and Recall

I wish to associate with these terms the overt manifestations of <u>results</u> of certain operations, and I wish not to <u>confuse</u> the results of these operations with either the <u>operations themselves</u> or the <u>mechanisms</u> that implement these operations. Example: After arrival from a flight I am asked about the food served by this airline. My answer:
"FILET MIGNON
WITH FRENCH FRIES AND SOME SALAD,
AND AN UNDEFINABLE DESSERT."
My behavior in response to this question--I believe--appears reasonable and proper. Please note that nobody expects me to produce in response to this question a real
filet mignon
with french fries and some salad
and an undefinable dessert.
I hope that after my previous discussion of storage and retrieval systems it is clear that my verbal response cannot be accounted for by any such system. For in order that the suspicion may arise that I am nothing but a storage and retrieval system first the <u>sentence</u>:
"FILET MIGNON
WITH FRENCH FRIES AND SOME SALAD,
AND AN UNDEFINABLE DESSERT."
had to be "read in" into my system where it is

stored until a querier pushes the appropriate retrieval button (the query) whereupon I reproduce
with admirable invariance of quality (high fidel-
ity) the sentence:
"FILET MIGNON
WITH FRENCH FRIES AND SOME SALAD,
AND AN UNDEFINABLE DESSERT."
However, I must ask the generous reader to take
my word for it that nobody ever <u>told</u> me what the
courses of my menu were, I just <u>ate</u> them.

Clearly something fundamentally different
from storage and retrieval is going on in this
example in which my verbal behavior is the result
of a set of complex processes or operations which
transform my experiences into utterances, i.e.,
symbolic representations of these experiences.

The neural mechanisms that perform the operations which permit me to identify experiences and
to classify these with other earlier experiences
determine my faculty to recognize (Re-cognition).
Those mechanisms and operations which allow me
to make symbolic representations of these experiences, say, in the form of utterances, determine
my faculty to recall (Re-call).

The hierarchy of mechanisms, transformational
operations and processes that lead from sensation
over perception of particulars to the manipulation
of generalized internal representations of the
perceived, as well as the inverse transformations
that lead from general commands to specific actions, or from general concepts to specific utterances I shall call "Cognitive Processes". In the
analysis of these processes we should be prepared
to find that terms like "recall" and "recognition"
--as convenient as they may be for referring quickly to certain aspects of cognition--are useless as
descriptors of actual processes and mechanisms
that can be identified in the functional organization of nervous tissue.

It could be that already at this point of my
exposé the crucial significance of cognitive pro-

cesses may have become visible, namely, to supply
an organism with the operations that "lift"--so to
say--the information from its carriers, the signals, may they be sensations of external or internal events, signs or symbols (11), and to provide
the organism with mechanisms that allow it to compute inferences from the information so obtained.

    To use more colloquial terms, cognition may
well be identified with all the processes that establish "meaning" from experience.   I may mention
that a somewhat generalized interpretation of
"meaning as "all that which can be inferred from
a signal" leads to a semantic rationale of considerable analytic power, independent of whether the
signal is a sign or a symbol. Moreover, I may add
in passing that this interpretation allows not only
for qualitative distinctions of "meaning" depending
on the mode of inference that is operative--i.e.,
the deductive, inductive or abductive mode (12)--
but also for quantitative estimates of the "amount
of meaning"--straightforwardly at least in the deductive mode (13)--that is carried by a given signal for a given recipient.

    I hope to make these points more transparent
later on in the development of my thesis which,
after these preliminary remarks, we are ready to
hear now.

II. Thesis

    In the stream of cognitive processes one can
conceptually isolate certain components, for instance

        (i) the faculty to perceive,
       (ii) the faculty to remember,
and (iii) the faculty to infer.

But if one wishes to isolate these faculties functionally or locally, one is doomed to fail. Consequently, if the mechanisms that are responsible
for any of these faculties are to be discovered,
then the totality of cognitive processes must be

considered.

Before going on with a detailed defense of
this thesis by developing a model of an "integrated functional circuit" for cognition, let me
briefly suggest the inseparability of these faculties on two simple examples.

First Example: If only one of the three faculties
mentioned above is omitted, the system is devoid
of cognition:

  (i) Omit perception: the system is incapable of
      representing internally environmental regu-
      larities.

 (ii) Omit memory: the system has only through-
      put.

(iii) Omit prediction, i.e., the faculty of draw-
      ing inferences: perception degenerates to
      sensation, and memory to recording.

Second Example: If the conceptual linkages of memory with the other two faculties are removed one by
one, nolens, volens "memory" degenerates first to
a storage and retrieval system and, ultimately, to
an inaccessible storage bin that is void of any content.

After these reductiones ad absurdum I shall
now turn to a more constructive enterprise, namely,
to the development of a crude and--alas--as yet in-
complete skeleton of cognitive processes.

III. Cognitive Elements and Complexes

I shall now develop my thesis in several steps
of ascending complexity of quality, rather than of
quantity, beginning with the most elementary case
of apparent functional isolation of memory but of
zero inferential powers, concluding with the most
elementary case of functionally unidentifiable memory, but of considerable inferential powers. Through-

out this discussion I shall use examples of minimal structural complexity for the sake of clarity in presenting the argument. I am well aware of many of the fascinating results that can be derived from a rigorous extension of these minimal cases, but in this context I feel that these findings may divert us from the central issue of my thesis.

My first case deals with the computation of concomitance. The detection of concomitances in the outside world is of considerable economic significance for an organism immersed in this world, for the larger an equivalence class of events becomes the fewer specific response patterns have to be developed by the organism. The power of inductive inference rests on the ability to detect concomitance of properties, and--as was believed until not long ago--the efficacy of the conditioned reflex rests on the ability to detect concomitance of events.

The principle of inductive inference is essentially a principle of generalization. It says that, since all things examined that exhibited property P_1 also exhibited property P_2, all as yet unexamined things that have property P_1 will likewise exhibit property P_2. In other words, inductive inference generalizes the concomitance of properties P_1 and P_2. In its naive formulation the "conditioned reflex" can be put into a similar logical schema which I will call "Elementary Conditioned Reflex" (ECR) in order to establish a clear distinction between this model and the complex processes that regulate conditioned reflexive behavior in mammals and other higher vertebrates. However, I shall return to these in a moment.

Part 1 of Figure 1 shows the minimal net capable of computing an ECR. Neurons A, B transmit the conditioning and conditioned stimulus respectively to the motoneuron with threshold $\Theta = 2-\varepsilon$, where $0<\varepsilon<<1$, which always fires when A fires, since the double excitation on its two synapses override its threshold of less than two units (A single sy-

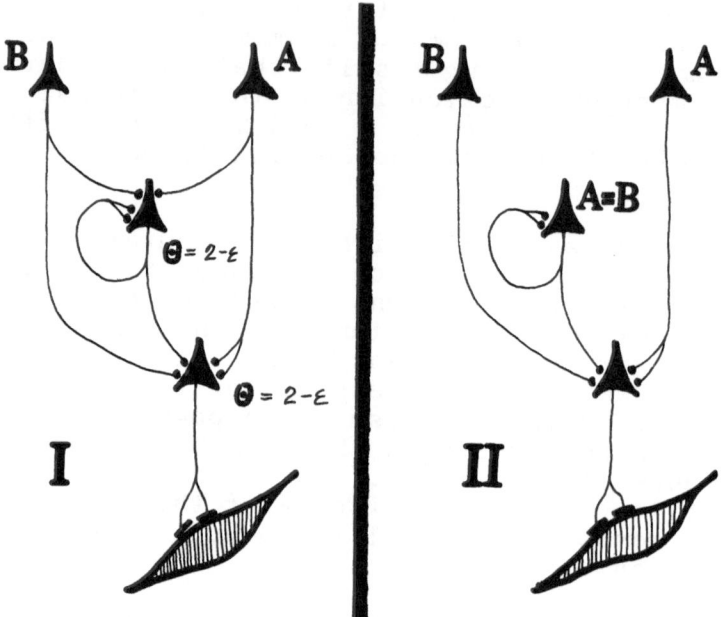

Fig. 1.   Minimal network for computing an elementary conditioned reflex (ECR).   (I) Before conditioning;   (II) After conditioning.

napse represented one unit of excitation).   The
motoneuron cannot be triggered by B only, for one
synapse is insufficient to override its threshold.

However, when first concomitance of A and B
occurs, the internuncial is activated and provides
sufficient facilitation for B to initiate the reflex.   The internuncial's recurrent collaterals
secure its permanent excited state and, hence-
forth, B only is sufficient to elicit a response.

In spite of the structural simplicity of this
four-element network, it exhibits some features
that are instructive in this context.   First it
should be noticed that it alters its function as a

consequence of the occurrence of certain stimulus
configurations: before concomitance of A with B,
the net is impervious to B, while afterwards it is
responsive to B as it is and was to A. Unfortunately, some authors seem to associate with this simple alteration higher mental functions by calling
this "learning through experience". Whether this
misrepresentation is caused by underestimating the
complexity of the processes that establish algo-
rithms for solving certain classes of problems--i.e.,
"learning" in its proper sense--or by overestimat-
ing the sophistication of this simple circuit, I
must leave to anybody's judgement. However, it
should not be overlooked that indeed a specific
external event caused this net to change its modus operandi, and that the occurrence of this event
is recorded in the reverberating loop of the self-
excitatory internuncial. This is particularly
clearly seen in Part II of Fig. 1 which despite
its degenerated internuncial afferents is a functional equivalent of the net in Part I after it
was modified by the specific event. The internuncial "holds" the equivalence relation A = B,
and one may be tempted to associate with this
store some form of elementary "memory" that indeed,
can be localized and functionally isolated. Alas,
this is not so, as I shall show in a moment, for
nothing can be inferred from this store except its
own truth: "A = B is the case".

However, note that the representation of the
concomitance of A and B is in form of a relation
between A and B in the sense of establishing equivalence between A and B. This may be interpreted
as an elementary representation of "meaning", for
the activity in the loop represents nothing more
nor less but "B means A", hence this network appears to be an elementary inductive inference computer. Alas, this is not so. In order for induction to be operative, inference about "as yet unexamined cases" has to be made. But this network
reiterates only the examined case, and more complex structures have to be considered to allow
for inductive inference. Can we get some leads as
to these structures from the examples of this net?
Perhaps, yes.

The answer may be gleaned from the fact that if equivalence of stimuli can be computed by the net, the general notion of "equivalence" must be somewhere "stored" in this net. Indeed it is, but not in a single element as one may be prone to believe, but in the whole functional and structural organization of the network before the event took place that caused it to become a record of the specificity of this event. We may conclude from this that an inductive net must keep its "equivalence structure" intact in order to be ready for every new case of "class B" to be classified as being also cases of "class A", and if this should prove to be false, to either drop this hypothesis or else switch to another one.

I would like to conclude the discussion of this simple net by citing a keen observation by Susan Langer (14) who considers the ontogenesis of mentation as being initiated by an ECR. In a passage devoted to the clarification of the distinction between symbol and sign she writes:

"There is a profound difference between using symbols and signs. The use of signs is the very first manifestation of mind. It arises as early in biological history as the famous 'conditioned reflex', by which a concomitant of a stimulus takes over the stimulus-function. The concomitant becomes a sign of the condition to which the reaction is really appropriate. This is the real beginning of mentality, for here is the birthplace of error, and herewith of truth."

As much as can be said about some features of this elementary four-element network, I wish to stress again the utter inadequacy of this net to account for even the most straight-forward cases of conditioned reflexive behavior in higher animals. The belief harbored perhaps by early reflexologists, that ultimately such behavior can be reduced to a logical or neural schema of the sort shown in Fig. 1 has--to my knowledge--been completely destroyed by the superb work of Jerzy Konorski (15) who showed that, for instance, in dogs

at the first application of the positive conditioned stimulus it elicits a quite distinct "orientation reaction", i.e., pricking up the ears, turning the head, etc., while salivation as response is
negligible. He goes on to demonstrate that in almost all experimental set-ups conditioned stimuli
"...do not usually possess a single modality, but
they supply a number of cues..." which the animal
utilizes and evaluates as to theis significance
in determining future action. Konorski reaches
the conclusion that essentially two principles
govern the acquisition of various types of conditioned reflexes, one, a principle of selection,
the other one a principle of inseparability of information from its utilization. Since I consider
these principles of considerable importance in my
argument, I shall state them more explicitly in
Konorski's own words:

(i) Selection

    "In solving a given conditioning problem the
animal does not utilize all the information supplied
by the conditioned stimuli, but it definitely selects certain cues, neglecting the other ones."

(ii) Inseparability

    "...it is not so as we would be inclined to
think according to our introspection, that receipt
of information and its utilization are two separate
processes which can be combined one with the other
in any way." Hence: "Information and its utilization are inseparable constituting, as a matter of
fact, one single process."

    If I may translate these observations into my
terminology of before, the principle of selection
becomes a "search for meaning" in the sense that
animal selects those cues--i.e., that information--
from which it can optimally draw inferences; while
the principle of inseparability becomes a "recourse
to self-reference" in the sense that the animal
evaluates the inferences drawn from that information always with regard to its utilization favorable

of its own self.

In search for a minimal network that would
exhibit these two principles of selection and of
inseparability of information from its utiliza-
tion--or of "search for meaning" and "self-reference"--I came across J.Z. Young's drawing of
a network representing a single memory unit or
a "mnemon" as he calls it (16). Although
Eccles' "The Cerebellum as a Neural Machine" (17)
abounds with examples of such networks, they exhibit many more functions than needed at this
moment and thus I cannot consider them to be
"minimal" in this context.

Figure 2 reproduces Young's drawing of the
organization of such a single memory unit. He
describes its general features as follows:

"...each unit consists of a classifying neuron that responds to the occurrence of some particular type of external event that is likely to
be relevant to the life of the species. The resulting impulse may initially activate either two
or more channels by branching of the axon. More
than one line of conduct is therefore possible.
The mnemon includes other cells whose metabolism
is so triggered as to alter the probable future
use of the channels on receipt of signals indicating the consequences of the actions that were
taken after the classifying cell had first been
stimulated."

From this it is easily seen that Young's
mnemon indeed incorporates minimally the two
principles mentioned earlier. The principle of
"selection" or of "search for meaning" of a particular stimulus is incorporated by the choice
of pathways that lead to different actions. What
that stimulus "means" becomes clear to the animal
of course, only after a test. "Attack" may under certain stimulus conditions mean "Pain",
under others "Pleasure". Note here the important
point that neither pain nor pleasure are objective states of the external universe. They are

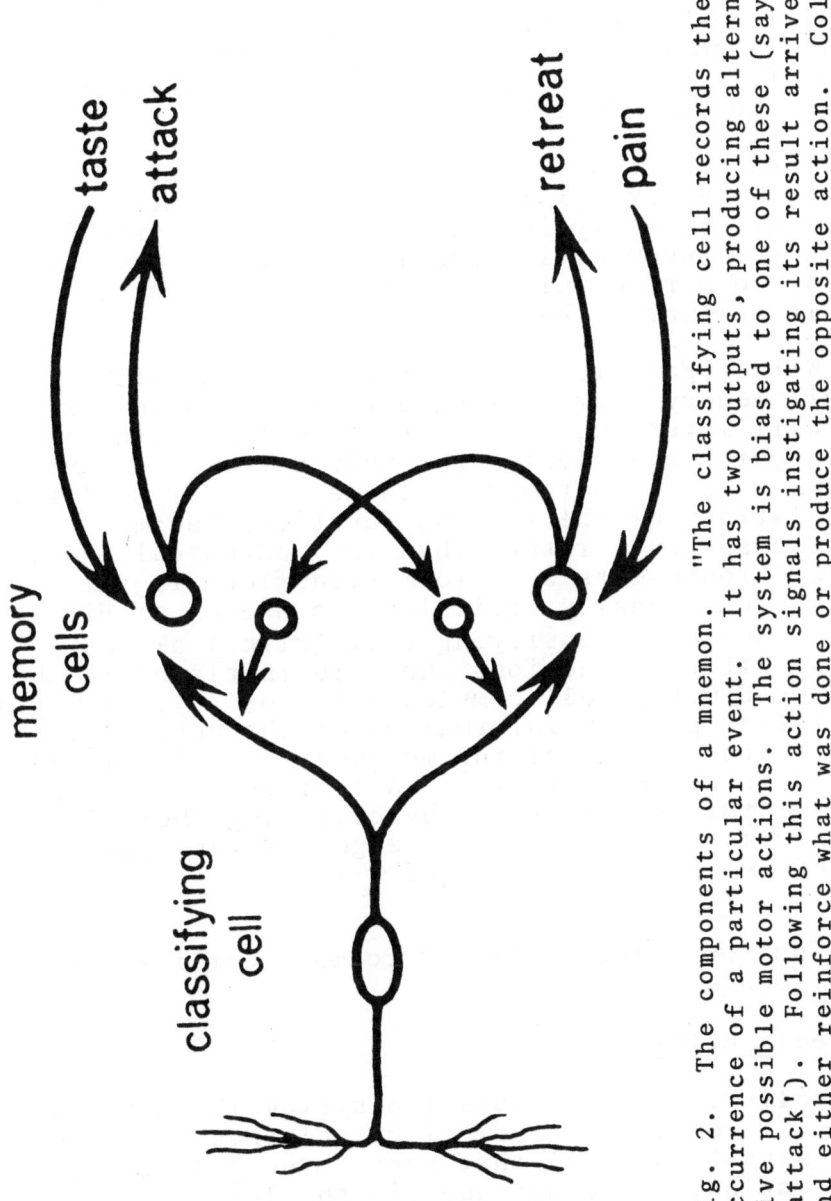

Fig. 2. The components of a mnemon. "The classifying cell records the occurrence of a particular event. It has two outputs, producing alternative possible motor actions. The system is biased to one of these (say 'attack'). Following this action signals instigating its result arrive and either reinforce what was done or produce the opposite action. Collaterals of the higher motor cells then activate the small cells, which produce inhibitory transmitter and close the unused pathway. These may be called "memory cells" because their synapses can be changed." (Reproduced with kind permission from J.Z. Young (16.)

states that are generated purely within the animal, they are "self-states" or--to use terminology
of physics--"Eigen-States" of the organism which
permit it to refer each incoming signal to its
own self, i.e., to establish self-reference with
respect to the outside world.

With this observation, the second principle
of inseparability of information and its utilization falls smoothly into place, for this system
checks the incoming information as to its usefulness by comparison with its eigen-states where
upon it initiates the appropriate actions.

With regard to the functional organization
of this memory element I wish to make two points
that will later become important in the synthesis
of a cognitive element. For this purpose, I have
redrawn Young's anatomical schema in order to let
the relations of the various functions become
more transparent, rather than the anatomical ones.
Figure 3 represents an information flow diagram
that is functionally equivalent to mnemon of Figure 2. Again a classifying cell (cl.c.) allows
for two alternate actions that are initiated in the
memory and motor cell complexes (A) or (B). Young's
collaterals pick up information of the action state
from the thick axon of the motoneurons (A) (B) and
feed it to comparators (A^+) (A^-) or (B^+) (B^-) which
evaluate the action states by comparing them with
the information of resulting eigen-states, either
desired (+) or else undesired (-).

The two points I wished to make earlier are now
as follows:

(i) Self-Reference

Self-reference enters the system through two
channels: one, via a priori established "good" or
"bad" signals (+) (-) that report the consequences
of an action; the other one via the loop $(A) \rightarrow (A+) \rightarrow A$
or, mutatis mutandis, via corresponding other loops
that report the states of its own actions.

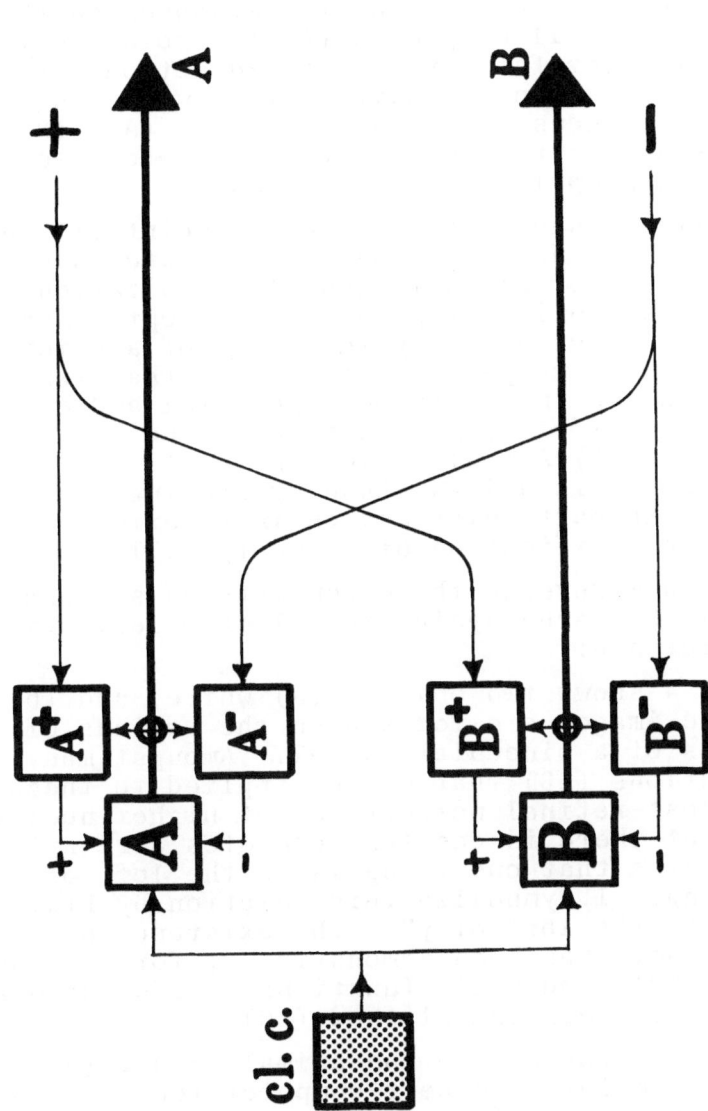

Fig. 3. Information flow diagram of a mnemon. (cl.c.) classifying cell; (A) (B) motoneurons and memory cell complex; (+) (-) information of eigenstates "good", "bad", or positive and negative internal reinforcement signals; (A+) (A-) (B+) (B-) comparators of action states with eigen-states.

(ii) Experience

Experience enters the systems through two operations: one, which modifies the synapses on the
memory cells in cell complexes (A) (B) so as to inhibit undesired or facilitate desired actions; the
other one, which compares past actions with its
present consequences in comparators (A^+) (A^-) (B^+)
(B^-) and transmits the results (+) (-) to complexes
(A) (B) for appropriate modifications.

I shall now show with respect to point (i) that
self-reference is an ubiquitous feature and is computed over and over again in neural organizations,
mostly by a resolution of paradoxes in representation, and not necessarily by reference of a priori
signals; and with respect to point (ii) that experience is gathered in a much more powerful and economic way by modifying the function of the recursive
loop $A(t-\Delta) \rightarrow A^+(t) \rightarrow A(t^+\Delta)$--t indicating time and Δ
cumulative synaptic delays--than by storing the
outcome of each particular action in a corresponding synaptic modification of a memory cell.

Let me now develop these comments in somewhat
more detail, by using again minimal examples. First
on self-reference:

Figure 4 shows two objects (a) white, and (b)
black, whose images are focused on the retinas of
the two eyes of a binocular animal. Amongst many
other operations (18) that may be applied to these
images by post-retinal networks or at higher nuclei,
I assume that there is one that computes a relation
which indicates that one thing is to the left of
another thing. I symbolize this relation by L(x,y),
read "x is to the left of y". The existence of such
anisotropic nets has been demonstrated, for instance
in pigeons (19), and their functional and structural
organization is well-established (20).

Since with respect to the animal object (a) is
behind (b), the left eye Left-computer reports L(b,a)
while the right eye Left-computer reports an opposite
state of affairs, namely, L(a,b). This apparent
paradox can be resolved by a computer $B(L_1, L_r)$
which realizes that the information L(b,a) is supplied by the left eye, L_1 (subscript 1=left), while

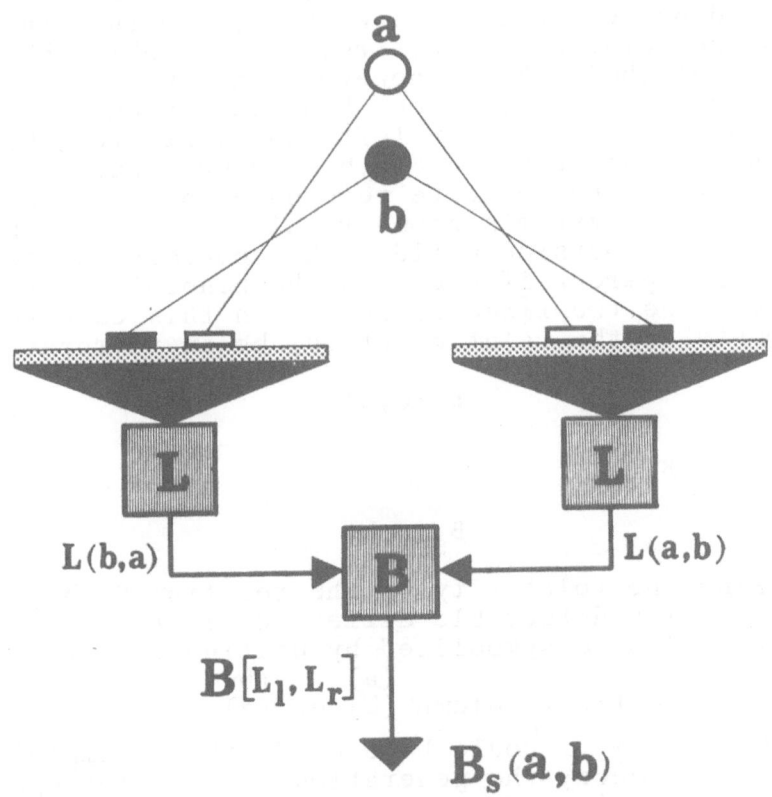

Fig. 4. Computation of "depth" by resolving a sensory paradox in binocular vision. (A) Networks computing the relation "x is left of y"; (B) Networks computing the relation "x is behind y".

L(a,b) is supplied by the right eye L_r (subscript
r=right). With this observation the paradox disappears, for the two apparently contradictory
results are in fact obtained from two distinct and
locally separated sensory systems which by no nec-
essity should deliver the same picture of the outside world. However, it is significant that a
consistent picture of the outside world can be
computed by generating a new space, "depth", in
which the relation B(a,b)--read "a is behind b" is
now established. Note, however, that this resolution could never have been obtained without reference to the animal's own left and right eye. Consequently, the relation B(a,b) too must carry a
subscript B_s, to indicate reference to "self" or to
indicate the system's geometrical relation to objects of the outside world. This becomes particularly transparent if one lets the binocular system
encircle the two fixed objects. In this case the
arguments in the relation B(a,b) begin to rotate

$$B_s(b,a)$$

$$\begin{matrix} (a) \\ B_s(b) \end{matrix} \qquad \begin{matrix} a \\ b \end{matrix} \qquad \begin{matrix} (b) \\ B_s(a) \end{matrix}$$

$$B_s(a,b)$$

mirroring the relativity of the relation "Behind".
(Absence of a detectable difference in the two L
computers I have symbolized by writing the arguments
of B in a vertical column $\begin{matrix}(a)\\(b)\end{matrix}$ or $\begin{matrix}(b)\\(a)\end{matrix}$.

Of course, I could have used other examples,
as, for instance, the generation of a "color space"
by the resolution of a triple-paradox which is produced by the divergent reports of the three types
of cones with different pigmentation regarding the
appearance of one and the same spot in the external
world. However, this case and other cases are not
minimal.

I shall now enlarge on my earlier brief comment regarding the use of recursive functions as a
more powerful tool in accounting for past experience
than simple storage of the outcomes of individual

acts. This comment was prompted by Young's observation of recursive loops that report back to a
central station via some synaptic delays. Turning
again to Figure 3 and following the arrows leading
from (A) to (A$^+$) back to (A), we realize that an
action A that took place in the past, say, one cumulative synaptic delay Δ ago, i.e., A(t-Δ), is evaluated by (A$^+$) at time t, i.e., A$^+$(t), which in
turn modifies the cellular aggregate in (A) that
will at best respond with new action after a cumulative synaptic delay Δ, i.e., with A(t+Δ).

I propose now to make changes neither in the
structure nor in the function of this subsystem,
but only in the interpretation of the modifications
that are supposed to take place. Instead of interpreting the synaptic modifications in the cellular
complex A as stores of the outcomes of various individual actions, I propose that these modifications
should be interpreted as a modification of the transfer function of the whole subsystem (A,A$^+$). Let me
demonstrate this idea again with a minimal example,
this time of recursive functions.

First, I have to point out that the term "recursive function" is a misnomer, for these functions
are like any other function, and it is only that
they are not as usual defined explicitly but are defined recursively. By this is meant that a function
which relates a dependent variable y to an independen
variable, say time t, is not explicitly given in
terms of this independent variable, say $y=t^2$,
y-sinωt, or in general y=f(t), but is given in terms
of its own values at earlier instances y(t) = F(y
{t-Δ}), where Δ expresses the interval between the
earlier instance and the instance of reference t.
A typical example of a recursive definition of a
function is, for instance, the description of growth
of a bacterial colony:

"The number of bacteria in a bacterial
colony at any time is twice the number
it was one generation ago."

If it takes on the average the time Δ for a bacterium to divide--i.e., one generation extends over a time interval Δ--then the recursive description of the size y of this colony is

$$y(t) = 2 \cdot y(t-\Delta)$$

I shall not discuss the mathematical machinery that "solves" these expressions, i.e., transforms them into explicit statements with respect to the independent variable t only. For instance, in the above case the "solution" is, of course,

$$y(t) = y(0) \cdot e^{\frac{t}{\Delta} \ln 2}$$

where y(0) is the initial size of the colony, i.e., its size at time t=0. The methods of solution are of no concern to us here, the point I wanted to make is only to assure you that a recursive definition of a function is as good as any other and, in some cases, may be even more powerful than an explicit expression (e.g., compare the terse recursive definition of above with the cumbersome explicit expression).

If we now go back to our original problem of finding an appropriate description for a system that acts according to the outcome of previous actions, then it seems--at least to me--that the conceptual tool of recursive function theory is just tailor-made for this purpose.

I shall now discuss the minimal case that corresponds to the "mnemonic" part of Young's mnemon. Figure 5 shows a three-element system (F,T, D) whose functional correspondence with the mnemonic features of the subsystem (A,A+) of Figure 3 will emerge in a moment.

Box F stands for the mechanism that computes the function*) $Y = F(X,Y')$ on its two arguments X

---

*) According to standard notation, capital letters X,Y represent a set of components $(x_1, x_2 ... x_n)$, $(Y_1, y_2 \cdots y_m)$, the value of each component representing, say, the stimulus or response intensity along a corresponding fiber. In other words, X and Y re-

and Y'. The argument X is an explicit function of
time X(t), and is called the "input proper". The
argument Y' is a representation of the "output" Y
of mechanism F at an earlier time, say t-Δ, and is
called the "recursive input". In order that F can
be informed about its previous output--or action--
the intensity of this action has to be measured by
an element, T, which translates this intensity into
a signal that is accepted ("understood") by F, and
feeds this information with a delay D back to F.

The functional correspondences of these elements with some of the physiological features in
Young's mnemon seem to be clear. D corresponds to
a cumulative synaptic delay that "holds" the whole
picture of this system's output activity for a while,
Δ, before it informs the cell aggregate in A of this
activity. T represents the motoneuron's collaterals
or terminations of sensory afferents that generate
the information of A's activity. F is, of course,
the aggregate (A,A^+), as yet without input of an
eigen-state (+) (-), but with an input proper, X,
which represents the signal from the classifying
cell (cl.c.).

Let us now watch this three-element system in
operation. Foremost, we wish to know its output
Y(t) at time t for a given input X(t) at that time.
Since F is given, we have

$$Y(t) = F\{X(t), Y(t-\Delta)\},$$

or, more conveniently, if we drop the reference to t
by merely writing Y and X, and for the previous
state Y' and X':

$$Y = F(X,Y')$$

However, this does not yet tell us the actual output of the system, because we do not know the value
of one of its inputs, namely Y', i.e., its previous
output. But this can be determined, using the recursive definition of Y:

$$Y' = F(X',Y'') \quad .$$

---

present the activity along whole fiber bundles and
not necessarily that of single fibers.

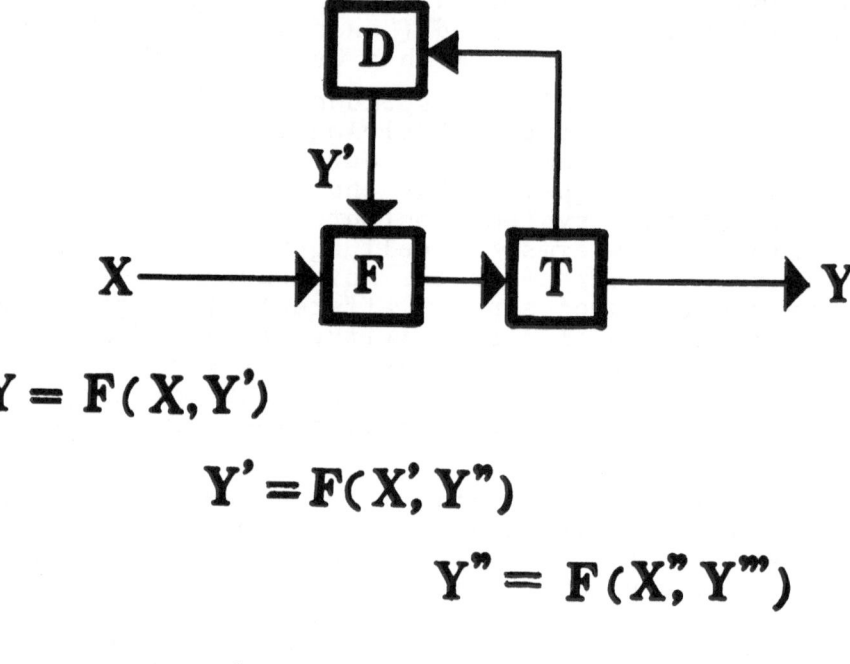

$$Y = F(X, Y')$$

$$Y' = F(X', Y'')$$

$$Y'' = F(X'', Y''')$$

$$Y = F(X, X', X'', X''', X'''', \cdots\cdots Y_0)$$

Fig. 5.   Circuitry and basic components of a recursive function computer.   F computing element;
X input proper; Y', recursive input of a previous
output Y' D, delay; T, translates action Y into a
representation of Y acceptable to F.

or in words: the previous output is a function of
the previous input and its previous-previous output.
Hence, by inserting the expression for Y' into the
earlier equation for Y, the present output becomes:

$$Y = F(X,X',Y'') \quad .$$

Again we may ask for the value of the previous-
previous output, and by applying again the recursion,
we will arrive at an expression which leads us three
steps back in time, and so on, until we arrive at
the "birth-date state" Y_0 of the system:

$$Y = F(X,X',X'',X''',\ldots,Y_0) \quad .$$

The remarkable feature of this expression is that
it clearly shows the dependence of the present output of this system on the history of the previous
inputs, rather than on just its present input or
to put this into more poetic terms, this system's
present actions depend on its past experiences.

Two features should be noted here. First, no
storage of representations of past events--save
for those traveling through the delay loop--take
place here. Reference to the past is completely
taken care of by the specific function F that is
operative. F is, so to say, the "hypothesis" that
predicts from previous cases future actions. Physiologically, F is determined by the functional organization of the cellular aggregate (A,A^+). Second, an external observer who wishes to predict the
behavior of this system in terms of its input-out-
put or stimulus-response pattern

$$Y = f(X),$$

and who has no access to its internal structure may
soon find to his dismay that he is unable to determine the elusive function "f", for after each experimental session the system behaves differently
unless--by lucky circumstances--he finds a repeated

sequence of inputs that will give him --by the very nature of that particular F--repeatedly a corresponding sequence of outputs. In the former case, this experimenter will in disgust turn away with the remark "unpredictable!"; in the latter case he will say in delight: "I taught it something!" and may turn around to develop a theory of memory.

Although such recursive function elements exhibit some interesting properties, in the restricted and isolated way in which I have discussed them, they are as yet incapable of responding to OK-signals (+), HANDS-OFF-signals (-), and any other signals that report eigen-states of affairs, or, in general, to self-referential information. Assume such information were available. The question is now to which of the elements in the three-element recursive function computer must this information go in order to modify its <u>modus operandi</u> in accordance with a desired eigen-state configuration? It seems to me that the question already carries its own answer: if such change is necessary at all, then the only effective way to modify the general properties of this computer is to change its "hypothesis" by which it computes future states from past experience, i.e., the recursive function F_1 which was operative until this moment must be altered to become, say, F_2, and perhaps later F_3, F_4 and so on, in order to achieve the properties that are commensurate with the system's eigen-state configurations. In other words, F itself has to be treated as a variable, as an element in a range of functions $\Phi(F)$, whose particular value F_i is determined by the eigen-states. Physiologically this means that the recurrent fibers that carry self-referential information are synapsing with cells in the (A, A^+) aggregate so as to change it from a computer that calculates F_i to one that calculates F_j. Mechanisms that achieve such modifications are well-known as, e.g., long range inhibitions and facilitations. However, I doubt that it will ever be possible to establish a detailed account of the

relations between individual synaptic changes and
the computational properties of the whole aggregate,
the main reason being that this is a problem that
does not have a unique solution, on the contrary,
it can be shown that with just a few cells making
up this computer, the number of different solutions
is, for all practical purposes, infinite. On the
other hand, I do not believe that such detailed
knowledge is of importance, as long as the principles are understood that make such modifications
possible.

Let me now briefly summarize some of the essential results of this discussion. Most of the
neural machinery is functionally organized to establish from sensory information--whether about
states of the outside world or about internal states
--relations between observed entities with respect
to the observing organism. This relational information modifies the modus operandi of a computer
system that computes new actions recursively on
the basis of the outcome of previous actions and,
hence, on the basis of the history of the stream
of external and internal information. Figure 6 is
a graphical representation of this summary in form
of a block diagram. I shall call this whole system
a "Cognitive Element", for it represents a minimal
case of a cognitive process, or a "Cognitive Tile",
for it may be used in conjunction with other such
tiles to form whole mosaics--or "tessellations"--
which, as a whole, permit the high flexibility in
representing relational structures not only of what
has been perceived but also of the symbols--the
"linguistic operators"--that ultimately are to con-
vey in natural language all that which can be inferred from what has been perceived.

The various components of this cognitive tile
are quickly explained. X stands for (external)
sensory input, and Y for the output of the system
as seen by an outside observer. Hence, this elementary component is a "through-put" system, as
suggested by the small inset, lower right. However,

because of its internal organization, this element
is quite a different animal from a simple stimulus-
response mechanism with fixed transfer function.

First, sensory information, X, is operated on
to yield relations $R_s(X)$ between observed activities
with respect to "self" (note subscript s), and is
then used as input proper for the recursive function computer which may be operative at this moment with any one of the functions F belonging to
range ϕ. Its output is fed back over two channels,
one being the recursive loop with delay D to allow
F to assess its earlier actions, the other carrying all the relational information of the system's
own actions $R_s(Y)$ as they refer to "self", and
operates on $\phi(F)$ in order to set the recursive function computer straight as to this tile's internal
goals and desires.

This element incorporates all those faculties
which I considered earlier to be necessary components of cognitive processes: to perceive, to remember and to infer. However, in this element none
of these faculties can be isolated functionally:
it is the interaction of all the processes here involved that "lift" the information from the input
signal and translate it into action meaningful for
this tile.

Nevertheless, if forced to interpret some of
this tile's functional components in terms of those
conceptual components I would reluctantly give the
following breakdown: (i) Perception is accomplished by the elements that establish self-referential relations in the spatio-temporal configurations
of stimuli and responses; (ii) Memory is represented by the particular modus operandi of the central
computer whose gross functional organization is determined and redetermined by evaluation of eigen-
states or relations; (iii) Inference in this tile
appears on three levels, depending on the type of
functions that are in range ϕ and on the type of
processes one wishes to focus on. Adductive in-

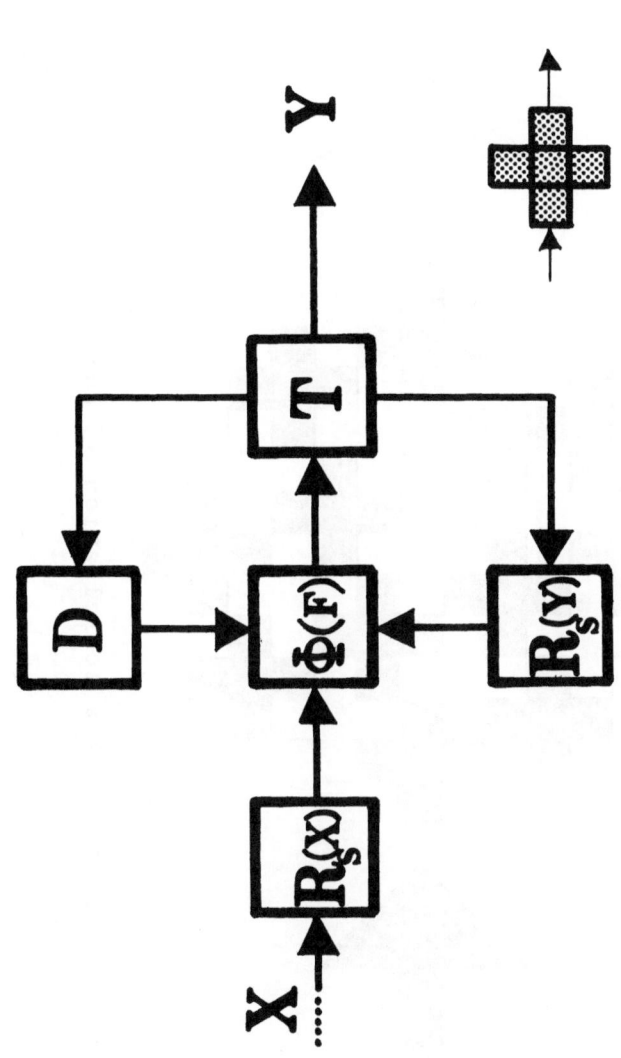

Fig. 6. Circuitry and basic components of a cognitive tile. Φ (F) general purpose computer for a range Φ of computable functions F; X input, Y output; $R_s(X)$, $R_s(Y)$ compute relations in the spatio-temporal configurations of input and output respectively, with reference to intrinsic properties of this particular tile; D a delay element; T translates Y into a representation of Y that is acceptable to this and other tiles.

Fig. 7.   Example of a tessellation of cognitive tiles.

ference is operative in the cumulative absorption
of comparisons of past external and internal experiences that give rise to the functional organization of the central computer. Inductive or deductive inferences are computed by the central system concurrently with any new signal, the inferential mode being solely dependent on strings of earlier failures or successes and of some of this tile's
internal dispositions to "disregard" false induc-
tions or to take them "seriously" by converting to
more stringent logical deductions.

I shall now conclude my thesis with only a
brief report on some properties as they may be relevant to this topic of aggregates of such tiles
or "tessellations" as they are usually referred to
in the literature. John von Neumann was the first
to realize the high computational potential of
these structures in his studies of self-reproducing
automata (21), and later Löfgren applied similar
principles to the problem of self-repair (22). We
use these, however, in connection with problems of
self-reference and self-representation.

Two features of cognitive tiles permit them to
mate with other tiles: one is its inconspicuous
element T which translates into a universal "internal language" whatever the "output language may
be; the other one is its essential character as a
"through-put" element. Consequently, one may as-
semble these tiles into a tessellation as suggested in Figure 7, each cross, white or black, corresponding to a single tile, while each square in a
cross represents the corresponding functional element as suggested in the earlier Figure 6. Information exchange between tiles can take place on
all interfaces, however, under observance of transmission rules implicit in the flow diagram of Figure 6. For instance, one tile may incorporate into its own delay loop preprocessed information
from an adjacent tile, but eigen-state information
of one tile cannot retroactively modify the operations of a "left" tile, although it can--via its
own output--modify that of a "right" tile, and so
on.

When in operation, this system shifts kaleido-
scopically from one particular configuration of co-
operating sets of adjacent tiles to other configurations, in an ever changing dynamic mode, giving
the impression of "clouds" of activity shifting,
disappearing and reforming as the task may demand.

We have studied such systems as yet only in
the "representative mode", i.e., in which these
tiles correspond to "linguistic operators" with
their multiple ramifications into various depth
of meaning. These systems are now simulated by
complex computer programs, one being a particularly interesting three-dimensional extension of the
two-dimensional scheme presented here and being
called "Cylinders" by its inventor Paul Weston (10,
23). These novel program structures represent at
this stage prototypes of systems which permit symbolic discourse between man and machine in the form
of natural language. We do not foresee fundamental
difficulties when switching to the "perceptive
mode", i.e., in which the inputs to some specified
"sensory tiles" are not symbols but signals from
some restricted, but relevant, environment.

We hope to provide with these studies the
foundation for a new architecture of future computers that may well serve as models for a cognitive
memory that has hindsight and foresight as well.

IV.   A Conjecture

Eccles' grand oeuvre The Cerebellum as a Neuronal Machine (17) is extremely encouraging to
look for small, highly organized, cell assemblies
that could be represented by the operational unit
I have developed above, namely, by a cognitive tile.
I have convinced--at least--myself that there are
numerous examples of networks whose actions can
be described by individual tiles or by smaller or
larger tessellations (24). The question however,
which plagues a theoretician is as to the minimal
physiological unit that could be described by the
corresponding minimal operational unit, i.e., a
single cognitive tile. Learning about the tre-

mendous complexity of a single Purkinje cell, with
its wide range of response activity and with a convergence of inputs up to about 200,000 synapses, I
believe most of my tile's functional properties can
be found in a single specimen of these cells, were
it not for one operational feature of the tile, its
computation of recursive functions, that would require the cooperation of at least one other cell
to form a single cognitive tile. However, I believe there is a way out of this dilemma by following the ideas proposed, for instance, by Holger
Hyden (25, 26), which suggest to look into the
cell, i.e., into modifications of the cell's molecular constituents, in order to account for some
mnemonic properties of a single cell.

The most pedestrian way to look at the potentialities of a complex molecule is to look at it
as a storage and retrieval device (27,28). This
possibility offers itself readily by the large number of excitable states that go hand in hand with
the large number of atoms that constitute such molecules. Consequently the chances are enhanced for
the occurrence of metastable states which owe their
existence to quantum mechanically "forbidden" transitions (28). Since being in such a state is the
result of a particular energy transaction, selective "read-out" that triggers the transition to the
groundstate--as in the optical maser--permits retrieval of the information stored in the excited
states.

There is, however, another way to allow for
information storage in macro-molecules where the
"read-out" is defined by structural matching (temp-
let). It is obvious that, m, the number of ways
(isomeres) in which n atoms with V valences can
form a molecule Z_n will increase with the number
of atoms that constitute this molecule. Each of
these configurations is associated with two characteristic energy levels (quantum states), one
that gives the potential energy of this configuration, the other one which is the next higher level
at which this configuration becomes unstable. Fig-

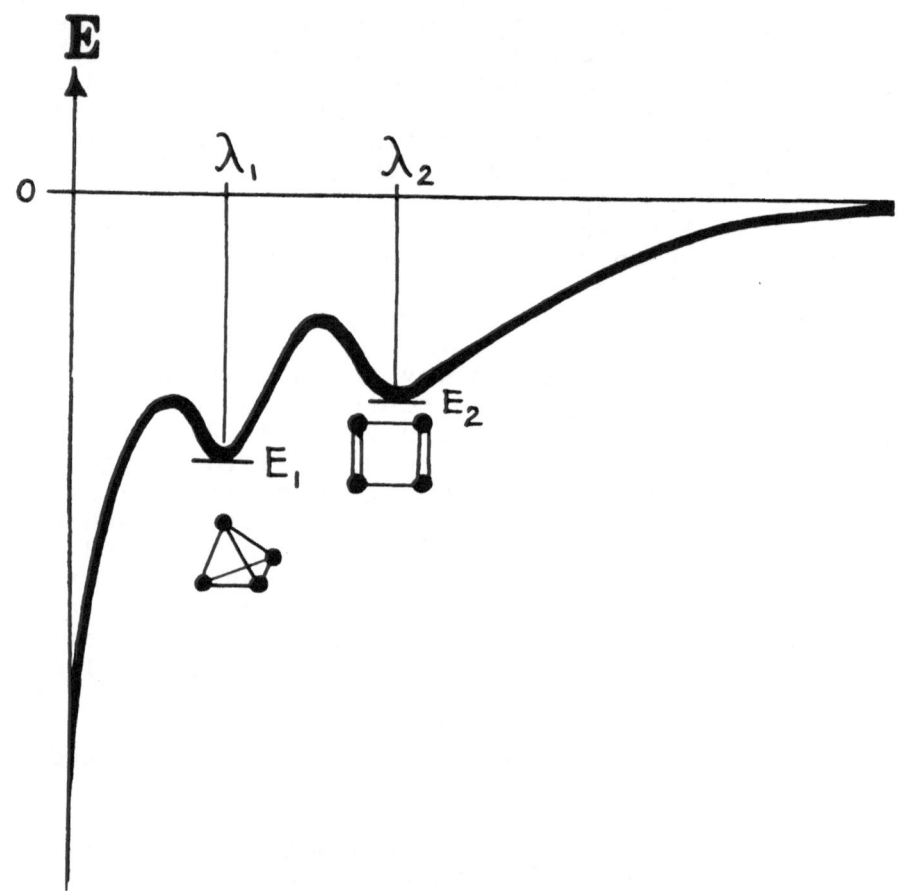

Fig. 8. Association of energy levels with the two
different configurations (isomeres) of a molecule
composed of four atoms, each having three valences
(n = 4; V = 3). λ_1 and λ_2 represent the eigen-values
in the solution of the Schrödinger wave equation.

ure 8 sketches this situation for the two isomeric
states of an hypothetical molecule Z_4 composed of
four 3-valence atoms Z. Simple considerations show
that the tetrahedral configuration is more stable
than the quadratic form, hence some energy must be
supplied to the tetrahedron to transform it into
the square. However, it will not stay indefinitely
in this configuration because of the quantum mechanical "tunnel effect" which gives each state a
"life-span" of

$$\tau = \tau_0 e^{\frac{\Delta E}{kT}}$$

where ΔE is the height of the energy "trough"
which keeps the configuration stable, k is Boltzmann's constant, T is the absolute temperature
surrounding this molecule and τ_0 is an intrinsic
oscillatory time constant associated with orbital
or lattice vibrations.

It is these spontaneous transitions from one
configuration into another one which tempt me to
consider such a molecule as a basic computer element, particularly if one contemplates the large
number of configurations which such macro-molecules can assume. Estimates of the lower and upper bounds of the number of isomeres are (29)

$$\underline{m} \approx \frac{5}{8} \cdot n$$

and

$$\overline{m} \approx \left(\frac{nV}{2p(V)} \right)^{p(V)}$$

where p(N) is the number of unrestricted partitions
of the positive integer N, and V and n are again

the number of valences and the number of atoms respectively.

Since each different configuration of the same
chemical compound Z_n is associated with a different
potential energy, the fine-structure of this molecule may not only represent a single energy transaction that has taken place in the past but may
represent a segment of the <u>history</u> of events during
which this particular configuration evolved. This
consideration brings me right to my conjecture,
namely to interpret the responses of such macromolecules to some energy transactions as those of
a recursive function computer element.

The idea to look upon various structural transformations which many of the macro-molecules per-
petually undergo as being outcomes of computations
is not at all new. Pattee, for instance, has demonstrated in a delightful paper (30) the isomorphism between the growth of some helical macromolecules with the operation of a finite, binary
autonomous shift-register. In his example the
recursive relation is only between a present state
Y and an earlier one Y':

$$Y = F(Y').$$

We, however, need to account of an "input proper"
X, in order to be able to interact with this system, i.e., to allow for a "read-in" and "read-out"
operation:

$$Y = F(X,Y').$$

Figure 9 sketches the four lowest energy
states numbered 1,2,3,4 of a molecule together with
the three energy thresholds ΔE_2, ΔE_3 and ΔE_4 which
keep the corresponding configurations stable at
least during the "life-span" of these states. For
simplicity, I assume these life-spans to be multi-
ples of the shortest life span $\tau*$, i.e., under nor-

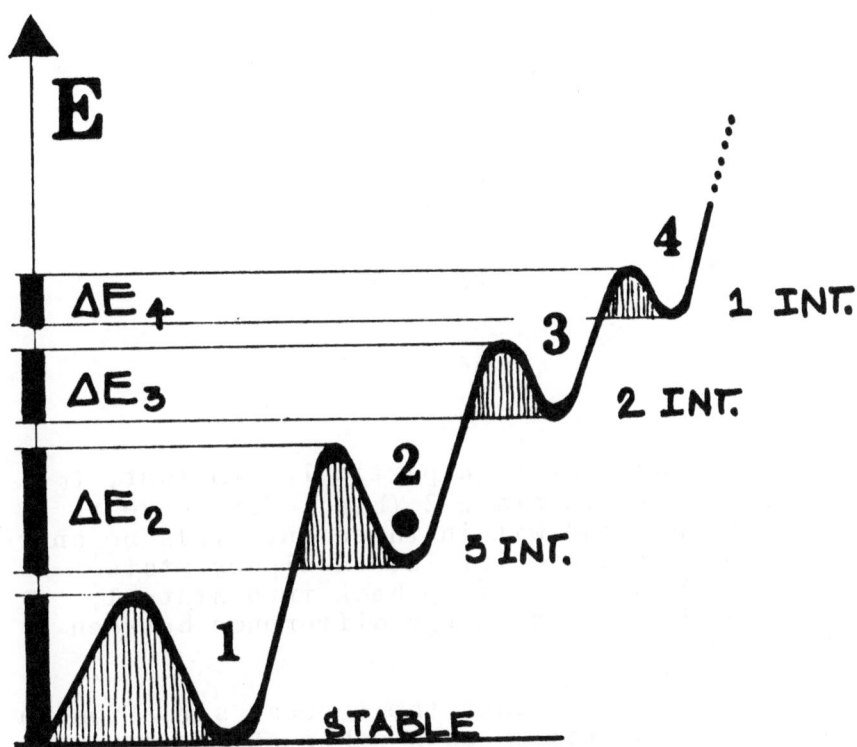

Fig. 9.   Four of the lowest energy levels corresponding to some molecular configurations together
with the threshold energies that keep these configurations stable within some definite intervals
of time.

mal temperature conditions ΔE_4 will give state #4 a lifespan of τ^*, and the others as suggested by Table I:

TABLE I

State	Threshold	Life Span
#1	large	∞
#2	ΔE_2	$3\tau^*$
#3	ΔE_3	$2\tau^*$
#4	ΔE_4	$1\tau^*$

Assume now that at a particular instant, t_0, this molecule is in state 2 (black dot suggests this position), and within three intervals no energy is supplied to kick it into a higher state. As a consequence, it will flip back into state 1, giving off the stored energy difference between state 2 and 1.

I shall now consider the general situation in which two events follow each other at times t_1 and t_2, each spaced approximately at intervals corresponding to τ^*, and each event either supplying (1) or else not supplying (0) the energy to lift the molecule into its next higher state.

Table II gives the result of these operations, indicating on the left whether or not the events at times t_1 and t_2 carried the required energies, and giving at the head of columns under t_0 the initial state of the molecule.

TABLE II*

t_2	t_1	t_0			
		1	2	3	4
0	0	1	2	2	3
0	1	2	3	4	4
1	0	2	3	3	4
1	1	3	4	4	4

Clearly, for each of the different initial conditions this molecule "computes according to the
four possible input-configurations (00) (01) (10)
(11) a different set of outcomes, in other words,
this computer changes its operations depending on
its initial state which is, of course, nothing
else but the result of previous operations.

It is easy to see how this idea can be extended to accommodate an arbitrary number of se-
quential events t_1, t_2, t_3...t_s and an arbitrary
number of molecular states 1,2,3,4,5,...m, and
thus gives rise to the possibility to interpret
the various induced and spontaneous states of a
macro-molecule as those of a recursive function
computer of considerable flexibility and latitude
of range.

However, there remains the question still
open as to the external mechanisms that will induce these changes. To this end we have to evaluate numerically the equation that was given earlier

*) The initial states are assumed to have been ac-
quired within one earlier interval. A more elaborate table is needed to indicate "aged" states.

and which relates the various quantities here involved, namely, the threshold energies ΔE, the
average life span of a state τ, and the other two
quantities, τ_0, an intrinsic time constant, and
the temperature T of the system. If we assume a
constant body temperature of 36.6°C, then T =
309.8° Kelvin and with the known value of Boltzmann's
constant there remain only the three variables τ_0,
τ and ΔE to be related. This is most clearly done
in form of a nomogram as given in Figure 10. Values along the three scales that are read off at
points which are connected by a straight line always represent a solution of our equation that
relates these quantities. The three scales represent the values of τ_0, the period of intrinsic
oscillations in seconds, of ΔE the energy threshold in electron volts, and τ the life span of a
state in seconds. Since a delivery of an energy
quantum of size ΔE is always associated with an
electromagnetic radiation of wave length λ, this
quantity is given along the middle scale in Ang-
strom units, the visible spectrum being represented by the heavy bar (4000 A to 8000A).

The numerical evaluation is now particularly
simple since there are essentially only two values
with small spread for τ_0, the intrinsic oscillation
period, to be considered. One is of the order of
3.10^{-15} seconds (31) and is associated with electron orbits within the crystal. Life spans that
are controlled by this time constant are those
of configurational change. The energy amount necessary to accomplish configurational change one can
calculate from the amount of kinetic energy per mole
that molecules must acquire before they can react.
This amount is well-established for proteins and
enzymes--it is the μ-value of the Arhenius equation
of reactions--and is found to be in the vicinity of
28,000 calories (32). Changing these thermal units
into electrical units we obtain a ΔE of about 1.1

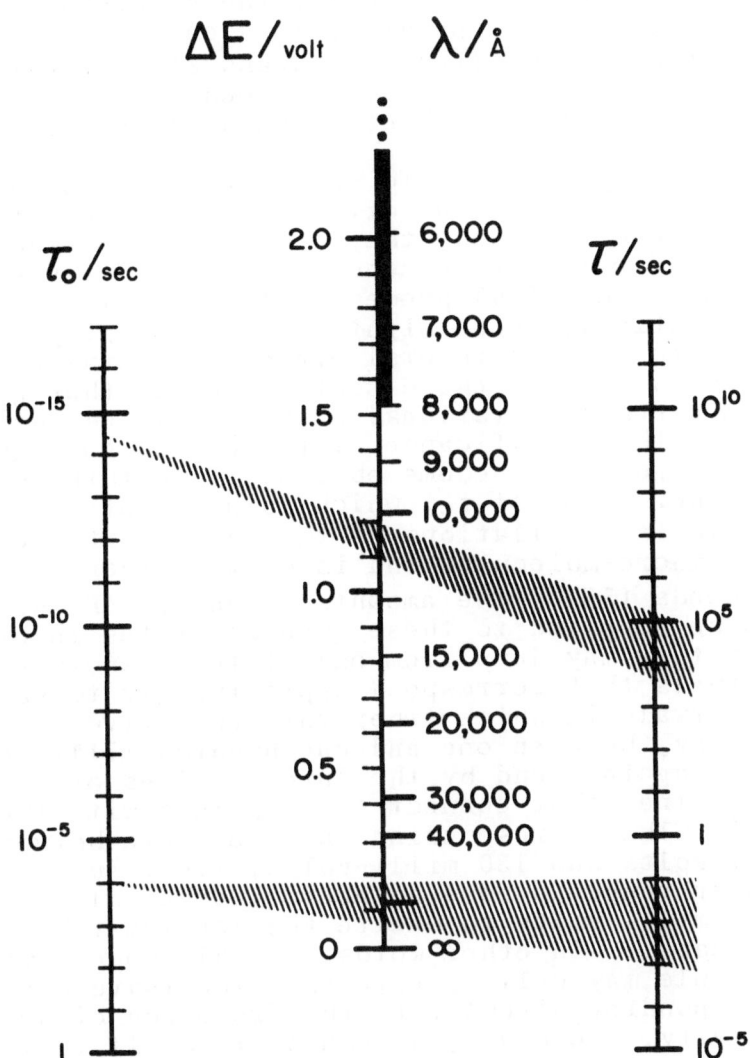

Fig. 10. Nomogram for τ, τ_0, ΔE, representing the

$$\tau = \tau_0 e^{\frac{\Delta E}{kT}}$$

for a fixed temperature of T = 310^0 K.

and 1.2 electron volts. Drawing the straight lines that connect the appropriate values on the τ_0-scale and the ΔE-scale, we find life spans of configurational changes on the τ-scale between 10^4 and 10^5 seconds, i.e., between about three hours and one day.

Apparently, these life spans are, on the one hand, too long to make an effective recursive element, on the other hand, they seem to be too short to account for long term memory traces. However, if one admits chemical processes to participate in these operations, these might be just the proper intervals to compute recursively over an arbitrary long stretch of time those configurations that give a neuron certain operational properties. Be that as it may, the significance of these slow configurational changes will become obvious if we turn now to the other value of τ_0, which is associated with the intrinsic oscillations of the lattice structure of these macro-molecules and is of the order of 10^{-4} seconds (33). The amounts of energy ΔE that have to be supplied to these quantum states in order that they may jump from one state to another at intervals that correspond approximately to various intervals in which fiber volleys arrive at a neuron, say, between one and one hundred milli-seconds, are again found by the intersections of straight lines that connect these points with the ΔE-scale. The corresponding ΔE values are between 50 milli-volts and 180 milli-volts, i.e., just in the proper range to have an action potential of about 80 milli-volts to excite the lattice vibrational states. In other words, in this mode a macro-molecule may well operate as a recursive element, responding directly to the frequency of neural activity. Moreover, as can be read off from the nomogram if a train of more than about 15 volleys of 80 milli-volts each and each volley following the other at intervals not longer than about 3 milli-seconds act on the molecule, then it will not have the time to go into lower energy states and will be "pumped" up into an energy level of about

1.2 volts which corresponds to levels in which configurational changes take place.

Now the game of recursion can be played including configurational changes whose relatively long
life spans allow us to make an almost unlimited
number of working hypotheses where only our imagination seems to be the limit.

I have presented this conjecture of molecular
computation only to suggest that there are avenues
open that point to a participation of molecules in
the grand spectacle of mentation in which they play
a dynamic rather than a static role.

ACKNOWLEDGEMENTS

Some of the ideas and results presented in this
paper grew out of work jointly sponsored by the Office of Education under Grant OEC-1-7-071213-4557,
by the Air Force Office of Scientific Research under Grant AF-OSR 7-68 and by the Air Force Systems
Engineering Group under Contract AF 33(615)-3890.

REFERENCES

1. Webster's New World Dictionary of the American
    Language.
2. Von Foerster, H., "Memory with Record" in: The
    Anatomy of Memory, D.P. Kimble (ed.), Science
    and Behavior Books, Palo Alto, pp. 388-433,
    1965.
3   Von Foerster, H., "Computation in Neural Nets",
    Currents Mod. Biol., 1, 47-93, 1967.
4. Von Foerster, H., "Time and Memory" in: Inter-
    disciplinary Perspectives of Time, R. Fisher
    (ed.), New York Academy of Sciences, New York,
    pp. 866-873, 1967.
5. Newell, A. and H.A. Simon, "The Logic Theory
    Machine" in: I.R.E. Transaction on Information
    Theory IT-2, pp. 61-79, 1956.
6. Minsky, M., "Steps toward Artificial Intelli-
    gence" in Proc. I.R.E. 49, pp. 8-30, 1961.

7.   Lindsay, R.K., Inferential Memory as the
     Basis of Machines which Understand Natural
     Language. In: Computers and Thought, E. Feig-
     enbaum and J. Feldman (eds.), McGraw-Hill,
     New York, 217-233, 1963.

8.   Raphael, R., A Computer Program which Under-
     stands. In: Proc. AFIPS, F.J.C.C., 577-589,
     1964.

9.   Von Foerster, H. and R.T. Chien, Cognitive
     Memory, Coordinated Science Laboratory, Univ-
     ersity of Illinois, Urbana, 1967.

10. Weston, P., Cylinders: A Data Structure Con-
     cept Based on a Novel Use of Rings. In: Accom-
     plishment Summary 1968, BCL Report 68.2, Bio-
     logical Computer Laboratory, Department of
     Electrical Engineering, University of Illinois,
     Urbana, 42-61, 1968.

11. Langer, S.K., Philosophy in a New Key, New
     American Library, New York, 1951.

12. McCulloch, W.S., Embodiments of Mind, M.I.T.
     Press, Cambridge, 1965.

13. Bar-Hillel, Y., Semantic Information and Its
     Measures. In: Cybernetics: Transactions of the
     Tenth Conference, H. Von Foerster, Margaret
     Mead, and H.L. Teuber (eds.), Josia Macy, Jr.
     Foundation, New York, 33-48, 1955.

14. Langer, S.K., op. cit., 30.

15. Konorski, J., The Role of Central Factors in
     Differentiation. In: Information Processing in
     the Nervous System, R.W. Gerard and J.W. Duyff
     (eds.), Excerpta Medica Foundation, Amsterdam,
     3, 318-329, 1962.

16. Young, J.Z., The Organization of a Memory
     System. In: Proceedings of the Royal Society,
     B, 163, The Croonian Lecture, 285-320, 1965

17. Eccles, J.C., M. Ito, and J. Szentagothai, The Cerebellum as a Neuronal Machine, Springer-Verlag, New York, 1967.

18. Lettvin, J.Y., H.R. Maturana, W.S. McCulloch, and W. Pitts, What the Frog's Eye tells the Frog's Brain. In: Proc. I.R.E., 47, 1940-1951, 1959.

19. Maturana, H.R., Functional Organization of the Pigeon Retina In: Information Processing in the Nervous System, R.W. Gerard and J.W. Duyff (eds.), Excerpta Medica Foundation, Amsterdam, 3, 170-178, 1962.

20. Von Foerster, H., Structural Models of Functional Interactions in Information Processing in the Nervous System, R.W. Gerard and J. W. Duyff (eds.), Excerpta Medica Foundation, Amsterdam, 3, 370-383, 1962.

21. von Neumann, J., The Theory of Automata: Construction, Reproduction and Homogeneity In: John von Neumann's Collected Works, A. Burks (ed.), University of Illinois Press, Urbana, 1964.22.

22. Löfgren, L., Kinematic and Tessellation Models of Self-Repair, TR 8, Contract NONR 1834(21) Electrical Engineering Research Laboratory, Engineering Experiment Station, University of Illinois, Urbana, 61, 1961.

23. Weston, P. and H. Tuttle, Data Structures for Computations within Networks of Relations. In: Accomplishment Summary 1967, BCL Report 67.2 Biological Computer Laboratory, Department of Electrical Engineering, University of Illinois, Urbana, 35-37, 1967.

24. See Reference 17, Figures 114, 115 and p. 311 ff.

25. Hydén, H., Activation of Nuclear RNA of Neurons and Glia in Learning. In: The Anatomy of Memory,

D.P. Kimble (ed.), Science and Behavior Books, Inc., Palo Alto, 1, 178-239, 1965.

26. Hydén, H., Studies on Learning and Memory. In: this volume.

27. Von Foerster, H., Das Gedachtnis; Eine Quanten-mechanische Untersuchung, F. Deuticke, Vienna, 40, 1948.

28. Von Foerster, H.,"Quantum Mechanical Theory of Memory. In: Cybernetics: Transactions of the the Sixth Conference, H. Von Foerster (ed.), Josiah Macy Jr. Foundation, New York, 112-145, 1949.

29. Von Foerster, H., Molecular Bionics. In: 1963 Bionics Symposium: Information Processing by Living Organisms and Machines, H.L. Oestreicher (ed.), Aerospace Medical Division, Wright-Pat-terson AFB, Ohio, 161-190, 1964.

30. Pattee, H.H., On the Origin of Macro-molecular Sequences, Biophys. J., 1, 683-710, 1961.
31. Schrodinger, E., What Is Life?, University Press, Cambridge, 1945.
Press, Cambridge, 1945.

32. Hoagland, H., Consciousness and the Chemistry of Time. In: Problems of Consciousness, H.A. Abramson (ed.), Josiah Macy Jr. Foundation, New York, 164-200, 1951.

33. Landau, L.D. and E.M. Lifshitz, Statistical Physics, Pergamon Press, London, 1958.

THE PHYSIOLOGY OF REMEMBERING

Karl H. Pribram

To those of us working at the brain-behavior
interface, and I am tempted to say to the scientific community as a whole, the overriding issue is
the one posed by Descartes at the beginning of the
Age of Reason. Are mind and body to be conceived
separately and thus studied by irreconcilably independent methods, or, can we come to understand
and overcome the implied dualism of the Nature of
Man? This issue is not an ethereal one--it is of
practical importance in the everyday course of man's
existence. We have seen that all the pills in the
world and the life-saving devices of medicine go
for naught if the minds of men do not keep pace
with the technical developments they can devise.

But my prediction is an optimistic one: By
the end of the twentieth century a fresh wind will
have swept away the remaining cobwebs that continue
today to divide the creative community.

Humanists and scientists will soon recognize
that despite the now obvious differences, their

methods as well as their aims display remarkable
similarities. The technique of the journalist in
making notes on the human scene is not altogether
different from that of the ethologist observing the
animal community, or of the pharmacologist charting
the effects of a drug on the nervous system or on
behavior. The re-enactment performed by the novel-
ist is not that foreign in approach to the re-enactment of the computer scientist simulating behavioral
processes or for that matter, to the in vitro re-enactment of life processes by the biochemist. Observation, analysis, reconstruction, test to meet an
esthetic[2] criterion--these are the procedures used
by creative men whether they be humanists or scientists because this is the way their brains work.

It is therefore to an understanding of man's
brain--the organ that regulates human affairs--that
we must look for this new view of man. And the
brain sciences at the present rate of progress are
not going to disappoint us. Already enough is known
which was not known only a quarter of a century ago
to mark clearly the outlines of the new view.

Let me take as my focus the study of memory as
so many others have done at this conference. Undoubtedly one of the prime functions of brain is to
allow us to re-experience and act on past and therefore future events. Yet only two decades ago the
foremost research scientist in the brain-behavior
field was able to state that what we then knew about
the brain made it inconceivable that either animals
or man could learn at all.

The winds of change began to blow less than
ten years ago. I predicted then that the 1960's
would be a decade of decision for the study of memory (22). And so it has turned out--old techniques
have been revived, refined and applied with new

---

[2]The esthetic criteria for "pure" science are truth,
evidence and elegance; those for the "purer" arts
are coherence, plausibility and impact. So differences there are, of course, but the human endeavor
is not always pure nor need it be.

vigor; new methods, especially those of information processing have been developed and focussed on the problem. Behavioral scientists have made especial progress in the analysis of verbal memory which has led to a beginning in an understanding of the processes of recognition and recall (11, 17). Biological scientists have, for the most part, addressed themselves, and successfully, to the issue of storage. Thus a major area of investigating memory has devolved on showing that biochemical and histological changes can come about in brain tissue as a function of experience; on showing how long such changes last; and, by imaginative use of the phenomenon of retrograde amnesia, how rapidly or slowly storage occurs (16)

But I want now to turn to another equally fascinating area, one explored in my own research--the question of what sorts of organizations of memory processes must exist in the brain to allow remembering to take place.

I use the word remembering advisedly because it gives a clue to what I consider to be the central issue in this area of research: literally, remembering or recollecting refers to the assembling of dismembered mnemic events. I propose, therefore, first to give the evidence for believing that mnemic events are distributed in the brain and then to describe experiments which tell us something about the way in which memory becomes usefully organized.

A Neural "Image"--The Spatial Distribution of Input Information

A recurrent puzzle in the study of brain has been the fact that a large portion of the input systems can be destroyed without impairing the organism's ability to make responses to highly organized patterned aspects of its environment. This observation is made daily in the neurological clinic and in the laboratory has been demonstrated in a variety of ways. Thus Lashley removed 80-90% of the striate cortex of rats without impairing

their ability to discriminate patterns (14). Galambos has cut up to 98% of the optic tracts of cats
and these animals could still perform skillfully
on tests necessitating the differentiation of
highly similar figures (18). Not only removals
but a variety of other methods for disturbing the
presumed organization of the input systems have
been tried to no avail: Sperry surgically cross-
hatched a sensory receiving area and even placed
mica strips into the brain troughs produced in order to electrically insulate small squares of tissue from one another (28). Lashley, Chow and Semmes
tried to short-circuit the electrical activity of
the brain by placing strips of gold foil over the
receiving areas (15). And I have produced multiple
punctate foci of epileptiform discharge within a
receiving area of the cortex by injecting minute
amounts of aluminum hydroxide cream (13). Such
multiple foci, although they markedly retarded the
learning of a pattern discrimination, do not interfere with its execution once it has been learned
(whether learning occurred before or after the multiple lesions are made).

These experiments have been interpreted to
show that each sensory system functions with a good
deal of reserve. Since it seems to make little

Fig. 1. Averaged recordings of electrical activity
obtained from the occipital cortex of monkeys performing a differential discrimination: circle as
opposed to vertical stripes. A standard 500 msec
of activity is represented in each trace; the amplitude represented is variable, however, and depends
on how many signals were averaged in order to make
the record; for example, many more signals were
obtained when the monkey made a correct response
than when he made an error during criterion performance. The records under STIM are the wave forms
evoked by a display lasting 1 msec; the records under RESP were generated just prior to the response;

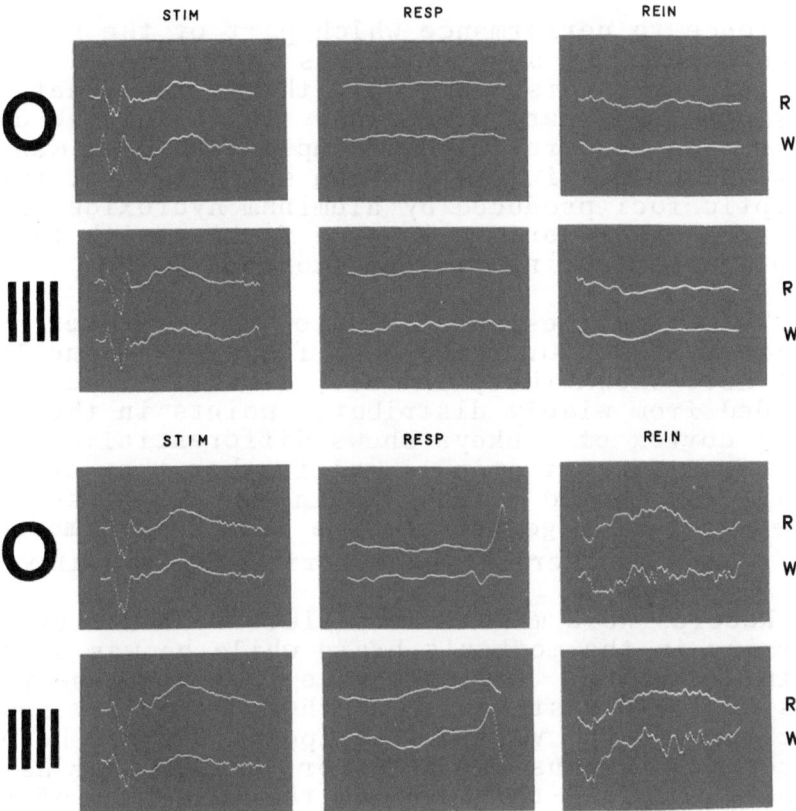

the records under REIN were generated after the response and during the period when reinforcing events
occurred. The upper six panels were made from records obtained while the monkey was performing at
chance; the lower six panels were made from records
obtained after the monkey attained an 85 percent
criterion (200 consecutive trials). The records in
line with R were made when the monkey performed correctly; those in line with W were made when the monkey was wrong. The waves generated just prior to
response (the intention waves) are similar whenever
the monkey is about to press the right half of the
panel, regardless of whether this is for the circle
or vertical stripes, and regardless of whether this
response proves to be correct or wrong.

difference to performance which part of the system
is destroyed, the suggestion has been offered that
this reserve is distributed in the system--that
the stored information necessary to making the discrimination is paralleled, reduplicated over many
locations. This interpretation suggests that the
epileptic foci produced by aluminum hydroxide cream
implantations interfere in some fashion with this
reduplication of information storage.

The correctness of this view has now been put
to direct test. Over the past few years Spinelli
and I have shown that, in fact, electrical activity
recorded from widely distributed points in the
visual cortex of monkeys shows differential patterns to different stimuli and further that other
widely distributed points within the system show
evidence of storage of response linked information
(26). Let me describe the experiment more fully.

Records were made of the electrical activity
occurring in the monkey's brain while he was solving this problem. From the wave form of these records we could distinguish whether the monkey saw
the circle or the vertical stripes; whether he made
the correct response or an error; and whether he
intended to press the right or the left half of the
panel once he knew the problem. All of these differential electrical responses occurred in the visual cortex (that part of the brain which also receives the visual input) though different electrodes recorded different events. Apparently experience and current input converge in the input
system (Figure 1).

Thus there is now, in addition to indirect,
some direct evidence that signals become distributed
within the input systems.[3] Further, this evidence

---

[3]The data presented describe the spatial distribution of input events. There is good reason to
believe that these events become temporally distributed as well. Evidence to this effect is also
beginning to accumulate (1).

shows that even within the input systems these signals reflect not only the occurrence of sensory
events but also provide response linked information: the intention to press one or another panel
and the outcome of this action.

What this amounts to is the production of a
neural representation initiated by the patterns
which excite the receptor--i.e. a neural "image"
triggered by an external world. This internal representation is, however, not a direct replica, or
as the cognitive psychologists are wont to call it,
an "iconic image".

Even the initiating events are expressions of
relationships which obtain between the effects of
excitation at one receptor point with that of its
neighbors. These relationships can be described
mathematically as convolutions. Thus, for instance,
the shape of the visual receptive field of a re-
inal ganglion cell represents the convolution of a
derivative of the shape of the retinal image produced at that point (27). Mathematical expressions of the type involved in these transformations
are called holographic because they are used to
make holograms, photographs of the interference
patterns produced when coherent light is split
to form a reference and a beam reflected from the
object to be imaged. By analogy, the neural image
is likely to be holographic in nature.

The question remains as to how interference
effects can be produced in the brain. Synaptic
events consequent on the arrival of nerve impulses
form wave fronts (9). Such arrival patterns can
interact with others and with wave forms produced
by the spontaneous potential changes which occur
in neural tissue. Immediate cross correlations
result and these can be the occasion for the generation of new spatial and temporal patterns of
nerve impulses. The assumption made here is that
the totality of this process can be conceived as
a neural "hologram", an "image" which is perceived.

Much work needs to be done to establish the

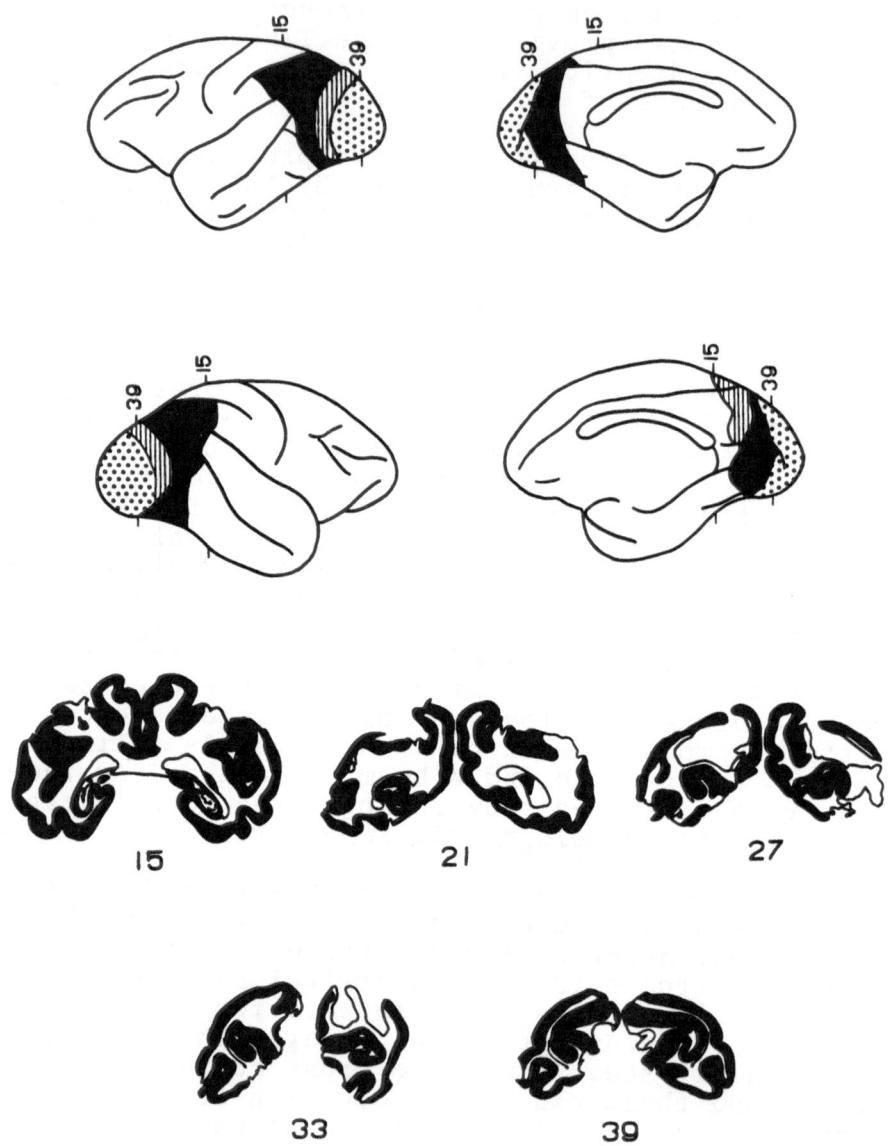

Fig. 2.   Disconnection of striate and infratemporal
cortex by extensive prestriate ablation.

limits over which this view holds. Do the mathematical expressions which interpret the shape of
visual receptive fields at the ganglion cell layer of the retina give equally useful interpretations at more central stations in the visual system? What sort of neural reference mechanism
corresponds to the coherent source in the holo-
gram--is it anatomical or is there a physiological spatial or temporal rhythm of neural firing
which serves this function? How is the effect of
the holographic cross correlation stored? By a
tuning of cell-assemblies or by a change in protein conformation, by both or some as-yet-undis-
closed possibility? These questions can be posed
in the laboratory with techniques available today.
Were there no other reason, this alone would make
the model of a neural holographic process an exciting one to pursue.

The Primate Sensory Specific Intrinsic Cortex

     This evidence for distribution of mnemic pat-
terns demands of remembering some sort of organizing process. Experimental data have accumulated
which make it likely that this process involves
the functions of the so-called association cortex
of primates. These regions of the brain are not
to be confused with the polysensory association
cortex which immediately surrounds the projection
areas which has been studied so extensively by
electrophysiologists in cats. Rather the primate
areas in question are located within the parieto-
temporal preoccipital convexity, somewhat remote
from the projection terminals. Further, they are
sensory specific. Sensory specificity was established by making lesions of various sizes in various locations and testing the ability to learn and
to retain discriminations in various modalities
(20. 23).

     These studies showed the parieto-occipital
area to be concerned in somesthetic choices (19,
33); the anterior temporal cortex to have something to do with taste (2, 24); the middle temporal region to be involved in auditory discrimina-

tions (8, 32); and the inferior temporal convolution important to vision (3, 6). By now these findings have been well established by a series of experiments involving over 950 monkeys.

The initial problem posed by these results can be stated as follows: Why can't, e.g. inferotemporally lesioned monkeys remember a + vs - or a 0 vs 1111 after surgery, when they do 90% or better before? This problem becomes compounded in the light of anatomical and physiological evidence that e.g. the inferotemporal cortex is afferently connected with the visual system only indirectly-- and that its function is not dependent on this input from the primary visual system.

For instance, removals of prestriate cortex which totally circumsect the primary visual cortex leave monkeys still able to perform visual discriminations at the 85% level when tested for the first time postoperatively (Figure 2).

What errors are made are within a field deficit produced by invasion of optic radiations as shown by subsequent histological analysis. Further, there is no effect on discrimination from cross-hatching the inferior temporal convolution, whereas undercutting this cortex produces a marked effect (25).

How the Brain Controls Its Input

To explain findings such as these I suggested some years ago the possibility that the inferotemporal cortex works through efferents to control-- organize--the input within the primary visual system. This possibility was entertained on the basis of neurobehavioral evidence (21) which showed that this cortex functioned to allow a rich sampling of the visual world to occur and these data have since been supplemented by further experiment (4,5).

Obtaining direct physiological evidence that this efferent control actually takes place has occupied my colleagues and me for the past years.

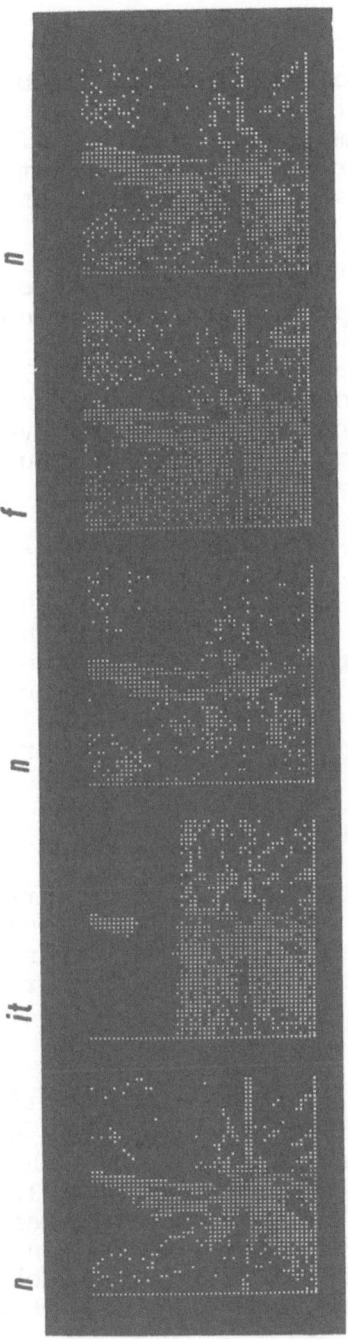

Fig. 3. Effects of stimulation of the posterior "association" cortex of a cat on a visual receptive field recorded from a neural unit in the optic tract. (These records are made by moving a spot with an X-Y plotter controlled by a small general purpose computer--PDP-8--which also records the number of impulses emitted by the unit at every location of the spot. The record shown is a section parallel to and 2 SD above the background firing level of the unit. Note the dramatic change in the configuration of the receptive field, especially after stimulation of the posterior "association" cortex--IT, inferotemporal.)

The following results have been obtained:

1. Changes in shape of and size of visual receptive fields of units in optic nerve can be produced by electrical stimulation of the inferotemporal cortex. Thus there is evidence that an effect can be registered as far peripheral as the retina (Figure 3).

2. Other evidence of the existence of efferent control as far peripheral as the retina has been obtained (29, 30, 31).

3. Changes are produced by electrical stimulation of the inferotemporal cortex in recovery functions of visually evoked responses. (Figures 4, 5).

Finally, the pathways over which this efferent control is exerted are being delineated. One outflow is to the superior colliculus. Another, much to our surprise, is a sizeable one to the putamen, one of the basal ganglia. Less dramatic are connections to the anterior commisure; to the n. medialis dorsalis and pulvinar of the thalamus; to the pons; and to other parts of the temporal lobe.

Receptor Control by Central Motor Structures

The meaning for behavior of these results can be best illustrated by the effect of inferotemporal lesions on the recording of stimulus, intention and reinforcement waves recorded from the visual cortex, as described in the first experiment discussed.

I fully expected that an inferior temporal lesion would selectively affect one of these wave forms only, leaving the others unchanged. Thus, we would have been able to say whether stimulus differentiation or some response-linked process were primarily involved. This did not happen. Instead, those electrodes from which we had obtained the best differential recordings now show-

ed no such differences and other electrodes which
had been relatively unimpressive showed some re-
liable differences. However, these differences
were now mostly response-linked in a peculiar
fashion which we have not as yet been able to de-
cipher clearly. It seemed as if the whole "frame
of reference" within which the brain activity had
been working prior to the lesion was now shifted--
and, in fact, shifting from time-to-time. Be-
haviorally, the monkeys appeared as surprised as
we by the effects of the surgery: they went into
the task situation fully confident of their ability, only to find that they made errors. This resulted in spurts of performance, hesitations, and
much trial and error to test out what had been
changed in the situation. It seemed as if they
were completely baffled for a while--not realiz-
ing, of course, as we did, that it was the in-
sides of their heads, their brains, we had altered
and not the situation.

The inferior temporal visual "association"
cortex (and as already noted, parallel findings
are available in the auditory system) thus seems
to work via structures to which motor functions
have usually been attributed. But recently, it
has become evident that even these very motor
functions are largely controlled via efferents to
their receptors--the gamma fibers to the muscle
spindles. By changing the bias on the spindle,
movement is induced. So perhaps it is not too
surprising if structures such as the corpora qua-
drigemina (colliculi) and the basal ganglia should
be shown to work via changes they effect in biasing input.

In summary, I have presented evidence that
information becomes distributed in the sensory
projection systems and that the functions of the
so-called posterior association areas of the primate are to organize input so as to make it memorable. This organizing process does not appear
to occur within these "association" areas per se,
however. Rather, their effect is exercised down-
stream via structures heretofore labelled as mo-

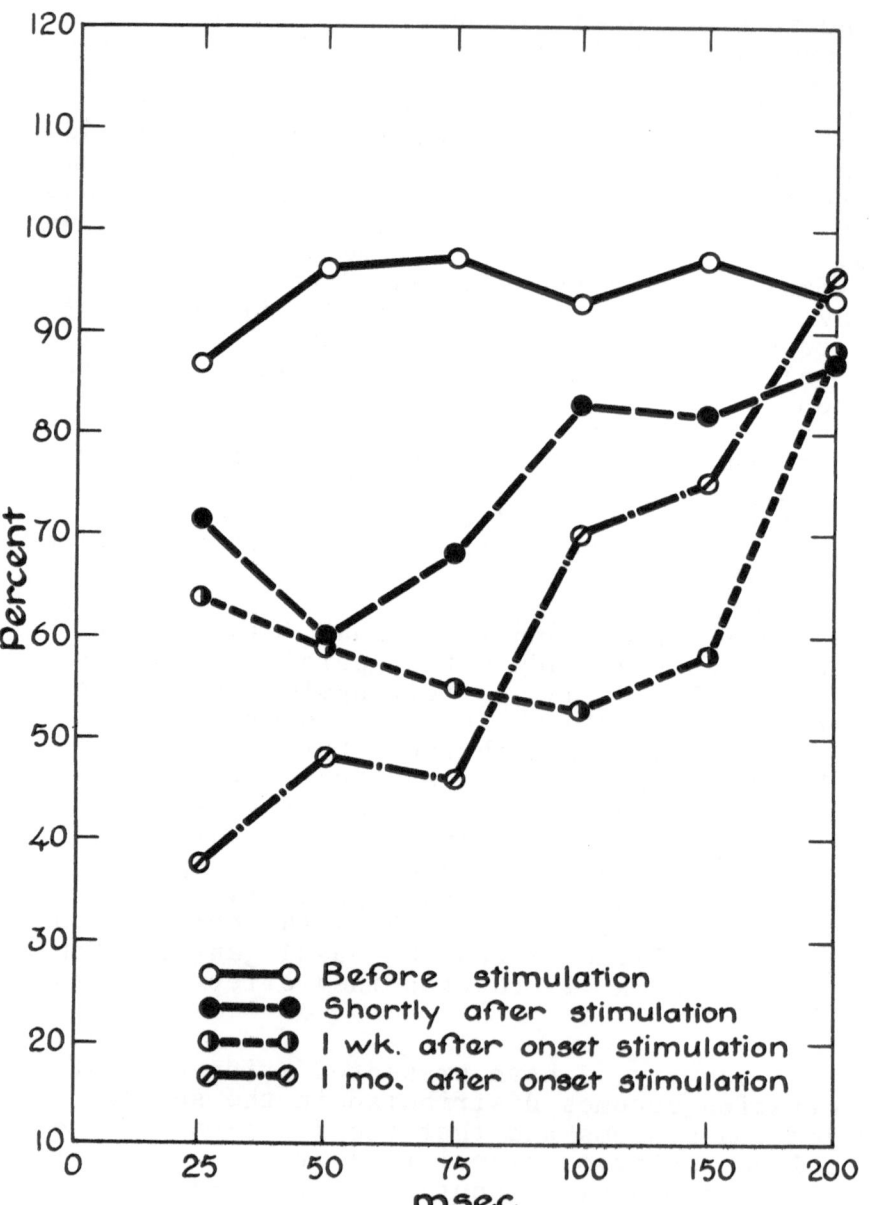

Fig. 4.   A plot of the recovery functions obtained
in one monkey before and during chronic stimulation
of the inferotemporal (IT) cortex.

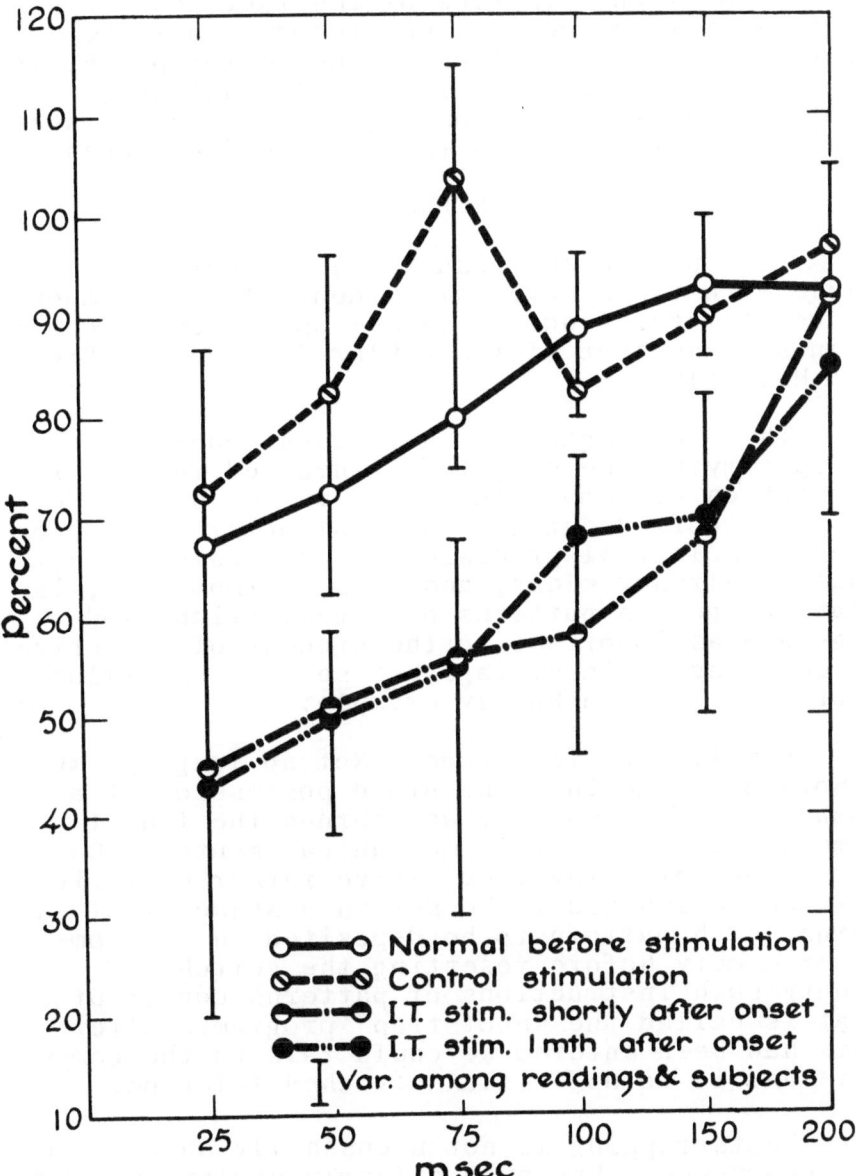

Fig. 5. A plot of the recovery functions obtained in 5 monkeys before and during chronic cortical stimulation.

tor in function. In view of the fact that such
control over movement is now recognized to be
largely due to regulation of muscle receptors, it
is perhaps not altogether unlikely that other receptor mechanisms such as the auditory and visual
are also controlled via these "motor" pathways.

CONCLUSIONS

This unforeseen result is, of course, compatible with the behavioral evidence for the importance of motor function in perception and recognition such as that of Ivo Kohler (12) and of Richard Held (10).

Thus the beginnings have been made--at least
in specifying the structures involved in the organization of memory functions in the brain. Next
we need to find out how these structures produce
the psychological processes we call remembering.
What is already clear, and this is important, is
that changes in patterns of organization--cod-
ing--are as important to the process of effective
remembering as is storage per se. Let me illustrate this with a homely example:

First, what is a code? Not so long ago my
laboratory came into the proud possession of a
computer. Very quickly we learned the fun of
communicating with this mechanical mentor. Our
first encounter involved twelve rather mysterious
switches which had to be set in a sequence of patterns, each pattern to be deposited in the computer memory before resetting the switches.
Twenty such instructions or patterns constituted
what is called the "bootstrap" program. After
this had been entered we could talk to the compu-
ter--and it to us--via an attached teletype.

Bootstrapping is not necessarily an occasion-
al occurrence. Whenever a fairly serious mistake
is made--and mistakes were made often at the begin-
ning--the computer's memory is disrupted and we
must start anew by bootstrapping.

Imagine setting a dozen switches twenty times and repeating the process from the beginning every time an error is committed.

```
U U D D U U U D U D U D

U U U U U U U U U D D D

U U U U D U D U D U U U

U U D D U U U U U U U U

U U D D U U U D U U D D

D D U D D D U D D U U U

U D U D D U D U U D U D

U D U D U U U U U U D U

D U D D U U U U U U U U

D U U U U U U U D D U U
```

And so on.

Imagine our annoyance when the bootstrap didn't work because perhaps on setting the 19th instruction an error was made in setting the eighth switch. Obviously, this was no way to proceed.

Computer programmers had early faced this problem and solved it simply. Conceptually, the twelve switches were divided into four triads and each combination of up down within each triad given an Arabic numeral. Thus

D D D became 0

D D U became 1

D U D became 2

D U U became 3

U D D became 4

U D U became 5

U U D became 6

U U U became 7

Conceptually, switching the first toggle on the
right becomes a one, the next left becomes a 2,
the next after this a 4 (and the next an 8 if
more than a triad of switches had been necessary,
i.e. if for instance our computer had come with
sixteen switches we should have conceptually divided the array into quads). Thus the bootstrapping program now consisted of a sequence of twenty patterns of four Arabic numerals:

e.g. 3 7 2 2

0 0 1 4

3 4 5 6

2 2 1 5

1 0 3 7

etc.

and we were surprised at how quickly those who
bootstrapped repeatedly, actually came to know
the program by heart. Certainly fewer errors
were made in depositing the necessary configura-
tions--the entire process was speeded and became,
in most cases, rapidly routine and habitual.

Once the computer is bootstrapped it can be
talked to in simple alphabetical terms: e.g. JMP
for jump, CLA for clear the accumulator, TAD for
add, etc. But each of these mnemonic symbols
merely stands for a configuration of switches.
In fact, in the computer handbook the arrangement
for each mnemonic is given in Arabic notation:
e.g. CLA = 7200. This in turn, is easily trans-

lated into U U U D U D D D D D D D should we be
forced to set the switches by hand because the
teletype has gone out of commission.

Programming thus is found to be in the first
instance the art of devising codes, codes that
facilitate learning, remembering and reasoning.
The logic of a computer is primarily a code, a
set of signals or signs which allows ready manipulation. The power of a program lies in the
fact that it is a useful code. If you doubt
this, try next month to check your bank statement
against your record of expenditures and do it all
using Roman rather than Arabic numerals. Can you
imagine working out our national budget in the
Roman system?

Coding and recoding are thus found to be es
sential operations of the brain. Recoding maintains some sort of representational identity[4]
among event structures much as such identity ob-
tains between the sequences of openings and closings of switches in a computer and the programs
which provide these openings and closings. Codes
are languages, however, and various languages
have greater or lesser powers and efficiency of
control. The evidence I have presented suggests
that one of the principal functions of the brain
is to construct languages although these languages have, by contrast with those used in computer science, built into them a good deal of ambiguity. From this ambiguity, of course, stem
the opportunities for construction of creative
novelties. If indeed it turns out that such languages are at least in part hierarchically related as there is good reason to suspect, we would
be a long way toward understanding in depth the
mind-brain issue. For it is through action that
these brain codes become externalized representations, e.g. as culture. Thus, it is turning out

---

[4]And hereby hangs the crucial problem: just how
isomorphic is this identity and by what means is it
achieved when isomorphism does not hold?

to be our understanding of coding, the language
of the brain, through which unity is affected in
the nature of man.

REFERENCES

1. Bagshaw, M. M. and Coppock, H. W., Galvanic
    skin response conditioning deficit in amyg-
    dalectomized monkeys, Exptl. Neurol., 20,
    188-196, 1968.
2. Bagshaw, M. M. and Pribram, K. H., Cortical
    organization in gustation (macaca mulatta), J.
    Neurophysiol., 16, 499-508, 1953.
3. Blum, J. S., Chow, K. L. and Pribram, K. H.,
    A behavioral analysis of the organization of
    the parieto-temporo-preoccipital cortex, J.
    Comp. Neurol., 93, 53-100, 1950.
4. Butter, C. M., The effect of discrimination
    training on patter equivalence in monkeys
    with inferotemporal and lateral striate les-
    ions, Neuropsychologia, 6, 27-40, 1968.
5. Butter, C. M., Mishkin, M. and Rosvold, H. E.,
    Stimulus generalization in monkeys with in-
    ferotemporal and lateral occipital lesions,
    D. I. Mostofsky (ed.), Stimulus Generalization,
    Stanford, Stanford University Press, 119-133,
    1965.
6. Chow, K. L., Effects of partial extirpations
    of the posterior association cortex on vis-
    ually mediated behavior, Comp. Psychol. Mono-
    gr., 20, 187-217, 1951.
7. Dewson, J. H. III, Noble, K. W. and Pribram,
    K. H., Corticofugal influence at cochlear nu-
    cleus of the cat: some effects of ablation
    of insular-temporal cortex, Brain Research,
    2, 151-159. 1966.
8. Dewson, J. H. III, Pribram, K. H. and Lynch,
    J., Effect of midtemporal lesions on auditory
    discriminations in monkeys. In preparation.
9. Eccles, J. C., The physiology of imagination,
    Scientific American, 199, 135-146, 1958.
10. Held, R., Experience and Capacity, D. P. Kim-
    ble (ed.) (Fourth Conference on Learning, Re-
    membering and Forgetting. New York, The New
    York Academy of Sciences. In press.

11. Kimble, D.P. (ed.), The Organization of Recall (Second Conference on Learning, Remembering and Forgetting). New York, The New York Academy of Sciences, 1967.

12. Kohler, I., The Foundation and Transformation of the Perceptual World, New York, International Universities Press, Inc., 1964.

13. Kraft, M.S., Obrist, W.D. and Pribram, K.H., The effect of irritative lesions of the striate cortex on learning of visual discriminations in monkeys, J.Comp. Physiol. Psychol., 53, 17-22, 1960.

14. Lashley, K.S., The problem of cerebral organization in vision, Biological Symposia, Vol. VII, Visual Mechanisms, Lancaster, Jacques Cattell Press, 301-322, 1942.

15. Lashley, K.S., Chow, K.L. and Semmes, J., An examination of the electrical field theory of cerebral integration, Psych. Rev., 58, 123-136, 1961.

16. McGaugh, J., The Anatomy of Memory, D.P. Kimble (ed.), First Conference on Learning, Remembering and Forgetting). Palo Alto, Science and Behavior Books, 1965.

17. Neisser, U., Cognitive Psychology, New York, Appleton-Century-Crofts, 1967.

18. Norton, T., Frommer, G. and Galambos, R., Effects of partial lesions of optic tract on visual discriminations in cats, Fed. Proc., 2168-1966.

19. Pribram, H. and Barry, J., Further behavioral analysis of the parieto-temporo-preoccipital cortex, J. Neurophysiol., 19, 99-106, 1956.

20. Pribram, K.H., Toward a science of neuropsychology: method and data, Current Trends in Psychology and the Behavioral Sciences, R.A. Patton (ed.), Pittsburgh, University of Pittsburgh Press, 115-142, 1954.

21. Pribram, K.H., The intrinsic systems of the forebrain, Handbook of Physiology, Neurophysiology, Vol. II, J. Field and H.W. Magoun (eds.), Washington, American Physiological Society, 1323-1344, 1960.

22. Pribram, K.H., Regional physiology of the
     central nervous system (the search for the
     engram-decade of decision), Progress in Neu-
     rology and Psychiatry (E.A. Spiegel, ed.),
     N.Y., Grune & Stratton, 45-57, 1961.
23. Pribram, K.H., Memory and the organization
     of attention, Brain Function, Vol. IV, D.B.
     Lindsley and A.A. Lumsdaine (eds.), Berkeley
     and Los Angeles, University of California
     Press, 79-122, 1967.
24. Pribram, K.H. and Bagshaw, M.H., Further
     analysis of the temporal lobe syndrome utili-
     zing fronto-temporal ablations, J. Comp. Neur.,
     99, 347-375, 1953.
25. Pribram, K.H., Blehert, S. Reitz and Spinelli,
     D.N., The effects on visual discrimination of
     crosshatching and undercutting the infero-
     temporal cortex of monkeys, J. Comp. Physiol.
     Psychol., 62, 358-364, 1966.
26. Pribram, K.H., Spinelli, D.N. and Kamback,
     M.C., Electrocortical correlates of stimulus
     response and reinforcement, Science, 157, 94-
     96, 1967.
27. Rodieck, R.W., Quantitative analysis of cat
     retinal ganglion cell response to visual stim-
     uli, Vision Research, 5, 583-601, 1965.
28. Sperry, R.W., Miner, N. and Meyers, R.E.,
     Visual pattern perception following subpial
     slicing and tantalum wire implantations in the
     visual cortex, J. Comp. Physiol. Psychol.,
     48, 50-58, 1955.
29. Spinelli, D.N., Pribram, K.H. and Weingarten,
     M., Centrifugal optic nerve responses evoked
     by auditory and somatic stimulation, Exptl.
     Neurol., 12, 303-319, 1965.
30. Spinelli, D.N. and Weingarten, M., Afferent
     and efferent activity in single units of the
     cat's optic nerve, Exptl. Neurol., 3, 347-
     361, 1966.
31. Weingarten, M. and Spinelli, D.N., Changes
     in retinal perceptive field organization with
     the presentation of auditory and somatic
     stimulation, Exptl. Neurol., 15, 363-376, 1966.
32. Weiskrantz, L. and Mishkin, M., Effect of
     temporal and frontal cortical lesions on audi-
     tory discri

tory discrimination in monkeys, Brain, 81, 406-414, 1968.

33.   Wilson, M., Effects of circumscribed corti-cal lesions upon somesthetic discrimination in the monkey, J. Comp. Physiol. Psychol., 50, 630-635, 1957.

IDEA, IMAGE AND MEMORY

DR. VON FOERSTER: I agree very much that a
clear distinction should be made between an idea
and memory, if memory is thought of as "storage".
I hope that I have been able to demonstrate in my
paper that there is indeed a genuine distinction
between a storage device and memory. The best one
can ever expect from a store is to get out what
one put in, while memory gives an <u>interpretation</u>
of that which came in. When one recalls a trip
to Europe one does not deliver to one's curious
listener the pattern of nerve impulses that came
through one's optic tract. One may say, perhaps:
"It was cold and rainy." This statement is clearly
an interpretation of one's experiences, and in this
case is given in a highly abstract form, namely in
form of symbols used to communicate this experience. When such a statement is uttered, the sound
pressure wave that reaches the listener's ear is
now, in turn, re-interpreted and regenerates "ima-
geries"--or whatever one may call these internal
representations--that are conforming to earlier
interpretation of experiences by the listener
that refer to "cold" or "rainy" without experi-
encing cold or being wet. This suggests to me
that those neural processes that generate concepts
like "cold" or "rainy" are in essence not too
different from those that generate ideas like
"Temperature" or "Climate". In other words, mem-
ory--as I see it--is the capacity to regenerate--
or recompute--interpretations of earlier situations. Of course, the same mechanism may be
used to compute situations not as yet experienced which may or may not be in the future.
Systems, neural or artificial, that perform such
interpretations are clearly fundamentally different from those from which one may retrieve
signals that were stored at an earlier occasion.

DEVELOPMENTAL AND CHEMICAL SPECIFICITY AND THE
COGNITIVE TILE

DR. BOGOCH: I would like to ask Dr. Von
Foerster whether he has any thoughts on the work
of Sperry and others on the developmental specificity that seems to be emerging in terms of the
connection of individual units within the nervous
system; and how this may relate to chemical specificity as it may be manifested later in life for
connections between these same units in learning
processes.

DR. VON FOERSTER: Yes, I am aware of Sperry's
work and of his fascinating results which link
chemical specificity with connectivity of neural
units. I do not see any difficulties to interpret
my tesselation model in those terms. In fact, my
functional unit, the "cognitive tile" has such
general properties that at this stage many interpretations could be given. This may be seen as an
advantage, but also as a disadvantage of my model.
As I confessed in my presentation I am not even
sure whether a single tile may represent a single
neuron or several neurons working together. And
similarly whether a single neuron may be represented by a single tile or an "elementary tesselation" of several tiles acting in concert. However,
there are no difficulties in introducing specific
constraints into a set of tiles such that they establish preferred connections amongst themselves
over those with others. In fact, one may let a
tile degenerate so far that it cannot refer to any
events in the past and act only as a transducer
of instantaneous signals. This will be immediately
the case when one cuts the recursive loop. However, the tranfer function of this transducer can
still be manipulated from the outside and these
manipulators may be chemical as well as electrical.

The important feature of this tesselation model is -- as I see it -- not so much that the general properties of a tile can be trimmed to correspond to certain neural function of which we know,
but rather that tesselations do the job in cases
where we do not know how in detail these functions

are accomplished.

SPATIAL AND TEMPORAL RELATIONSHIPS AND THE COGNI-
TIVE TILE

DR. FREED: I would like to ask about spatial
and temporal relationships in the interaction between these proposed cognitive tiles.

DR. VON FOERSTER: Spatial relationships in
a tesselation are most naturally realized by the
specificity of location of each tile. As I tried
to point out in my example of the binocular animal, self-reference enters essentially by reference to its own left eye and right eye. Similarly, signals origniating from various tiles never
lose the tag from where they come. In other
words, a tile responds to a spatio-temporal configuration in its "signal space" and may degenerate into a purely spatial animal, if for instance
the recursive loop is cut, or may turn into a purely temporal animal by selecting a function F which
does not distinguish between signals that arrive
over channel x_1 or x_2, that is $F(x_1,x_2) = F(x_2,x_1)$.

DR. MANDEL: I am very happy with Dr. Von
Foerster's conclusion.

We heard this morning several millions of synapses might be involved. It means that we will
have several millions of these macromolecules
which are able to work as computers in distances
which are, on the level of molecules, extremely
high. The question arises, how to bring together
the work of an extremely great number of macromolecules working at a great distance.

DR. VON FOERSTER: This is, I believe, a very
important question but, alas, it is also very difficult to answer. What you essentially ask is to
give you my ideas as to the sites of these molecular computers within the confines of a single
neuron. At present I can only answer this by constructing an even more complex conjecture out of
my elementary conjecture of computing molecules.
As I see it, the question of sites will be answered

differently for various types of neurons if one
makes the distinction I suggested earlier between
spatio-temporal neurons and their degenerated
forms, the purely spatial neuron and the purely
temporal neuron. Take for the moment a purely
spatial neuron. Since it has to conserve integrity as to whence the signal comes, and this is
in the neuron directly coupled with the location
of the synapse that receives the signal, it seems
to me that these molecular computers must be located near those synapses. Moreover, these molecules will operate in the mode of configurational
changes which is an affair involving hours, and
thus they will supply little or nothing of the
information that may arrive at the synapse in
terms of fast changing pulse patterns to the overall response of the neuron.

Take now a purely temporal neuron. Here
spatial information is completely disregarded,
that is it computes a function whose value is
independent of which one of its independent
variables supplies a certain amount of the overall
stimulus. A simple additive function, say $\sum x_i$,
will do this, since it is irrelevant in this
operation whether x_3 carries 12 units and x_9 carries 2 units, or reversely $x_3=2$ and $x_9=12$, for in
both cases the sum remains the same $x_3 + x_9 = 14$.
Here these computers may be located near the site
which initiates an action potential, that is near
the axonal region of the neuron. Hence, I would
predict that chemical variety in dendrites will
be found in neurons that preserve spatial information, which neurons operating on purely temporal
information will show chemical variety in the cell
body. Moreover, a temporal neuron with a large
number of afferents will have its large number of
molecular computers operate in the fast mode of
lattice vibration quantum jumps until signal over-
load may force them to go into the slow mode of
configurational changes. Temporal neurons with
few afferents seem to operate mostly in the fast
mode on temporal sequences in the range of several
seconds.

Spatio-temporal units will be a combination of these mechanisms. I am afraid that this is not an exact answer to your question; but does it give an indication?

DR. MANDEL: It gives an indication.

FEEDBACK LOOPS AND THE TESSELATION MODEL

DR. ECCLES: I was particularly interested in the Tesselation Model of Professor Von Foerster. I think that, when attempting to model the nervous system, it is important not to overlook the fact that the nervous system has an immense organized complexity.

I also liked the feedback loops which are put in so that you have circulating impulses, continuously feeding back information. There is already anatomical and physiological evidence that such loops are immensely developed in the nervous system. Not only do these feedback loops occur but also they involve much more wide-spreading pathways. You have heard already this morning that there is a feedback control to receptor organs from the central nervous system modifying the input from receptors. The central nervous system also can inhibit synaptic relays in the spinal cord, and this explains, of course, the many kinds of anesthesias that have a central so-called psychic origin.

An additional point is that feedback loops extend still further peripherally. For example in the control of movement there is feedback from a wide variety of relationships of the body not only one part to another within the body such as is given by muscle and joint receptors; but also the body in relationship to the external world, such as is provided by receptors of touch and pressure, of vision, of hearing and of labyrinthine sense. All of this diverse sensory information is continually being fed back during every movement, giving what may be called the dynamic loop

control of movement. All movements are as it were
provisional, and during their course are subject
to continual revision by these feedback loops.
These concepts are inherent or implicit in Dr.
Von Foerster's presentation, and of course he
would agree that we are just at the beginning of
understanding the operational complexity of these
postulated neuronal mechanisms.

SOME FUNCTIONAL SPECIFICATIONS WHICH MUST BE MET BY THE BIOCHEMICAL MECHANISMS WHICH STORE MEMORY

DR. JOHN: I would like to summarize some of
our neurophysiological observations which show
some of the characteristics of the mechanism which
is facilitated in McGaugh's experiments, is interfered with by Agranoff, or produces the manifestations that the other speakers have described.

If an animal learns that a stimulus has meaning, a number of different parts of the brain display similar electrical responses when that stimulus is subsequently presented (1). The previously
different electrical responses from certain anatomically different regions have become similar.
If an animal is differentially conditioned to make
one response to flicker at one frequency and a
different response to flicker at a second frequency
and a test stimulus is presented midway between
the two conditioned stimulus frequencies, the
animal will sometimes treat this test stimulus as
though it meant frequency one, and sometimes the
same physical event is responded to as though it
meant frequency two.

The shape of the evoked potential caused by
such a test stimulus has one form, when the animal
processes it as though it had one meaning, and
a different form when the same physical stimulus
is processed as though it had a different meaning.
This differential form is displayed in many different regions of the brain (2).

Using chronically implanted movable microelectrodes, we have studied the relationship between the evoked potential elicited when a stimulus produces generalization and multiple neuronal discharges.

The results show clearly that when a novel stimulus produces generalization from an animal, not only does the evoked potential to the new stimulus look like the usual response to the familiar stimulus (3), but the multiple unit histogram has exactly the same shape as when the usual conditioned stimulus for that behavioral response is present (4,5).

The anatomical distribution of these neuronal patterns was mapped by moving the microelectrodes very slowly (about 500 microns a week) through vast distances on the cellular scale. Some of these maps extend 6000 microns. Neurons responsive to the conditioned stimuli were found throughout these domains. Wherever responsive neurons were found, the pattern of response to stimulus one was characteristic for that stimulus. In those responsive neurons, stimulus two, the differential stimulus, elicited quite a different temporal pattern of firing. Throughout these enormous domains, the characteristic differential responses of these neuronal ensembles were quite constant. The number of responsive cells varied between regions. In all these ensembles, a particular signal elicited its characteristic pattern as long as the animal retrieved the stored information, operationally defined as performance of the correct conditioned response.

When performance fails to occur, a set of components disappears from both the evoked potential and the post-stimulus histogram. Those components usually occur at about 40 milliseconds, often following an inhibitory period which markedly diminishes neural discharge.

In all this work, we have only once found a single cell, studied for six days, which fired to stimulus one and didn't fire to stimulus two.

In these chronic studies, responsive cells usually respond to both conditioned stimuli. They responded with one pattern of firing to one signal, a different pattern of firing to the other signal. They did not respond reliably to any input and showed great variability to all inputs. Invariance was only observed in the statistical characteristics of the ensemble.

These various data indicate that conditioned stimuli elicit differential firing patterns from neurons which are distributed throughout extensive anatomical regions. Basically similar evoked potential waveshapes and neuronal firing patterns are displayed in different anatomical regions. Differences between stimuli seem to be represented in the different firing patterns which they elicit in responsive neurons rather than by the specific neurons which are caused to fire. Individual cells show highly variable and unreliable responses to conditioned stimuli, while the average firing patterns of neuronal ensembles reveal impressive constancy as long as appropriate behavior subsequently ensues. These average firing patterns are not solely determined by the physical effect of the present stimulus but partially reflect the release of information stored in memory, since they are precisely reproduced when novel test stimuli elicit behavioral generalization.

Such data make it unlikely that memory is stored in discrete connections or modifications of synaptic excitability which make this cell belong to the neuronal system representing a particular memory. No evidence has been found which supports the idea that the readout of a specific memory involves the firing of particular cells while others remain silent. Rather, our results indicate that the activation of a memory involves the release of a specific firing pattern in extensively distributed neuronal ensembles, in which invariant characteristics are achieved statistically by the average behavior of the ensemble in spite of the highly variable and unreliable behavior of individual cells.

If we are right in our interpretation of these data, we have described the functional specifications which must be met by the biochemical mechanisms which store memory. It might further our understanding of these processes to consider how a neural ensemble might acquire such properties.

(1) John, E.R., Mechanisms of Memory, 468 pp.,
     Academic Press, New York, 1967.
(2) John, E.R., Bartlett, F. and Shimokochi, M.,
     Electrophysiological correlates of the engram.
     Manuscript in preparation, 1969.
(3) Ruchkin, D.S. and John, E.R., Evoked potential
     correlates of generalization, Science, 153,
     209-211, 1966.
(4) John, E.R. and Morgades, P.P., Chronic micro-
     electrode studies of conditioned responses,
     Fed. Proc., 27, 277, 1966.
(5) John, E.R. and Morgades, P.P., Neural correl-
     ates of conditioned responses studied with
     multiple chronically implanted moving micro-
     electrodes, Exptl. Neurol., in press, 1969.

ON THE ROLE OF NUCLEIC ACIDS IN COMPUTER THEORIES

DR. GLASSMAN: One of the things that kept
coming into my mind as I sat here this morning is
that we are not taking into account the distinction
between the intraneuronal and the interneuronal
events in the nervous system. Interneuronal events
involve the reactions between neurons and lead to
the establishment of networks and circuits which
most of us feel are responsible for memory. The
intraneuronal events are part of this process but
are concerned mainly with whether a neuron will
participate in a particular network. The remarks
concerning DNA do not really apply to Dr. Von Foerster's "computer molecules" because their purpose is
to act like a switch to determine whether a nerve
cell will participate in a network and not directly with the memory itself. Am I incorrect?

DR. VON FOERSTER: No. Absolutely correct,
perfectly correct. Only that, you see, one has the
possibility of computing recursive functions, which
does away with the storage problem.

If you calculate Young's model of the possibility of storing all the pleasure action associations, then it will be necessary, if you make
brutal numerical calculations, that even if the
octopus would have had the brain of Professor Young,
you know, in order just to retract an arm, it
wouldn't be able to do that. Numerically, it's
just out. While with a recursive function calculator you could do this with the greatest ease.
You would need three neurons for that. It's funny
how powerful and efficient these fellows are.

But, you see, the moment you have this concept
of a recursive function computer, a relational signal that comes in and reports about relations,
about entities with respect to the system, there
is a triple relation. There is a relationship between, let's say, my eyes that are behind with
respect to the position of the observer. You must
have the relation with respect to the observer,
otherwise you will be forced to assume A is always

behind B, which is false because behind is always in relation to the system.

So you have incoming signals representing relations of entities to themselves with respect to the observer. Then you have the recursive operation which makes the function compute what may be the future states of affairs. Then they are corrected by the prior perceptive system or by just watching what it has done.

DR. GLASSMAN: Thus, in your concept of the computer molecule, the computer molecule is part of an event in the cell which merely decides whether a cell will join in these feedbacks and loops, and so on?

DR. VON FOERSTER: Yes.

DR. GLASSMAN: But the real memory is stored in the pathways and loops between cells and their relationships; is that true?

DR. VON FOERSTER: May I just insist, for argument's sake, that I wouldn't talk about storage at all. I would say there is a computer cell here which is doing some computations and functions, does predictions, just telling what's supposed to be done next and not only next and next, after next and next after next. It is only the question of the richness of the recursiveness which is involved. You can predict I don't know how long in the future with such things.

My computer molecule here was only to do the following thing. If you take one of these cognitive tiles as a fundamental entity, you may have the group of cells and you may interpret now the interaction between these molecules by loop actions, as indeed a recursive function, as a computer would indeed operate because it has to need the information of the preceding state or the preceding states. You can take a group of neurons to do that job.

Now, you have to answer the question: "Dear Dr. Von Foerster: Where is your recursive function operator?"

I say, well, if I am squeezed into that corner, I give you molecules which may do that. But I don't say, "Look here, it all works through the molecules. It works with everything."

If it can do it on the very local specific level, it will do it on the very local specific level. I think it's open to the cleverness of the system, however it's developed, either this way or that way. But I don't say it must be this. I leave it completely open because the model is so general.

DR. GLASSMAN: Let me just add this to further clarify this discussion. Dr. Sacks said that the short time sequences do not allow for the synthesis of molecules. But I would point out that in your model systhesis is not involved. These are molecules that are present all the time. So the events can occur as fast as you would like them to because you do not have to wait for synthesis to occur.

I want to make another comment on the amount of DNA required for this model. One does not have to postulate that there are millions of different kinds of "computer" molecules. You can have a very few of these involved in the process, and by just making permutations of ten or twenty of them, one could really make a lot of decisions about firing without having the problem of whether the DNA can code for them or not.

DR. WELSH: I have been fascinated by your computer model Dr. Von Foerster, and I would like to hear you and Dr. Eccles and Dr. Mandel put this together in such a way that we might know from Dr. Eccles' point of view, for instance, is there a reverberating system of neurons; from Dr. Mandel's system, could it be that the proteins that are being synthesized are themselves used in this system, or perhaps used to produce transmitter sub-

stances, and that it is these transmitter substances
that are released in split second timing which then
cause a neuron to either fire or not to fire. Possibly you might get all kinds of feedback loops in
which transmitter substances control the firing or
the non-firing of specific cells within the central
nervous system.

DR. ECCLES: I am not going to answer the
question directly but I will raise some oblique
suggestions.

There seems to be some misunderstanding about
the relationship of this highly specific DNA or
RNA store to the neuronal mechanisms of memory
coding. I believe that we do require the operation
of RNA in the protein synthesis involved in the
laying down of long-term changes in the nervous
system. I would postulate that there is a micro-
growth of one kind or another giving for example
bigger and better synapses and on the other hand
that there are atrophic changes.

The operation of the nervous system at any
one time is I think not dependent upon what immediately can be produced by these systems of RNA
and protein synthesis. But all the time behind
this ongoing activity there are the slower, more
plastic changes dependent on these chemical mechanisms, so that the system is always changing.
These plastic changes of neuronal connectivity form
the basis of storage and retrieval. Thus the system is always changing. These plastic changes of
neuronal connectivity form the basis of storage
and retrieval. Thus the system is changing by
virtue of the biochemical events and the coding
of the information at the chemical level. However
at any one instant it is responding in the specific
manner given by its already preformed connectivity,
and by the dynamic operation of impulse traffic
that derives from the immediately preceding inputs and that form the basis of a short-term memory. At any one instant there is an impulse
traffic deriving from all of the on-going immense
input and the past immediate history, forming the
so-called dynamic memories. But the patterned

operation of all these impulse discharges in the thousands of millions of neurones is of course dependent on the neuronal connectivity that in part at least is the basis of long-term memory.

DR. WELSH: Is the memory then in the connect-ions between neurons or is it in specific molecules such as biochemists have been looking for?

Could memory be in new synaptic connections between specific neurons or should biochemists continue to look for a memory molecule?

DR. ECCLES: I don't want to answer these questions directly!   I don't want to say what is memory.   In the first place it is something that we discover in ourselves or in animals under ex-perimental situations.   For example in response to particular sensory inputs we get experiences, conscious experiences, which link our immediate situations together in some way with the past ex-periences.   This involves mechanisms both of stor-age and of retrieval of information.

But memory is more complicated than any of those things because it also involves conscious-ness.   Memory involves a whole immense system of structure and dynamic operation that is dependent on the metabolic events of biochemistry, the micro-structural growths of neurogenesis and the pattern-ed operations of the impulse traffic in the organ-ized complexity of the brain.

ON SEARCHING FOR ENGRAMS

DR. GLASSMAN: Since all of the presentations of this meeting are devoted to the vertebrate brain, it should be pointed out that many of the problems associated with localizing the engram might be lessened if we look at some of the simple invertebrate preparations now under study. We do not seem to be able to deal in depth with the kind of information processing going on in the vertebrate brain, and I think that the exciting work going on in Aplesia and in the headless cockroach will help a great deal along these lines. In particular, if we could identify and focus on a synapse that was involved in learning, we might have a chance to discover the kinds of changes that occur when transmission across a synapse is facilitated, as is now required by many theories of learning and memory storage.

DR. PRIBRAM: I am afraid I agree somewhat with Dr. Eccles on the problem of finding the so-called engram. Not that it is impossible. But let's think about a computer for a moment and the problem of coding I was discussing earlier.

We purchase a new computer and are not very skilled at putting together an efficient program. So we write a very long program to do something rather simple.

We then try to find out whether the computer has learned to use that program: i.e., at some stage of processing we kill the computer and look at all of the switches and make a record of how many of them are open and how many of them are closed.

After the computer has been in the laboratory for a number of years, and we are sophisticated programmers, we write another program which is very efficient. What before taxed the computer's capacity is now contained in a sub-routine which occupies only a part of the computer's capacity.

We now again kill the computer and compare the early and late switching patterns.

It could be that we find a quantitative change of some sort that has taken place. But this change might even show a diminution of activity at later stages of learning. And, of course, to try to discern anything more from the patterning would not be easy. That is why we may be close to the limit of what kinds of questions we can ask about storage.

In other words, what we can ask and receive an answer to is: does anything at all happen in the brain as a function of experience? Let us not demand a search for specific engrams.

I have been delighted with the progress made during this decade, that in fact such changes do take place. Of course, it would be nice to correlate the types of changes with some of the specific types of experiences. But to push it much further would be like looking for the difference between a jazz recording and one of a symphony by trying to analyze the individual bumps on the record.

ON RECOGNITION MOLECULES IN CIRCUITRY
(PIGEON BRAIN MUCOIDS IN TRAINING, LEARNING
AND MEMORY)

DR. BOGOCH: The terms training, learning and memory are used independently in our studies to indicate different levels of certainty about the nature of an operational laboratory event, and different levels of complexity of concept. Each of these three is different from the general term "activity" in that a higher degree of ordered activity of the nervous system is implied. Each of the three is also different from the other two. Thus, when we state that an animal is training, or even more carefully, in the training situation, we imply nothing about learning or memory except that these may or may not occur. When the term learning is used, it requires evidence of a changed response to a stimulus, with elaboration of concomitant variables, as in several useful definitions cited in

the recent monograph, THE BIOCHEMISTRY OF MEMORY:
WITH AN INQUIRY INTO THE FUNCTION OF THE BRAIN MU-
COIDS, Oxford U. Press, 1968 (4). And when the
term memory is used, it implies a much more comprehensive, indeed higher hierarchical level of orga-
nizational change in the nervous system. Learning
is a process which utilizes old memories and con-
structs new ones; memory is a state which contains
learning in its past. Further, in memory, biochemical changes are possible at one or all of sensory
receptor, transmission, encoding, storage, retrieval, association and discharge levels of function,
together with their separate supporting reactions
(4).

   The hypothesis which we have advanced (1-6),
that the brain mucoids, aminoglycolipids, and glycoproteins perform recognition functions, requires
that there be structural specificity in their carbohydrate units. For example, in the proposed
"sign-post" mechanism (4), these mucoids would determine through recognition functions the degree
of synaptic contact and thus of facilitation of impulse transmission between any two nervous system
cells.

   The uniqueness of this hypothesis rests not
upon the specificity of the amino acid portion of
the molecule, which is of course a property shared
by many unconjugated proteins, but rather in the
specificity conferred by the conjugated carbohydrate
component. Similar bacterial glycoproteins are
functionally related to the behavioral properties
of these bacteria. For example, their contact properties, aggregation, motility and virulence, are
shown to be closely related to the nature and configuration of the terminal carbohydrate units in
the bacterial outer cell membranes (23). It was
this fact which prompted the initiation of our work
on the brain mucoids (12,1,4) 13 years ago, with
the thought that they might perform analogous functions in the nervous system (4).

   Our previous work has demonstrated that the
brain glycoproteins may be considered in terms of

various possible information processing roles, because of their heterogeneity (1,3-5,7-9), location in membranes and at synapses (3-5,10), development (4), pathology (9,11,4), recognition functions (12-15), biosynthesis (4,5), variation in different behavioral states (16-22), and concentration in different training states (2,4,5).

If glycoproteins serve recognition and contact functions in the nervous system, it may be possible to detect structural differences in the carbohydrate components of the brain glycoproteins as a function of nervous system activity in general, of training more specifically, and eventually, specifically in relation to learning. The present communication reports the comparison of the carbohydrate components of three brain glycoproteins, which have previously been shown to vary in the concentration of their protein components in the training as compared with the resting state. In passing, it may be noted that this new series of studies on training pigeons confirmed the previous observation (4) that determined through their protein moieties, in training there is a decrease in the concentration of 10B, an increase in the concentration of 11A, and a consequent change in the ratio of 11A to 10B. The stepwise acidic hydrolytic cleavage and determination of the carbohydrate units of these three glycoproteins indicates considerable structural differences, especially in the composition of their most readily hydrolyzable carbohydrate units.

Methods

Male white Carneaux pigeons were trained in the standard operant conditioning apparatus as previously described (4) in a non-stressful, non-aversive food reward paradigm, first to peck each of two keys for food reward, and then to perform discriminations between two keys, performance on one only being rewarded. The programmed instruction and automaticity of recording of all responses are advantages in this learning system, as is the ability to state quantitatively at any given time in the training process the amount of the training

task which has been accomplished. All pigeons had
the same number of one-hour training sessions, 13
for the pre-experimental training where both keys
were equally reinforced, and 21 for the experimental
sessions where only one key was reinforced. The total training period extended over a three month period with rest days interspersed irregularly. In the
present experiment (B series), where a random distribution between right and left keys would be represented by a score of 50 (50% of pecks on each
key), the "amount learned" could be expressed by the
deviation from random, i.e., preference of the correct key over the incorrect key, expressed as deviation from 50. Thus complete "learning" of the task
would be represented by pecking only on the correct
key (a score of 50).

A double-blind procedure was again maintained,
whereby those performing the training of pigeons
had no information of the chemical studies being
undertaken, and vice versa. Pigeons were placed
in rank order according to the degree to which the
correct preference had been acquired, and their last
day score, as described above, was compared with the
absolute concentration of 10B and 11A glycoproteins,
expressed in mg/gm of whole brain.

The pigeons were sacrificed immediately after
the last session by dipping the head in dry ice-
acetone, and the brain extracted for glycoproteins
by the "modified procedure", all as previously described (4). In addition to the concurrent freez-
ing of the brain at sacrifice, the repeated homo-
genizations of brain, high speed centrifugation,
concentration, stepwise elution from purified DEAE-
cellulose (Cellex-D), and Folin-Lowry determination
of proteins in each of 350 tubes of eluate, all are
important components of the method and all as previously described (4). The total protein from each
individual pigeon brain was thereby separated by
combination of elution tubes into the 16 groups
previously referred to (4). 10BI and 10BII are sep-
arable by these methods into two quite sharp, discrete peaks, whereas 11A, occasionally a discrete
sharp peak, is more often continuous with the last

portion of the 10BII peak. These glycoproteins
were collected separately from each brain, then combined, concentrated by perevaporation to 5-10 cc
volume and dialyzed exhaustively in the cold against
0.005 molar phosphate buffer, PH 7.

Each glycoprotein was then subjected to stepwise acidic hydrolysis and dialysis under the following conditions: 1) 0.1N HCl hydrolysis, 80°C,
1 hour, followed by dialysis; 2) hydrolysis of the
residue from the 0.1N HCl hydrolysis with 2N HCl
hydrolysis, 4½ hours at 100°C, followed by dialysis; 3) hydrolysis of the non-dialyzable residue remaining after the 2N hydrolysis with 4N HCl, 100°C,
4½ hours, followed by dialysis. This resulted in
three dialyzates, representing the hydrolytic split
products which became dialyzable after increasing
acidic hydrolysis, and the final non-dialyzable residue remaining after 4N HCl hydrolysis. The analysis of each of these fractions in terms of hexose,
hexosamine and neuraminic acid, expressed as per
cent of protein in the original unhydrolyzed fraction, is shown in Table 1. Protein was determined
according to the Folin-Lowry method (24), the hexose determinations were the orcinol-H_2SO_4 method of
Sorensen-Haugaard (25) with 50% galactose: 50% man-
nose as standard; the hexosamine determination was
a modified Elson-Morgan method (26) with galactos-
amine as standard. Neuraminic acid was determined
two ways, according to the thiobarbituric acid method (27) and by means of the Bial's orcinol reaction
(28). All determinations were repeated on all fractions between two and five times. On the 0.1N dia-
lyzate, there was close agreement between Bial's
orcinol and thiobarbituric acid methods for neuraminic acid, the former reading slightly higher, as
expected (for example, 0.69 and 0.59% respectively,
and 1.02 and 0.92% respectively). The true values
probably lie between those obtained with each method (4).

Results

The data from a previous series (EPII and CP, Table
XVI in ref. 4) have been fully discussed elsewhere

(4). While the experimental conditions for the training of pigeons were different in that situation, and the pigeons were allowed to rest for three to eleven months after training had been completed before they were sacrificed and the glycoproteins determined, the relationship of 10B to 11A in untrained pigeons (CP) as compared to those who have been trained (EPII) was demonstrated: the concentration of 10B was decreased, of 11A increased, and the ratio of 11A to 10B increased in the EPII group. The present (B) series (32) again showed the relationship of the concentration of 10B and 11A in that there was an increase in 11A concentration, a decrease in 10B concentration, and an increase in the ratio of 11A to 10B. These changes accompanied training which was comparable in many ways but not identical to that observed in the previous experiment.

The data in Table 1 demonstrate the composition of the carbohydrate units determined in isolated 10BI, 10BII and 11A from Series B. Some comparative indication is given of the ease or difficulty with which these different constituents can be hydrolytically released. With 0.1N HCl hydrolysis at 80°C, as expected, the neuraminic acid which is present is liberated, together with some hexose components, but no hexosamine. With 2N HCl, more of the hexose components and the bulk of hexosamine components are liberated. With 4N HCl, most of the remainder of the hexose and hexosamine components are liberated. 10BI, 10BII, and 11A are seen to differ in terms of the total concentration of hexose: 10BI and 11A being close, whereas 10BII contains some 50% more hexose than the other two. With 0.1N HCl hydrolysis, the readily hydrolyzable carbohydrate units are released. 10BII has 50% more neuraminic acid than 10BI, but 11A has no demonstrable neuraminic acid. In addition 10BII has four times the amount of hexose in this (0.1N) fraction than is derived from 10BI, and some five times the amount seen in 11A. The total amount of hexose in the combined, less readily hydrolyzed fractions, that is those released by 2N and 4N HCl hydrolysis, together with the small amount in the non-dialyzable re-

Table 1.

CARBOHYDRATE CONSTITUENTS OF PIGEON BRAIN 10BI, 10BII and 11A,

AS % OF PROTEIN

Hydrolyzed Fraction	HEXOSE			HEXOSAMINE			NEURAMINIC ACID		
Dialyzates	10BI	10BII	11A	10BI	10BII	11A	10BI	10BII	11A
0.1N	1.9	7.5	1.4	0	0	0	0.64	0.97	0
2N	6.8	5.7	6.6	14.6	16.2	3.3	-	-	-
4N	2.0	2.8	2.0	1.0	1.1	1.3	-	-	-
Non-Dialyzable	0.4	0.7	0.9	0.6	0.6	0	-	-	-
TOTALS	11.1	16.7	10.9	16.2	17.9	4.6	0.64	0.97	0

maining residue, are seen to be very similar in all three glycoproteins, that is 9.2%, 9.2% and 9.5% respectively for 10BI, 10BII and 11A. There is slightly more hexosamine in 10BII than in 10BI, but much less hexosamine in 11A than in the 10B glycoproteins.

10BI and 10BII can therefore be distinguished from each other by the above quantitative analytical data as two discrete glycoproteins. Furthermore, each of these is distinct from 11A, another glycoprotein with considerably less hexosamine and no neuraminic acid. The identification of the individual sugars in each of these glycoproteins has been partially reported (4) and in each of these hydrolytic fractions is now in progress. Antisera prepared in rabbits (29) separately to 10BI and 10BII, and to 11A, suggest the presence of two shared antigenic components in all three, and the possibility of additional individual antigenic sites for each. Antisera prepared earlier in this laboratory (30) to Tay-Sachs' Disease whole 10B fraction have given at high dilution immunofluorescent staining of astrocytic glia only (31).

Both the protein data and the carbohydrate data of this series of experiments again support the conclusion that these glycoproteins are involved in biochemical reactions occurring during the training process (2-5). With the definition of these distinguishing carbohydrate compositions as shown in Table I added to the chromatographic separation of these substances, and with the additional evidence of 10B decrease and 11A increase accompanying the training procedure, the glycoproteins involved have been named Trainein I and II (10BI and 10BII respectively) and Learnein (11A). While the change in concentration of Trainein and Learnein is further evidence that these glycoproteins are in some way related to training and learning processes, their exact function, and whether of a primary or secondary nature (4), remain to be elucidated. The biosynthetic data for the incorporation of 14C-glucose into these glycoproteins (3-5) support an extremely rapid and sizable synthetic potential commensurate with their proposed function. The fact that both Trainein and Learnein may contain even greater amounts of glycosidically bound carbohydrate constituents during the earlier phases (indeed minutes) of the training process (see EPI in reference 4) taken together with the possibility of microheterogeneity within each of these glycoproteins, points to considerable remaining complexity in the resolution of their exact functions.

Nonetheless, an important point has now been clarified. If the carbohydrate constituents of each of these glycoproteins showed no appreciable structural differences there would be reason to question the validity of the recognition aspects of the mucoid-memory hypothesis (4). With the shift from Trainein to Learnein supported, and the present data indicating marked distinctions between them in terms of the structure of their carbohydrate components, the possibility that carbohydrate specificity underlies specificity of transmission at junctures between cells is further supported. The recent electron microscopic evidence by Droz and associates (33,10) that glycoproteins are heavily concentrated at synaptic junctures is certainly relevant to our

earlier subcellular fractionation studies (4,5).
Kanfer & Richards' (34) demonstration that puromycin
blocks ganglioside synthesis is also relevant. Thus
it remains plausible that a glycoprotein-"sign-post"
mechanism (4) represents the (or one) chemical basis
of the establishment of the cluster of specific circuits in the nervous system which could constitute
a memory record. Some ways in which both genetic
and experiential influences might operate under
these conditions, together with additional theoretical considerations have been recently discussed (4,
32).

REFERENCES

1.   Bogoch, S., Nature, 190, 153, 1960.
2.   Bogoch, S., Neurosciences Res. Prog. Bull.,
      3, 38, 1965.
3.   Quamina, A. and Bogoch, S., Proc. XIIIth Col-
      log. Protides of Biological Fluids, Brugge,
      H. Peeters (ed.), Amsterdam, Elsevier, 211,
      1966.
4.   Bogoch, S., The Biochemistry of Memory: With
      an Inquiry into the Function of the Brain Mu-
      coids, Oxford University Press, N. Y., 1968.
5.   Bogoch, S., Belval, P. C., Sweet, W. H.,
      Sacks, W. and Korsh, G., Proc. XVth Collog.,
      Brugge, Protides of Biological Fluids, H.
      Peeters (ed.), Amsterdam, Elsevier, 129, 1968.
6.   Bogoch, S., Nervous System Proteins, Ch. in
      Handbook of Neurochemistry, Vol. I, A. Lajtha
      (ed.), Plenum Press, N. Y. In press.
7.   Bogoch, S., J. Biol. Chem., 235, 16, 1960.
8.   Bogoch, S., Nature, 185, 392, 1960.
9.   Bogoch, S., Cerebral Sphingolipidoses, S. M.
      Aronson and B. W. Volk (eds.), New York, Acad-
      emic Press, 249, 1962.
10.   Rambourg, A. and Leblond, C. P., J. Cell.
      Biol., 32, 41, 1967.
11.   Bogoch, S. and Belval, P. C., in Inborn Dis-
      orders of Sphingolipid Metabolism, S. M.
      Aronson and B. W. Volk (eds.), New York, Per-
      gamon Press, 273, 1966.
12.   Bogoch, S., Virology, 1, 458, 1957.
13.   Bogoch, S., Lynch, P. and Levine, A. S., Viro-

logy, 7, 161-9, 1959.

14. Bogoch, S. and Bogoch, E. S., Nature, 183, 53, 1959.
15. Bogoch, S., Paasonen, M. K. and Trendelenburg, U., Brit. J. Pharmacol., 18, 325, 1962.
16. Bogoch, S., Am. J. Psychiat., 114, 122, 1957.
17. Bogoch, S., AMA Arch. Neurology and Psychiatry, 80, 221, 1958.
18. Bogoch, S., Am. J. Psychiat., 114, 1028, 1958.
19. Bogoch, S., Dussik, K. T. and Conran, P., New Eng. J. Med., 264, 521, 1961.
20. Bogoch, S., Dussik, K. T., Fender, C. and Conran, P., Am. J. Psychiat., 117, 409, 1960.
21. Bogoch, S., Biological Treatment of Mental Illness, M. Rinkel (ed.), New York, Farrar, Straus and Giroux, 406, 1966.
22. Campbell, R., Bogoch, S., Scolaro, N. J. and Belval, P. C., Am. J. Psychiat., 123, 952, 1967.
23. Nikaido, H., The Specificity of Cell Surfaces, B. D. Davis and L. Warren (eds.), Prentice-Hall, N. J., 3, 1967.
24. Lowry, O. H., Rosebrough, N. J., Farr, L. and Randall, R. J., J. Biol. Chem., 193, 265, 1951.
25. Sorensen, M. and Haugaard, G., Biochem. Z., 206, 247, 1933..
26. Elson, L. A. and Morgan, W. T., J. Biochem. 28, 988, 1934, as modified in S. Bogoch, J. Biochem., 68, 319, 1958.
27. Warren, L., J. Biol. Chem., 234, 1971, 1959.
28. Bogoch, S., Biochem. J., 68, 319, 1958.
29. Bogoch, S. and Das, B. R.   Unpublished data.
30. Bogoch, S. and Rajam, P. C.   Unpublished data.
31. Mori, K. and Sweet, W. H. Personal communication.
32. Bogoch, S., Brain Mucoids in Training Pigeons, Ch. in Symposium on Protein Metabolism, A. Lajtha (ed.), Plenum Press. In press.
33. Droz, B., Ch. in Symposium on Protein Metabolism, A. Lajtha (ed.), Plenum Press. In press.
34. Kanfer, J. N. and Richards, R. L., J. Neurochem., 14, 513, 1967.

MORE ON RECOGNITION MOLECULES

DR. UNGAR: I am in general agreement with most of what Dr. von Foerster said, especially with his plea not to consider memory as an abstract entity separated from the other mental functions. I should, however, like to say a word about storage versus computation. I do not think one can deny that a calculator operates only because of the information stored in its design. Even a simpler machine, like an automobile, can move only because a certain amount of information is built into its structure. Where it moves and how fast, will depend on the conditions of the environment and essentially on the driver.

Now, the brain has a vast store of information built into its organization, far vaster than any computer yet devised. This store is genetically determined and exists before any learning has taken place. In fact, no learning could take place without it, just as no computing could be done by a calculator that is not wired for that purpose.

Computers are certainly better models for the brain than anything previously proposed, provided that one keeps in mind that, according to the best evidence, the brain is not an electronic but a chemical computer. Impulses pass from one unit to the next by chemical transmitters and the connections are established by molecular recognition. Computation has, no doubt, an important part in the selection of the connections and the routing of the impulses through them.

The main problem is really to find out what takes place in the neurons during learning. We can take the well-known Pavlovian paradigm as an example. A dog will salivate when it is presented with food. This is an unconditioned, innate response which is elicited by nerve impulses travelling along a prewired pathway whose

connections are genetically determined, probably
by means of a chemical recognition mechanism.

Let us now see what happens when a conditioned stimulus is given. If you ring a bell,
for the first time, the dog will prick its ear
or bark or whatever dogs do when they hear a bell
but it will not salivate. However often you ring
it, the bell will never elicit the flow of saliva
unless it is presented together with the unconditioned stimulus. Under proper conditions, this
will cause the neurons corresponding to the unconditioned pathway and the conditioned stimulus
to fire simultaneously. By the permeability
changes that this firing implies, an exchange of
"recognition molecules" takes place between two
hitherto unconnected pathways.

DR. BOGOCH: Could these be glycoprotein
molecules?

DR. UNGAR: It may very well be so but the
important fact is that from now on the pathway
of the unconditioned response will recognize impulses arising from the conditioned stimulus and
will be fired by them. Consolidation of the
conditioned response probably requires that the
same "recognition molecules" be synthesized on
both sides of the critical synapses at which new
connections have been created. This implies in
some of the neurons activation of DNA and RNA
sequences that were repressed. Although, in the
strict sense, no new molecular species is involved, the synthesis of messenger RNA which is new
to the given neuron, may produce changes in the
base ratios.

DR. HYDEN: I would like to answer Dr.
Ungar's question in the following way. At birth,
there exists a great number of fully developed
pathways in the brain, phylogenetically given,
without which the organism could not function.
There also exists certain genetically programmed
behavior which only requires the stimulation by
key-factors in the environment to be elicited.

The neurons and glia building up these pathways and providing these functions are highly differentiated. Cytogenetically speaking, they have each only a small percentage of their genes activated. On top of this organization, one has reason to assume the existence of a mechanism which serves experiential learning and memory during a life-time. A retrieval of a memory may require a differentiated response from millions of neurons. These neurons have specific proteins in common which lead all the neurons to respond on a certain input and activate the transmittor substances. We have interpreted our data on synthesis of brain muclear RNA and acidic protein during learning to mean that they reflect an additional activation of gene areas in neurons as a response to specific environmental factors. This should be part of the mechanism which serves experiential learning.

So in this respect I agree with Dr. Ungar. Existing "pathways" in the brain need to be combined, but I see the creating of the new combination as a gene-RNA-protein problem.

THE MOLECULAR STRUCTURE OF MEMBRANES

Fritiof S. Sjöstrand

The discovery by means of electron microscopy that membranes are the most common supramolecular structures of cells has made it justifiable to ascribe important functions to this
type of structure. That membranes are involved
in other functions than those of a boundary de-
limiting the territory of a cell or a cell nucleus
became obvious with the discovery of membranes as
the dominating structural components of energy
transducing structures like the outer segment of
retinal photoreceptor cells (Sjöstrand, 1949,
1953a), followed by the demonstration of a similar
structural organization of the chloroplast grana
(Steinmann, 1952, Steinmann and Sjöstrand, 1955)
and of the membrane components of mitochondria
made independently and simultaneously by Palade
(1952, 1953) and Sjöstrand (1953b, 1953c).

With this characteristic association of membranes and complex metabolic functions involving
a number of different types of highly specialized
molecules, it appears justifiable to ascribe to
the membranes important metabolic functions.

A number of different types of membranes in
the cytoplasm of cells, like the rough surfaced
membranes, which are associated with protein synthesis, the Golgi membranes as well as various

types of smooth surfaced membranes of cytoplasmic vesicles and granules, form the boundaries of closed compartments in the cytoplasm. In these cases we are concerned about both the specific metabolic function of the membranes themselves and the functional significance of the compartmentalization of the cytoplasm.

When considering the plasma membrane we are inclined to focus our interest on the permeability properties of this type of membrane and to try to ascribe the permeability characteristics primarily to certain physical properties of the membranes, although active transport of sodium ions from the inside to the outside involves an energy requiring metabolic process in the membrane. With more precise knowledge of the function of the plasma membrane it is likely that also this type of membrane will appear as a metabolically rather active and complex structure. Particularly in the nervous system we are inclined to ascribe rather specialized functions to those areas of the plasma membrane which form the synaptic contacts between neurons. These areas must be equipped with specific receptor sites for transmittor molecules, and the number and efficiency of these sites might well change dynamically to allow those modifications of the activity patterns of the nervous system that are the basis for learning and memory functions.

The mitochondrial membranes are the membranes we know most about from a biochemical point of view. It is rather well established that the respiratory chain is associated with the mitochondrial membranes and that these membranes are the "structural factor" that the pioneers in the study of tissue respiration assumed to be responsible for the fact that these components could not be brought in solution in an active form.

An understanding of why complex metabolic functions are associated with structural organization at a supramolecular level is most important in order to appreciate some of the very basic properties of living systems. Such an understanding requires a precise knowledge of the molecular or-

ganization of the membranes. From such a knowledge
concepts can be deduced regarding the mode of interaction between the components in membrane bound
multienzyme systems.

When applying electron microscopy using conventional techniques, we have inherited the methodology from light microscopy and, when considering
the molecular structure of membranes we have inherited from the Danielli-Davson model for the
plasma membrane. The latter model was deduced
from rather primitive data relating to some physical properties of that particular type of membrane.
Such inheritance can impose and has imposed considerable inhibitions and bias when exploring the
new situation created by the developments of new
instruments for direct and indirect studies of
membrane structure. Let us try to release ourselves from this burden of inheritance and become
intellectually free.

Fig. 1 shows an electron micrograph from 1952
in which the periodic structure of the myelin
sheath was discovered and where the regular periodicity was demonstrated to consist of two sub-
periods (Sjöstrand, 1953d). The main period we
can consider to extend between the thick opaque
layers. This period is subdivided into two equal
halves by an intraperiod layer.

This pattern supports the concept regarding
the molecular structure of the myelin sheath that
was developed on the basis of X-ray crystallographic data by F. O. Schmitt and collaborators
(1941). Due to the large dimensions of the radial
periodicity observed in the X-ray diffraction patterns these investigators assumed that the period
contained two bilayers of lipids. To ascribe the
periodicity to a layering of lipids was justifiable
because W. J. Schmidt (1936) had demonstrated by
means of polarization optical analysis that the
lipid molecules were oriented radially in the
sheath while the proteins presumably were arranged
in layers oriented perpendicular to the long axis
of the lipid molecules.

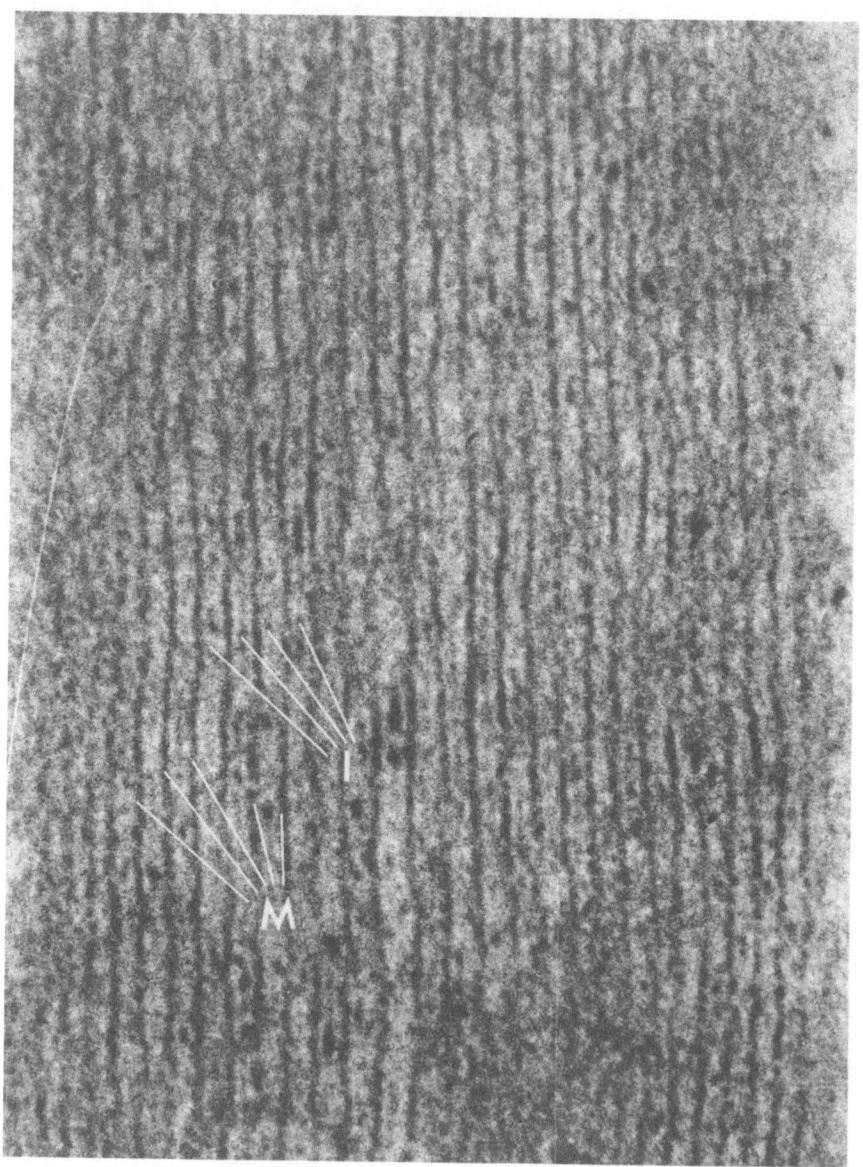

Fig. 1.   Myelin sheath of axon in the sciatic nerve
of mouse, showing the layered structure with the
main period layers (M) and the intraperiod layer
(I).   Magnification 340,000 X.   (From Sjöstrand,
1953d.)

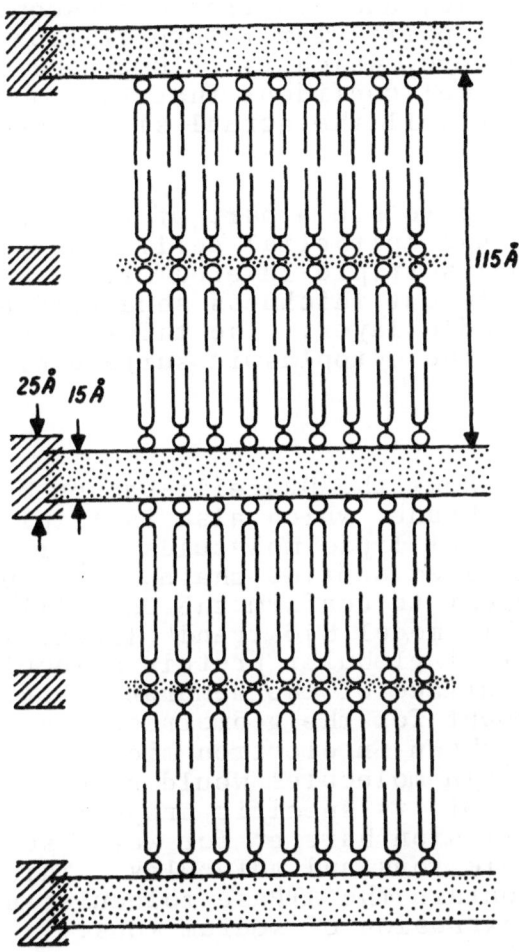

Fig. 2. Schematic presentation of the molecular structure of the myelin sheath as deduced from X-ray diffraction data by Schmitt et al. (1941), adjusted to dimensions found in electron micrographs of the myelin sheath. The opaque layers observed in the electron micrographs were interpreted to correspond to concentrically arranged layers of protein. The polar ends of the lipid molecules were also assumed to contribute to the staining and to the width of these layers. (From Sjöstrand, 1957.)

In this case polarization optical, X-ray crystallographic, and electron microscopic data agree with one particular interpretation of the molecular structure, which is illustrated in Fig. 2. We therefore feel confident that this model describes rather well the actual situation.

In Fig. 3 we see a part of the brush border in an intestinal epithelial cell. The plasma membrane appears triple-layered with two opaque layers of different thickness separated by a lightly stained middle layer. The thicker opaque layer is located at the cytoplasmic surface of the plasma membrane.

Fig. 4 shows the Danielli-Davson model for the plasma membrane, which also is triple-layered with a bilayer of lipid molecules separating two layers of unrolled peptide chains. The pictures therefore appear to confirm the Danielli-Davson model. To this model Sjöstrand (1960a, 1960b) added a layer of globular protein molecules at the cytoplasmic surface of the membrane (Fig. 5) in order to account for the geometrical asymmetry of the membrane shown in electron micrographs. These globular protein molecules would give to the plasma membrane some of its specific properties other than that of a diffusion barrier for water soluble and lipid insoluble ions and molecules. This layer would thus contain the ATPase molecules associated with active extrusion of sodium ions.

When the plasma membrane is symbolically represented by such a mixed structure we emphasize its function as a diffusion barrier. The mitochondrial membranes, on the other hand, are more likely predominantly metabolically active structures and their barrier function might well be adjusted to their metabolic function. Fig. 6 shows a survey picture of mitochondria in a proximal convoluted tubule cell in the mouse kidney. It clearly demonstrates that the mitochondrial membranes appear in pairs, "double membranes" (Sjöstrand,

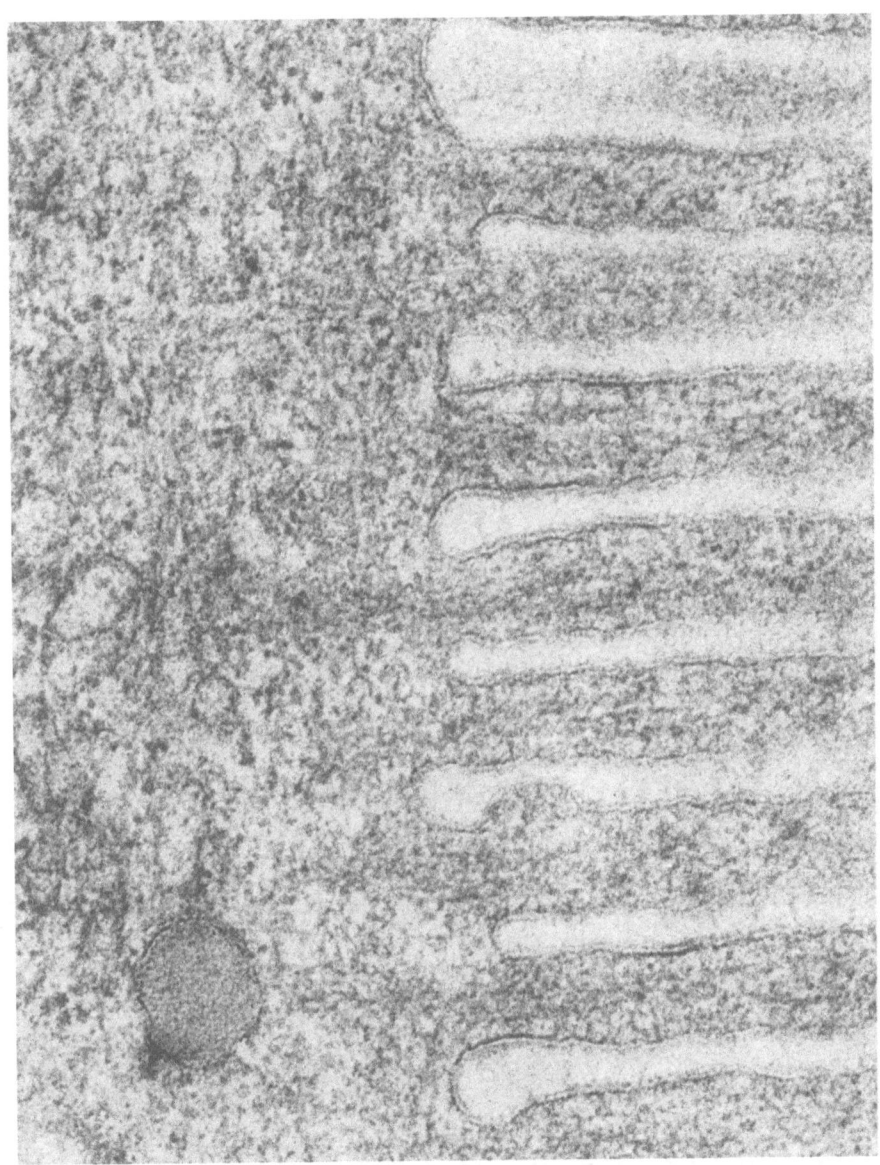

Fig. 3. Basal region of brush border in columnar cells of the mouse small intestine showing the bases of the brush border processes. The plasma membrane appears triple layered and geometrically asymmetrical. Magnification 140,000 X. (From Sjöstrand, 1963d.)

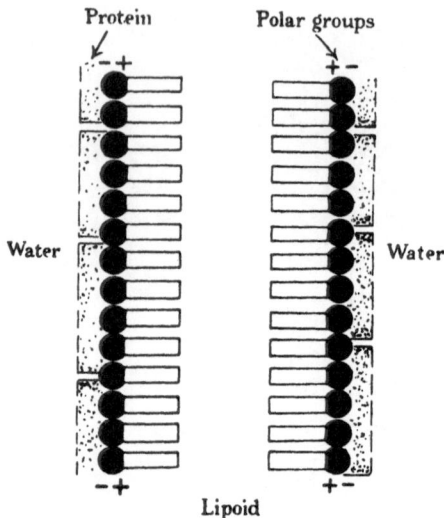

Fig. 4.   The Danielli-Davson model for the plasma
membrane in which a layer of oriented lipid molecules is sandwiched between two layers of rolled
out, denatured protein molecules.   Secondarily
globular proteins could be assumed to associate
with the layers of proteins that had become denatured at the lipid-water interface.   (From Danielli and Davson, 1952.)

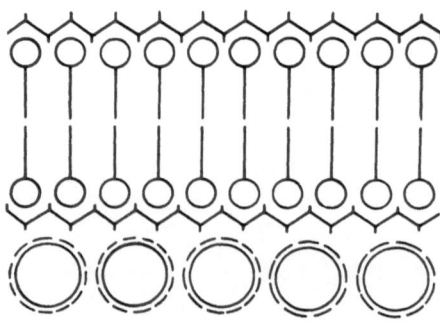

Fig. 5.   Modified Danielli-Davson model for the
plasma membrane in which a layer of globular
proteins is assumed to represent a major structural component of the plasma membrane adding enzymes like ATPase to the membrane.   (From Sjöstrand,
1960a.)

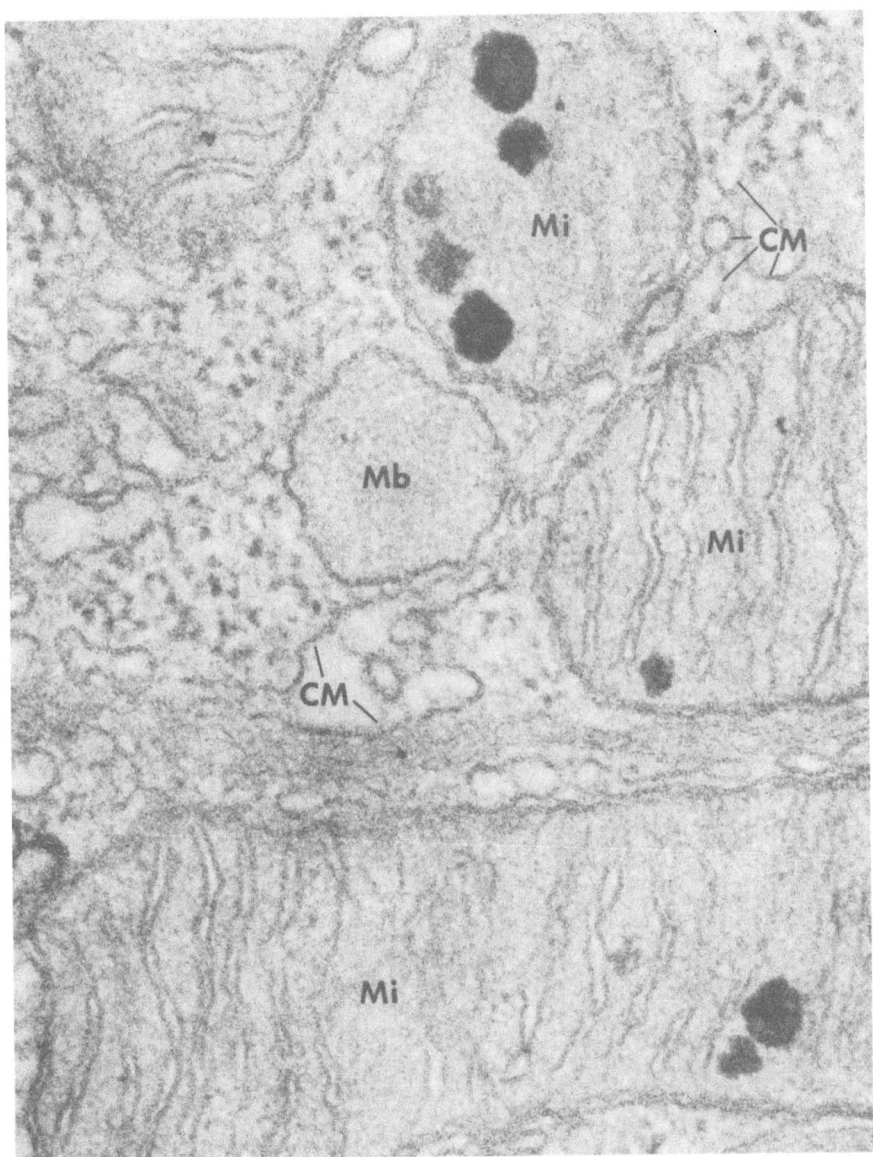

Fig. 6. Survey picture of a region in a proximal convoluted tubule cell in the mouse kidney showing mitochondria (Mi), a microbody (Mb), smooth surfaced cytomembranes (CM) and free ribosomes. The mitochondrial membranes appear to consist of two components or membrane elements separated by a space which presumably is due to artifactual swelling. Magnification 120,000 X.

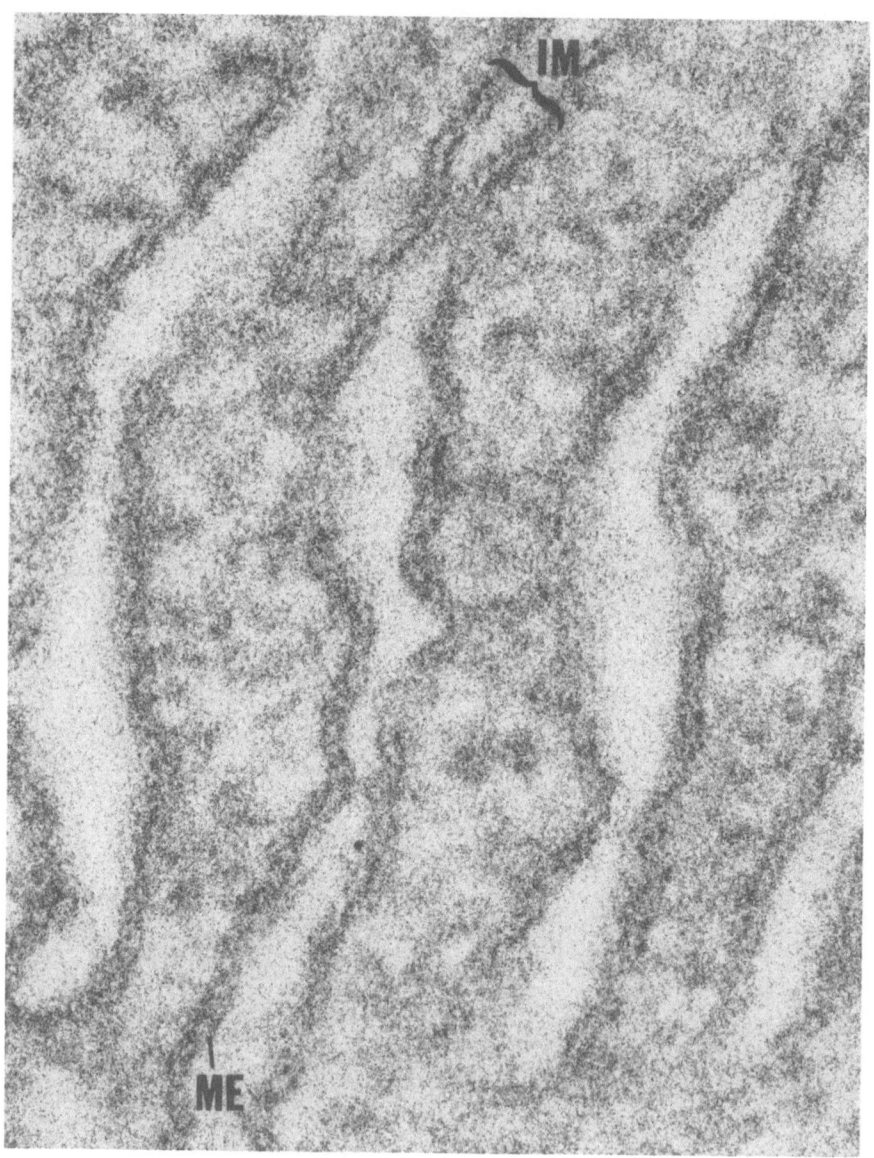

Fig. 7. High resolution close-to-focus picture of mitochondrial membranes in osmium-fixed kidney tissue (mouse). The inner mitochondrial membranes (IM) oriented vertically consist of two membrane elements (ME) which appear triple layered although the staining is not confined entirely to two sur- face layers but also includes regions in the middle of the elements. Magnification 600,000 X.

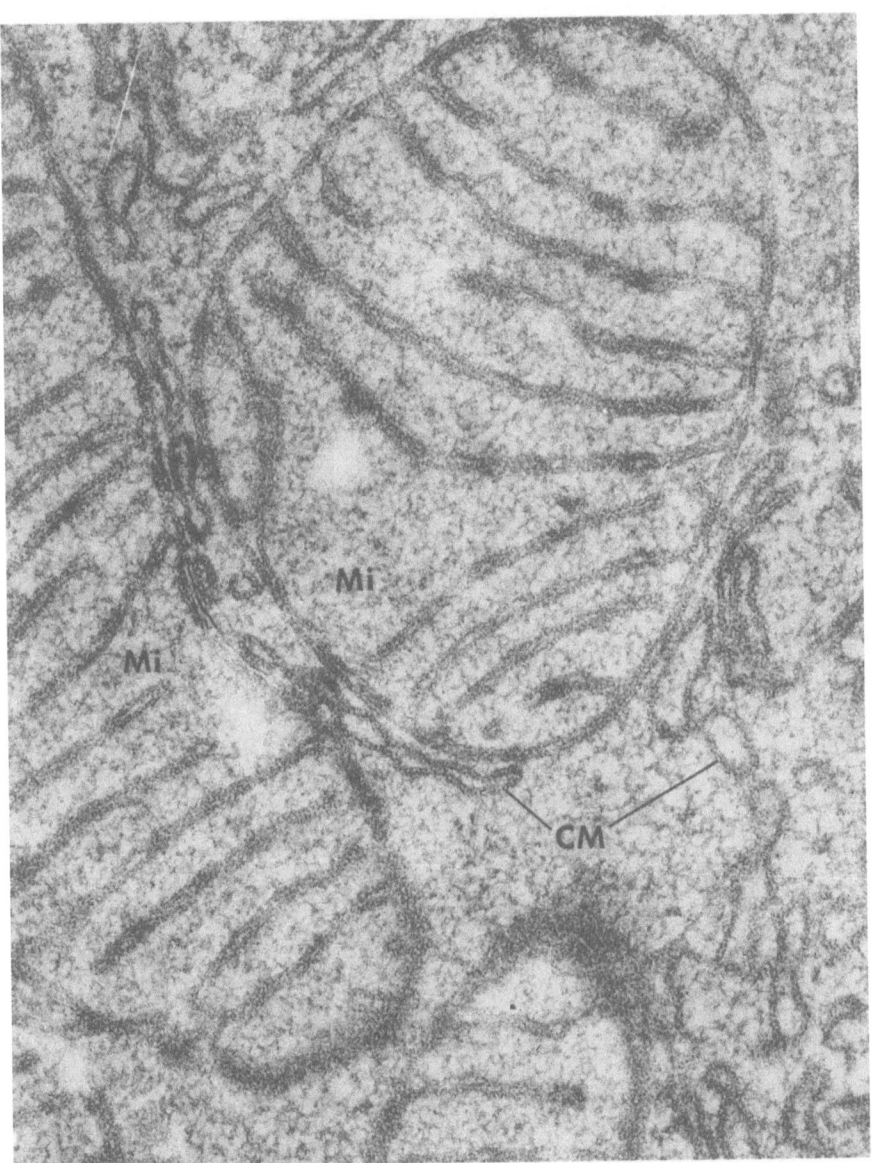

Fig. 8. Mitochondria in proximal convoluted tubule cell of the mouse kidney fixed for 15 minutes
in vivo by dripping a 3% solution of potassium
permanganate onto a small exposed area of the
kidney surface. Short dehydration time. The mitochondrial membrane elements and the smooth surface
cytomembranes (CM) appear to be sub-divided into
globular sub-units. Mag. 130,000 X.

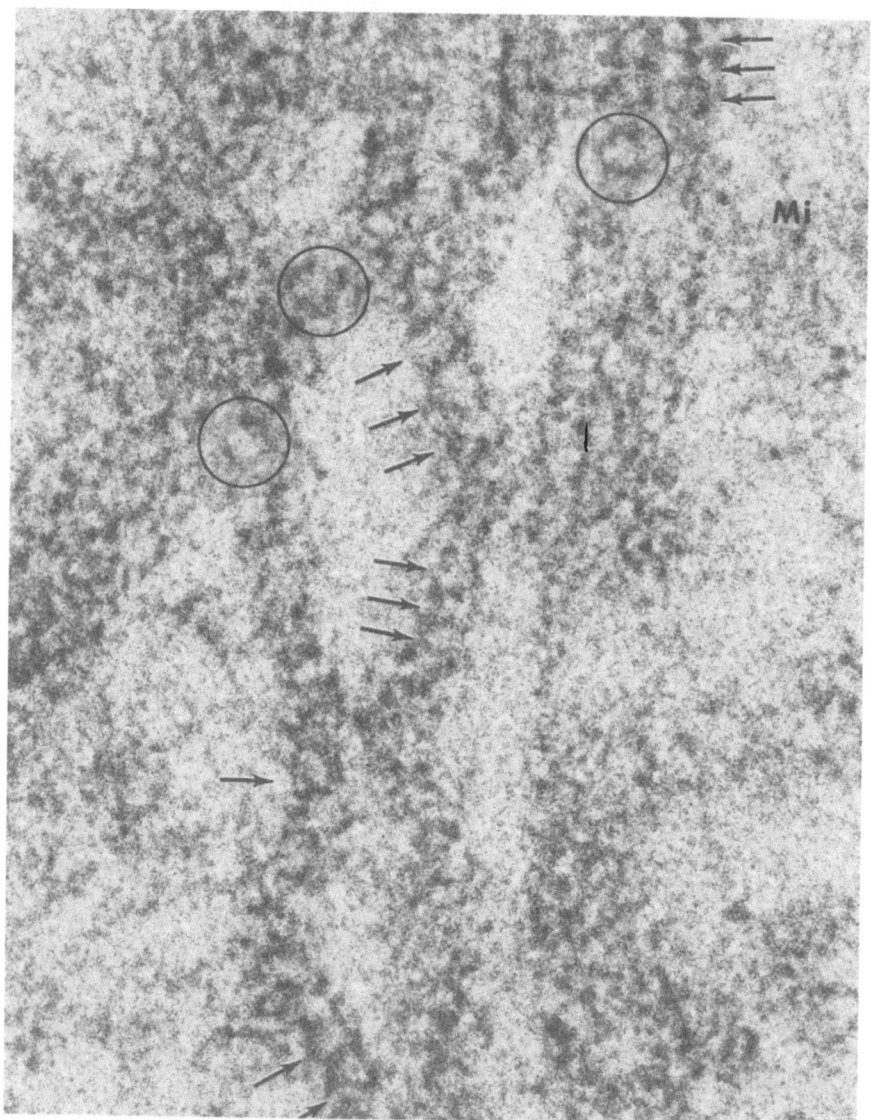

Fig. 9. High resolution (5-10 Å) slightly under-
focussed electron micrograph of smooth surfaced
membranes and part of outer mitochondrial membrane
(Mi), showing the globular substructure of the
membranes. The globules consist of a lightly
stained area delimited by stain grains measuring
5-10 Å in diameter. The globules are indicated
by arrows and some are encircled. Mag. 1,000,000 X.
(From Sjöstrand, 1968a.)

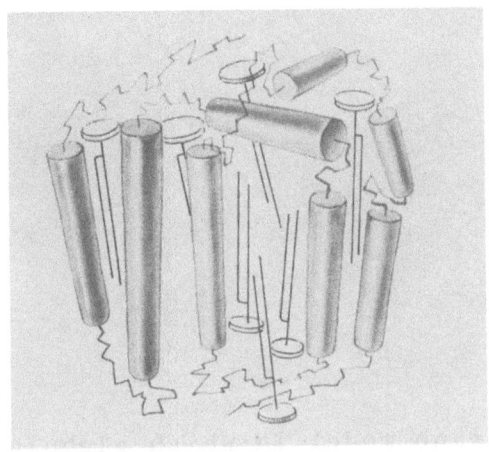

Fig. 10.   Schematic representation of a hypothetical lipo-protein complex.   The α-helix
parts are indicated by cylindrical sections.   Due
to a preferred orientation of the α-helices, the
lipid molecules are considered to assume a similar preferred orientation.   For schematic simplicity the two hydrocarbon chains of the lipid
molecules are drawn straight.   (From Sjöstrand,
1968b.)

1953b, 1953c).   The mitochondrion is bounded by
a pair of membranes, as originally shown by
Sjöstrand (1953b, 1953c) and in the interior we
find pairs of membranes with very narrow stalk-
like connections with the outer membrane pair.
These pairs of membranes can either be considered as menbranes bounding different compartments
within the mitochondrion, or they can be conceived of as two components of a composite membrane.   In the latter case the separation of
the two components is considered artifactual.
In the undamaged mitochondrion the two "membrane
elements" are assumed to be closely opposed to
form the composite membrane.   Such a close
apposition of the membrane elements is characteristic for mitochondria that have been preserved
without swelling of the mitochondria, like after
freeze-drying preservation (Sjöstrand and Baker,
1958).

Fig. 11. Membrane model in which globular lipo-
protein complexes form the backbone structure
of the membrane. The lipid molecules are associated with the protein molecules fo form lipo-
protein complexes and in addition to fill in
the holes between the globular lipo-protein
complexes. (From Sjöstrand, 1968b.)

        The two components of the double membranes
we refer to as the membrane elements of the mitochondrial membrane. They appear triple layered
in Fig. 7 and measure about 50 Å in thickness
in material that has been prepared according to
conventional procedures. In an effort to expose
the tissue to less modifications of the structure
by the fixative and by the dehydration medium
than the conventional preparatory procedures,
experiments were carried out in which the kidney
tissue was fixed in vivo for only 5-15 minutes
and dehydrated very rapidly (Sjöstrand 1963a,
1963b, 1963c, 1965).

        In such material, shown in Fig. 8, the
membrane elements show a more complex structure
than in material fixed and dehydrated according
to conventional procedures. The membrane elements appear to be subdivided into globular components by minute grains of stain that are located both at the surfaces and the interior of

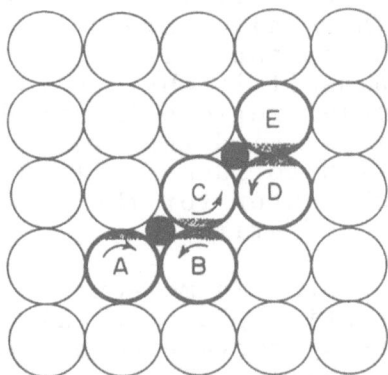

Fig. 12. Schematic presentation of how the active sites of a globular protein molecule in a membrane can make contacts with the active sites of adjacent molecules through thermal rotatory oscillations of the molecules. (From Sjöstrand, 1968b.)

the membrane elements. The latter grains appear to form septa extending across the membrane elements.

At higher magnifications, as shown in Fig. 9, the globular subunits of the membrane elements consist of a lightly stained center surrounded by a number of stain grains. The latter measure 5-15 Å in diameter and the resolution of these pictures of extremely thin sections is better than 10 Å. Such a high resolution, which once was considered not possible to achieve on positively stained sections, is explained by the favourable contrast conditions characteristically for such material. Each stain grain presumably consists of a compact aggregate of anhydrous stain ions. This assumption is necessary to make in order to explain the high contrast of the individual stain grains. When staining the sections further, for instance by applying a second electron stain, the grains of stain increase in size but their number changes usually very little. This supports further the concept that staining involves the formation of an aggregate of anhydrous stain ions or molecules at certain sites in the structures of the cells. These sites are primarily associated with the proteins of the cell. From the dimensions of the stain grains it is obvious that these sites represent particular regions in protein molecules which preferentially bind the stain. The first binding of a stain ion to such a site presumably acts as a "nucleation" center for the formation of a compact aggregate of stain ions.

The sites in protein molecules that could bind the stain preferentially are certain amino acid side chains, particularly certain polar and charged side chains, while non polar side chains are less likely to be involved. At a resolution in the range of 5-10A, intramolecular structural features will be shown in the pictures. A globular protein molecule with a molecular weight of about 40,000. A preferential staining of the periphery of such a globular structure would be reasonable if we assume that the globules are protein molecules since the hydrophilic, polar amino acid side chains are likely to be located preferentially at the surface of the molecule while the non polar hydrophobic amino acid side chains are accumulated in the interior of the molecule. X-ray diffraction studies have revealed such a tendency of amino acid side chain distribution in globular protein molecules.

A reasonable concept regarding the structure of lipo-protein complexes places the lipid molecules with their hydrocarbon tails buried in the interior, hydrophobic region of globular protein molecules, while their polar ends are assumed to be associated with the polar side chains at the surface of the molecules. Such an intimate association of lipid and protein molecules would account for the associated with proteins in membranes are of a rather precise composition. The restrictions of the space available for the hydrocarbon tails in the interior of the protein molecule would require a particular shape and dimensions of the hydrocarbon tails. Fig. 10 shows a schematic drawing of a lipo-protein complex in which the α-helix regions have been particularly emphasized.

We have discussed the globular structures
that we have observed in the mitochondrial membrane
elements and in cytoplasmic membranes as possibly
representing globular protein molecules. They
could also be assumed to represent globular lipid
micelles. One important observation by Fleischer
et al (1965) supports the former interpretation.
They found that the mitochondrial membrane elements
appeared identical in electron micrographs before
and after mitochondrial fractions obtained by tissue homogenization and differential centrifugation
had been extracted with lipid solvents to remove
more than 90% of the lipids. This observation
clearly indicates that the lipids are not essential
for the triple-layered structure observed in electron micrographs of conventionally prepared material.

The tendency of certain protein molecules to
aggregate in membranous structures is easy to account for by assuming that these proteins contain
a large proportion of non polar amino acid residues and that non polar side chains occupy positions not only in the interior of the molecules
but also at certain areas at the surface of the
molecules. Such hydrophobic areas would give to
the molecules a tendency to aggregate to cover up
these areas to exclude them from contact with the
aqueous medium. The difficulties in solubilizing
the membrane proteins can be understood on the basis that these proteins are hydrophobic. The X-ray
data of Dickerson et al (1967) show that in the
cytochrome c molecule hydrophobic regions extend
all the way from the surface through the interior
of the molecule.

Let us accept the interpretation that the basic structure of the mitochondrial membranes consists of globular protein molecules and lipo-protein complexes (Sjöstrand, 1966, 1967a, 1967b).
Let us also assume that lipid molecules fill out
the holes between the globular components. This
way a predominantly hydrophobic region will occupy

the middle part of the membrane, both lipid molecules and protein molecules contributing to the
hydrophobic character of this region.

Such a structure imposes restrictions on the mobility of the protein molecules. The molecules thus
could be confined to certain positions in the membrane securing a certain topographic relationship
between molecules in, for instance, the respiratory
chain and other multi-enzyme systems. It is, however, reasonable to assume that the molecules are
not frozen in these positions but retain a certain
mobility and are subject to thermal oscillations.
Through the thermal movements active sites of a
component in a multi-enzyme system could be brought
in contact with the active sites of other components
of the complex. One component could be conceived
of as oscillating between alternate contact rela-
ents. This way electron transfer through the respiratory chain could be associated with alternat-
ing contacts between active sites of the cytochrome
molecules.

This model of a membrane is illustrated in
Figs. 11 and 12. Fig. 12 shows schematically the
dynamic aspects of such a membrane model. It also illustrates the possibility of collisions between active sites of two molecules with a third
molecule, for instance, a substrate molecule attached at the active site of one component. In
this case a substrate molecule could be handed
over from the active site of one component to the
active site of the following member of a multi-
enzyme complex without being released to the medium.

Such a model has a justification if it stimulates to new efforts in collecting data regarding
the actual structure of membranes. It is obvious
that this model emphasizes the proteins as the most
important components determining the basic structure of membranes. To test this model and to obtain, on the whole, information regarding the arrangement of the metabolically most important components of the membrane, the protein molecules, it

is necessary to analyze material in which the native conformation of the protein molecules has not
been to drastically changed. A denaturation of the
proteins with an unfolding of the peptide chains
could very well lead to the formation of a new and
artifactual structural arrangement of the proteins
satisfying the condition of not exposing the hydrophobic amino acid side chains to the aqueous medium. We face, in fact, a rather delicate problem
when tranferring the tissue to a plastic in order
to make it possible to prepare thin enough specimens for high resolution electron microscopy.

Changing the milieu of the proteins means that
we change a major factor determining the conformation of the molecules. The hydrophobic interaction
within the molecules contributes greatly to the
globular shape of protein molecules. This interaction is likely to be affected by a less polar solvent than water, with conformational changes as a
consequence. A series of experiments (Sjöstrand
and Barajas, 1968) was carried out to try out some
preparatory procedure that would be more satisfactory than conventional methods from the point of
view of preserving the native conformation of protein molecules. Both fixation and dehydration in
an organic solvent of the tissue was avoided by
air-drying of thin slices of the tissue and infil-
trating them directly by the plastic. The mitochondrial membranes then showed such a varied and
irregular structure that it appeared obvious that
this treatment did not prevent conformational
changes from occuring. It therefore seemed important to find some method to stabilize the native
conformation of the proteins before the tissue was
transferred to the embedding medium.

Glutaraldehyde has been used in electron microscopy as a fixative that is characterized by retain-
ing the enzymic activity of a number of cellular enzymes. This indicates that it affects very little
the native conformation of these enzyme molecules.
Glutaraldehyde acts as a cross-linking agent and has
been used by, for instance, Quiocho and Richards
(1964, 1966) to stabilize protein crystals mechanically. When cross-linking carboxypeptidase A crys-

tals the crystals became mechanically very stable
and the enzymic activity of the crystals was retained to 50% after one hour cross-linking in 1%
glutaraldehyde.

Short time cross-linking with glutaraldehyde
through in vivo transfusion of kidney tissue and
air-drying of the tissue did not lead to better
preservation of mitochondrial membranes than air-
drying of unfixed tissue. It therefore appeared
important to subject the tissue to dehydration by
some proper medium after cross-linking.

As dehydration medium, ethylene glycol was
chosen because it has been shown through the
studies of Tanford et al (1955a, 1955b, 1962) "that
ethylene glycol may be an unusually inert reagent
towards proteins in general" (Tanford et al, 1962).
This conclusion was based on optical rotatory dis-
persion studies of solutions of γ-globulin and β-
lactoglobulin in various concentrations of ethylene
glycol. In contrast to ethanol these proteins
showed no indications of conformational changes
even in high ethylene glycol concentrations. γ-glo-
bulin did not undergo conformational changes in
concentrations of up to 90% ethylene glycol. Sage
and Singer (1958) and Singer (1962) found that ribonuclease did not change with respect to its α-helix
content in pure ethylene glycol. They concluded that
the hydrogen binding capacity of ethylene glycol was
similar to that of water. The condition for intra-
peptide hydrogen binding would therefore be rather
similar in these two media.

Ethylene glycol has been used by Pease (1966)
in his "inert dehydration" preservation of tissues
which involves ethylene glycol dehydration of unfixed tissues. This technique results in excellent preservation at a low to medium resolution,
but was found unsatisfactory for high resolution
studies.

The technique we eventually arrived at as most
satisfactory at the present time involves a careful
selection of 1) a stabilizing compound, glutaraldehyde, which acts by extensive intermolecular cross-

linking of the proteins, and 2) a dehydrating
agent, ethylene glycol, which has been shown to affect the native conformation of a number of proteins less than most other organic solvents.   Furthermore, the time the tissue was exposed to glutaraldehyde and ethylene glycol was chosen very
short.   These times could be reduced to 3 to 15
minutes for cross-linking and 3 minutes for dehydration because the glutaraldehyde was applied by
perfusion of the tissue and tissue slices measuring only 50 μ in thickness were dehydrated.

The appearance of the tissue after preservation and embedding according to this technique was
compared with that after freeze-drying of partially dehydrated tissue which was embedded at -20°C.
In this case the temperature of the tissue had not
been raised above -20°C until after the plastic
had polymerized to a solid state.   The partial dehydration of the cross-linked tissue was achieved
by immersion of 50 μ thick tissue slices in 10-50%
ethylene glycol.   After partial dehydration the
tissues could be frozen without the formation of
ice crystals.   Both techniques gave similar results.

Fig. 13 shows a section through mitochondria
in a proximal convoluted tubule cell in a rat kidney which was cross-linked for 3 minutes by perfusion with a 1% glutaraldehyde solution and then
dehydrated in ethylene glycol.   The membranes of
the mitochondria appear lightly stained against
the more intensely stained matrix.   The two membrane elements of the mitochondrial membranes can
be distinguished.

At higher magnification shown in Fig. 14, the
lightly stained membrane elements appear to consist of particles of varying dimensions.   Each
particle consists of a lightly stained center delimited by stain grains.   The dimensions of these
particles can be measured where the mitochondrial
membranes are oriented obliquely in relation to
the plane of the section (Fig. 15).   At such orien-
tations of the membranes the particles are particularly clearly outlined because they are observed

Fig. 13. Survey picture of mitochondria in proximal convoluted tubule cell of the rat. The
tissue was cross-linked for 3 minutes through
perfusion with a 1% solution of glutaraldehyde
in phosphate buffer. Dehydration for 3 minutes
in ethylene glycol before infiltration with Vestopal W and embedding. The mitochondrial membranes appear as light cross-bands separated by
the darkly stained matrix. They are well preserved
with little indication of fragmentation. The two
membrane elements can be seen in places where the
membranes are oriented perpendicular to the plane
of the section. They are then separated by a
dark layer (arrows). They are closely apposed.
The mitochondrial membrane elements appear con
siderably thicker than in Fig. 8, where they
measure only 50 Å in thickness. They show a particular structure with the individual particles
appearing in negative contrast. Mag. 120,000 X.
(From Sjöstrand and Barajas, 1968.)

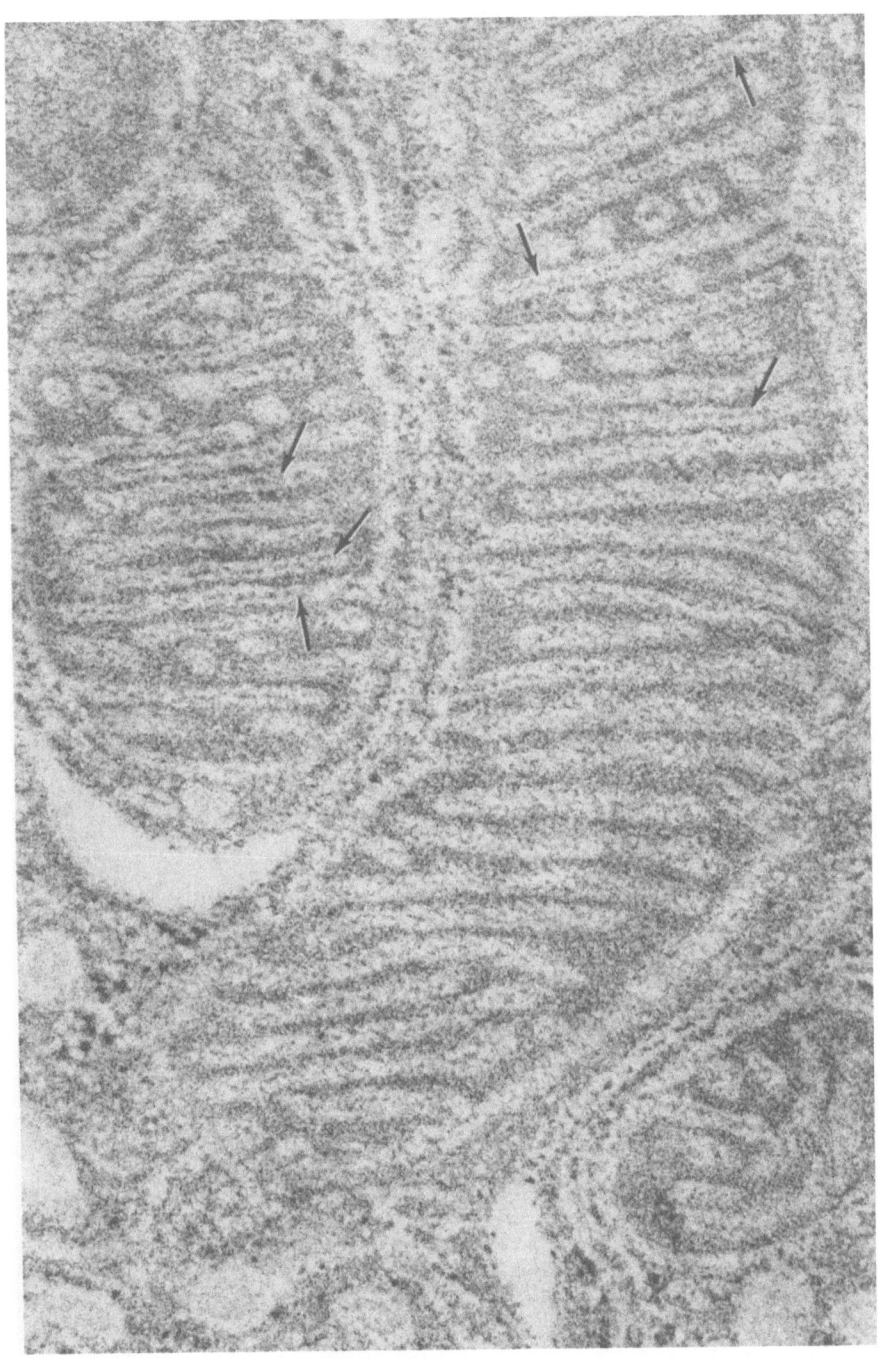

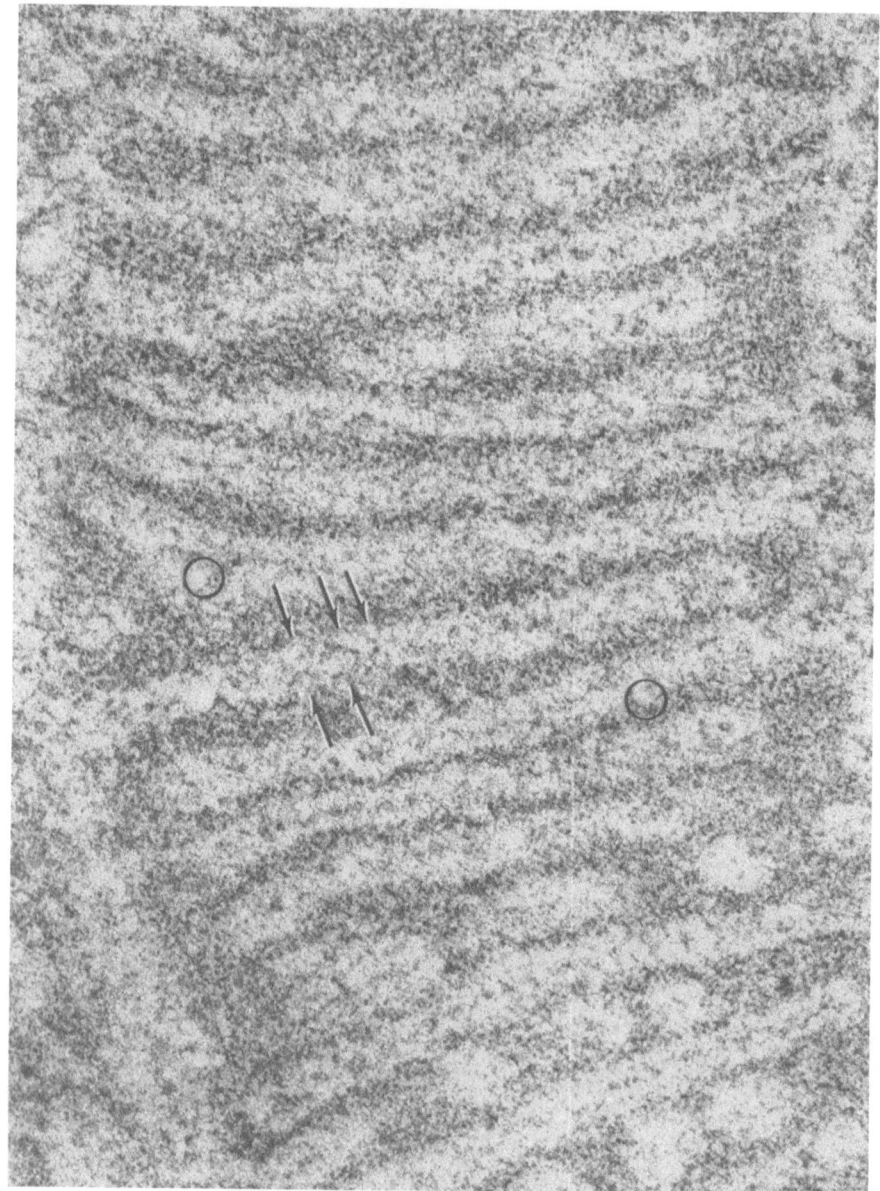

Fig. 14. Higher magnification of a region in Fig.
13 showing the particulate structure of the mitochondrial membranes. Magnification 240,000 X.
        (From Sjöstrand and Barajas, 1968)

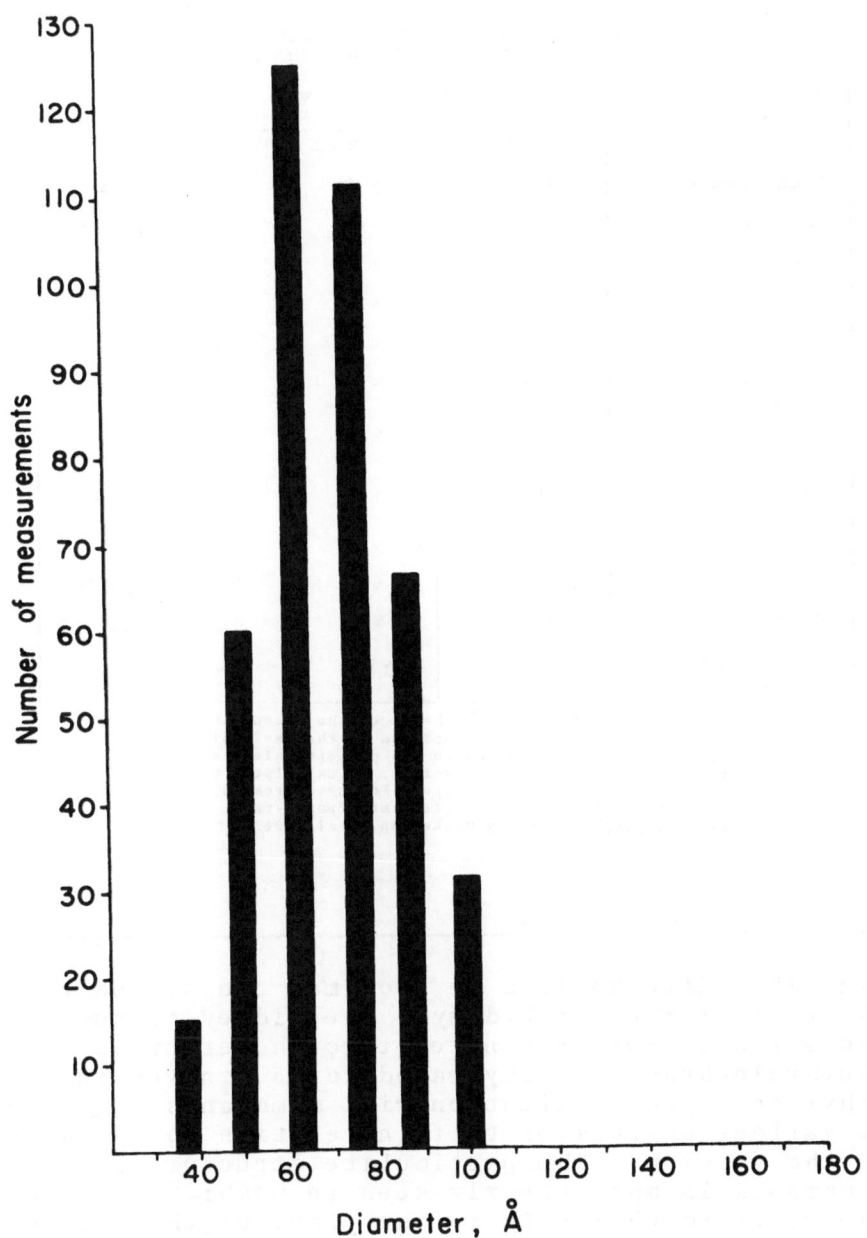

Diagram 1.

MOLECULAR WEIGHTS OF PROTEINS LOCATED IN MITOCHONDRIAL
MEMBRANES AND THEIR ESTIMATED DIMENSIONS

Compound	Molecular Weight	Diameter* (Å)	Source
Cytochrome c	12,400	25 x 25 x 37	Dickerson et al., 1967
Structural protein	22,000	~ 40	Criddle et al., 1962
Cytochrome b	30,000	~ 40	Goldberger et al., 1961
Succinate dehydrogenase	200,000 4 subunits 49,000	~ 70 ~ 50	Singer et al., 1956
Coupling factor F_1	280,000	~ 70	Racker, 1965
Cytochrome a	360,000 (pentamer) 72,000 (monomer)	~ 80 ~ 55	Ambe and Venkataraman, 1959; Criddle and Bock, 1959
Cytochrome c_1	360,000 subunits 51,000	~ 80 ~ 50	Criddle et al., 1962
Choline dehydrogenase	850,000	~ 90	Kimura and Singer, 1962
NADH dehydrogenase	1,000,000	~ 90	Singer et al., 1957
α-glycerophosphate dehydrogenase	2,000,000	~ 100	Ringler, 1961

*With the exception of cytochrome c, the dimensions
of the molecules have been assumed on the basis of
the dimensions that are known for protein molecules
of corresponding molecular weight, or calculated on
the basis of a spherical shape. For cytochrome c
the dimensions were those determined from X-ray
crystallographic data by Dickerson et al (1967).

Fig. 15. Mitochondria in proximal convoluted tubule cell of the rat kidney. The kidney tissue was
cross-linked for 15 minutes through perfusion with
glutaraldehyde and dehydrated for 3 minutes in
ethylene glycol. Mitochondrial membranes are seen
at various degrees of tilt in relation to the plane
of the section. The particulate structure of the
membranes is most clearly seen in membranes which
are oriented obliquely to the plane of the section
(arrows) because superposition of structural components of the membranes interferes less under the
latter conditions. Magnification 100,000 X. (From
Sjöstrand and Barajas, 1968)

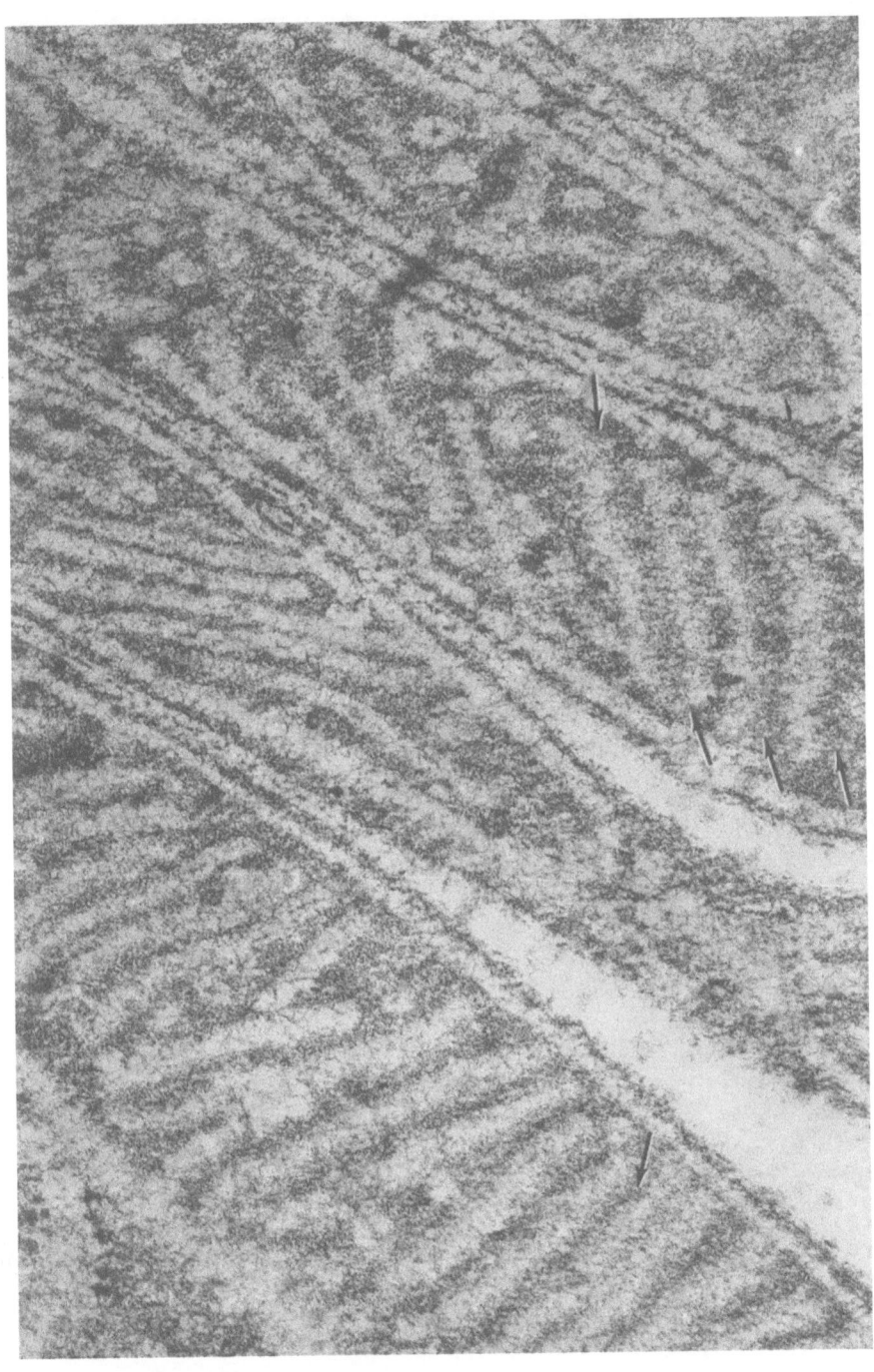

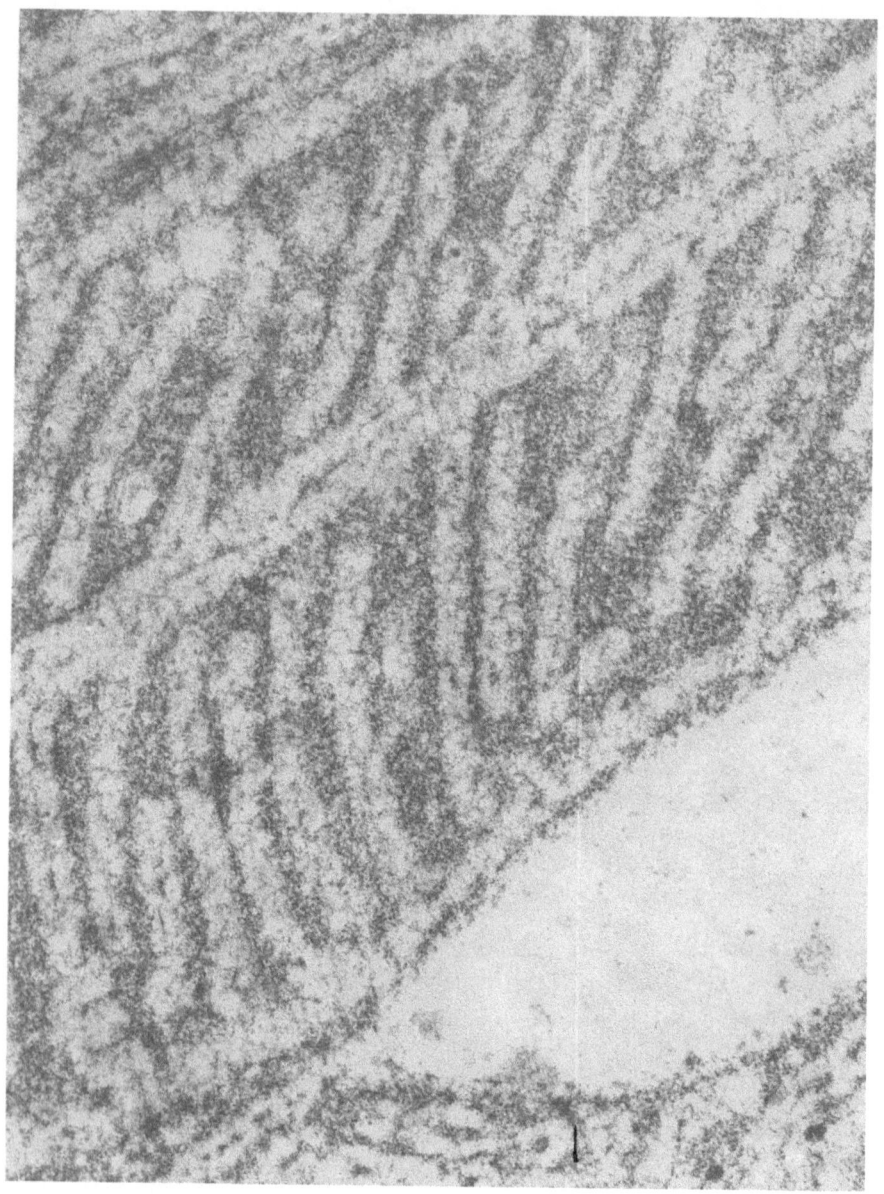

Fig. 16. Mitochondria in proximal convoluted tubule cell of the rat kidney, treated as described in
legend to Fig. 15. The mitochondrial membranes
show a particulate structure. Magnification,
170,000 X.

with less interference of superposition effects.
The diameters of the particles range from below
40 Å to about 100 Å as shown in Diagram I.

After 15 minutes cross-linking (Fig. 16) the
mean diameter of the lightly stained particles had
increased from 70 Å to 90 Å. Particles measuring
more than 100 Å in diameter could observed.

The range 40 to 100 Å of the dimensions of the
particles corresponds to that we would expect for
the respiratory chain components and other enzymes a
associated with the mitochondrial membrane. Some
membrane bound components of mitochondria are listed in Table I, and their molecular weights are
shown. The diameters of these molecules shown in
the table are, with the exception of cytochrome c,
reasonable assumptions made on the basis of the
molecular weights and a simple spherical shape of
the molecules. From this table it is obvious that
we are dealing with a complex mixture of components
of greatly varying dimensions. Such a mixture of
particles cannot be associated in a regular crystalline arrangement but are likely to form a membrane structure of rather irregular appearance.
This is characteristic for the membranes observed ·
after this type of tissue preservation.

The total mean thickness of the mitochondrial
membranes was calculated to about 300 Å and the
mean thickness of the individual membrane elements,
which are closely apposed, was found to be about
150 Å. This is about three times the thickness of
the membrane elements as observed after conventional preparation procedures.

One type of particle with a characteristic
geometrical shape was observed. Rectangularly
shaped particles measuring about 60 x 100 Å appeared strikingly frequently (Fig. 17).

The particles in the mitochondrial membrane
elements we interpret to be protein molecules.
This interpretation is based on the fact that
these particles in size correspond to the membrane

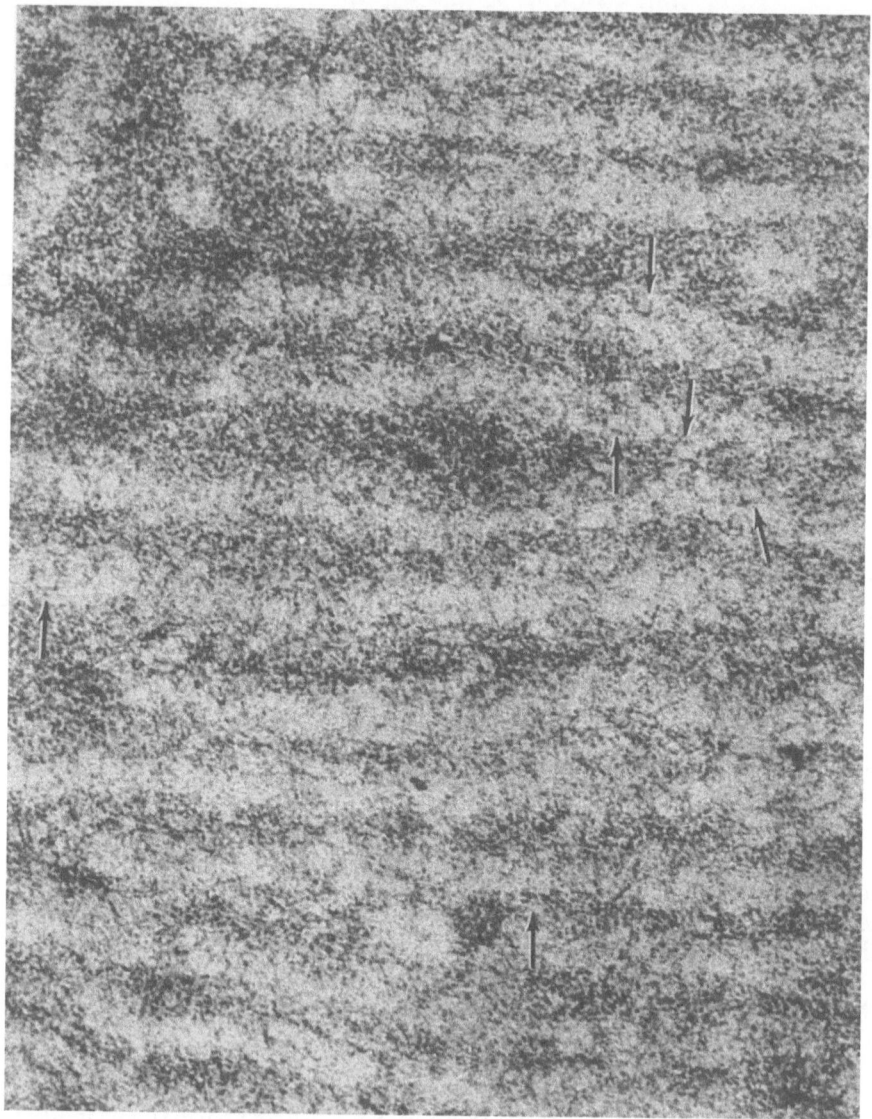

Fig. 17. Part of rat kidney mitochondrion. Cross-linking and dehydration as described in legend to Fig. 15. In these obliquely oriented inner mito-chondrial membranes several lightly stained parti-cles with a rectangular or square shape can be ob-served (arrows). Magnification 240,000 X.

proteins. That the particles appear in negative contrast as light particles against a dark background can be given the explanation presented earlier in this paper. If this interpretation were correct this particulate structure should be greatly modified by denaturation of the proteins in the tissues.

In order to achieve extensive denaturation of the protein, 50 μ thick slices of cross-linked kidney tissue were exposed to 60° C for 400 minutes before dehydration by means of acetone ane embedding in Vestopal. The membranes in denatured tissue appeared to consist of a light layer measuring about 50 Å in thickness (Fig. 18). Frequently a triple layered, "unit membrane" type of pattern could be observed. In mitochondria the membranes appeared as two lightly stained continuous layers separated by an opaque middle layer. This pattern can be interpreted to correspond to two triple-layered membrane elements in close apposition. The opaque surface layers facing each other will then fuse in the middle of the membrane to form the middle opaque layer. The opaque layers facing the matrix will not appear as separate layers because the staining of the matrix is as opaque as that of the surface layers.

It is obvious that the denaturation of the proteins leads to a destruction of the particulate structure of the cellular membranes and the appearance of a membrane pattern similar or identical to that observed after conventional preparation of tissues for electron microscopy. The triple layered, in the case of mitochondria, only about 50 A thick membrane profiles observed in electron micrographs of conventionally prepared tissues therefore represents a pattern caused by extensive denaturation of the membrane proteins. This is what could be expected since the tissue is exposed to denaturing agents during dehydration and also to denaturing fixatives over long periods of time.

After cross-linking the tissue with glutaraldehyde for 15 minutes and dehydration in acetone the membranes appeared triple layered, exhibiting

Fig. 18. Mitochondria in proximal convoluted tubule
cell of the rat kidney cross-linked for 15 minutes
through perfusion with glutaraldehyde. 50 μ thick
slices of the tissue were kept for 400 minutes at
65°C for heat denaturation before dehydration in
acetone and embedding.

The mitochondrial membranes appear to consist
of two continuous light layers separated by an
opaque middle layer. The light layers measure
about 50 Å in thickness. Compare the thickness of
the membranes and of the membrane elements in this
picture and in Fig. 13, which represents a slightly
lower magnification. Magnification 130,000 X. (From
Sjöstrand and Barajas, 1968)

Fig. 19. Mitochondrion in proximal convoluted tubule cell of rat kidney cross-linked for 15 minutes
through perfusion with glutaraldehyde and dehydrated in acetone.

The mitochondrial membranes appear similar to
those in Fig. 18. Magnification 120,000 X. (From
Sjöstrand and Barajas, 1968)

Fig. 20. Mitochondrion in proximal convoluted tubule cell of rat kidney treated like specimen described in legend to Fig. 19, with the exception
that the 50 μ thick tissue slices were stored in
buffer in a refrigerator overnight.

The mitochondrial membrane elements appear
either as about 50 Å thick light lines (arrows) or
as light layers several times thicker than these
lines and with a particulate substructure. Magnification 140,000 X. (From Sjöstrand and Barajas,
1968)

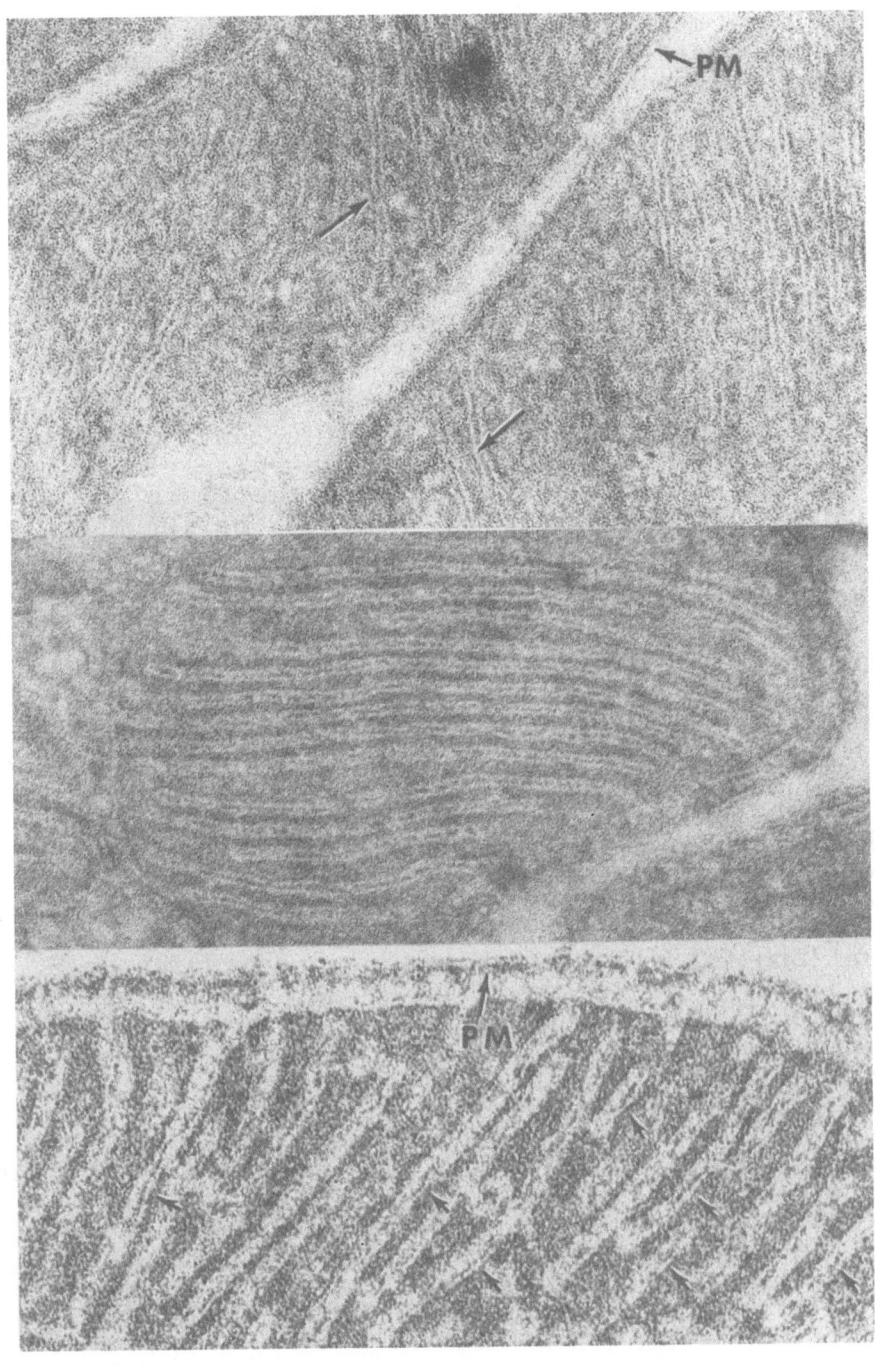

the "unit membrane" type of structure (Fig. 19).
In a few cases when the tissue had been exposed to
this treatment parts of the membrane appeared this
way, while other regions showed a particulate pattern similar to that observed after ethylene glycol dehydration (Fig. 20). This mixed appearance
of the membranes can be explained by a partial denaturation of the membrane proteins.

The outcome of the experiments in which we
tried to avoid an extensive denaturation of the
membrane proteins supports the concept that the
membranes consist of globular protein molecules of
different dimensions. It was not possible to find
any support for the existence of a continuous lipid layer in mitochondrial membrane elements or in
cytoplasmic membranes. There is therefore no justification to consider the Danielli-Davson model
for the plasma membrane as representing a general
feature of cellular membranes. The structural pattern which has been referred to as supporting this
concept, the triple layered pattern of the "unit
membrane," could be shown to develop as a consequence of extensive denaturation of membrane proteins. This pattern has earlier been shown to develop independent of the lipid content of mitochondrial membranes as shown by the experiments by
Fleischer et al (1965). It has also been shown
that proteins like cytochrome oxidase, with a lipid content as low as 8%, easily aggregate to form
membranes with a triple layered "unit membrane"
type of profile after being subjected to the conventional fixation and embedding procedures used
in electron microscopy (Sun et al, 1968). The
triple layered appearance of a membrane profile
can therefore not be interpreted exclusively in
terms of a lipid bilayer structure. More or less
pure protein membranes show the same profile pattern. According to the observations made on glutaraldehyde cross-linked and ethylene glycol dehydrated tissue, the thickness of the mitochondrial membrane elements (150 Å) exceeded the mean diameter
of the membrane particles (70 Å after 3 minutes
cross-linking). This means that the particles are

Fig. 21. Apical region of columnar cell in the epithelium of mouse small intestine. Below the brush
border processes a zone with no or only a few vesicles is seen. Below this zone the cytoplasm contains a large number of smooth surfaced vesicles
(V). Magnification 72,000 X.

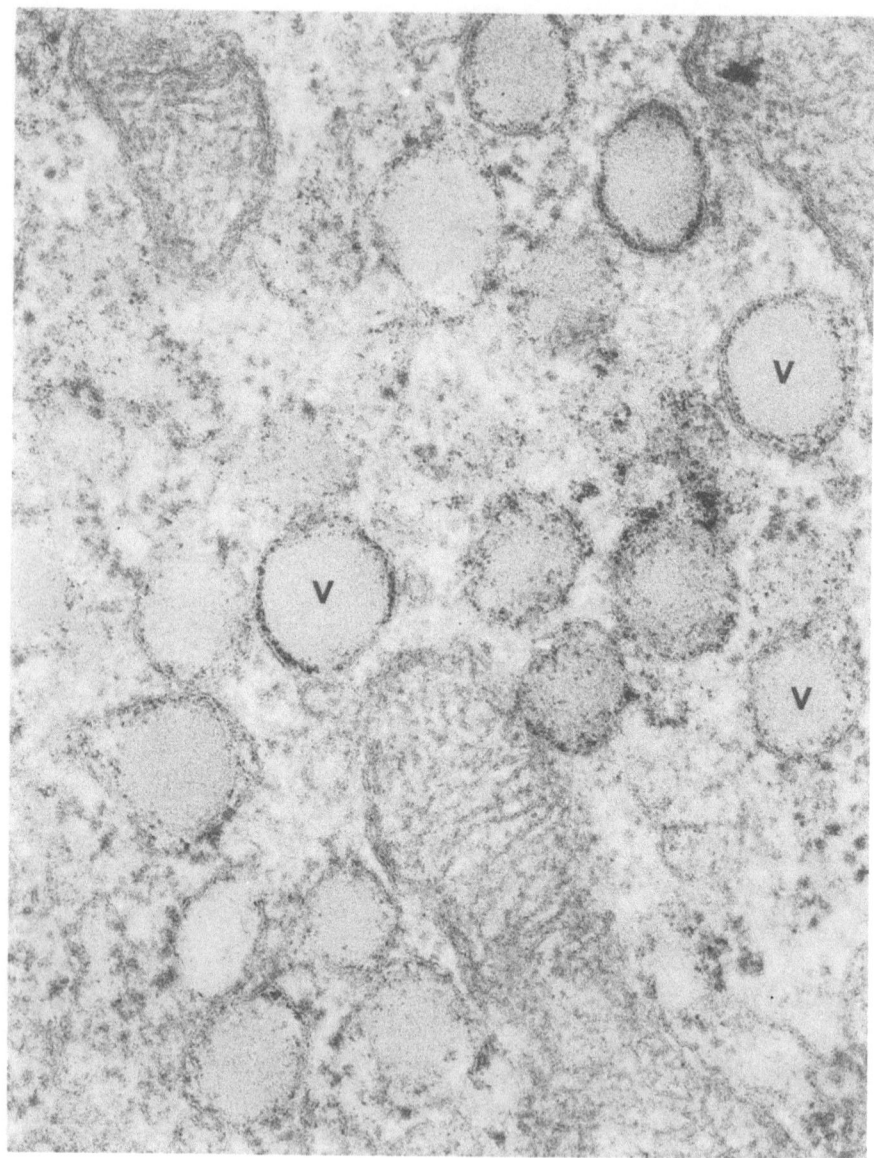

Fig. 22.   Apical region of columnar cell in the
epithelium of mouse small intestine after feeding
the animal a fatty meal.   Fat has accumulated in
closed vesicles (V) bounded by smooth surfaced
membranes.   Magnification 110,000 X.

not arranged in one plane but in a staggered way,
with the possibility of regions in which two layers of particles exist. Such an arrangement would
be expected if the enzyme molecules of the membranes are associated in aggregates or assemblies
like it has been assumed for the respiratory chain
components.

A particulate substructure was also observed
in the plasma membrane although this type of membrane has not yet been studied in a rigorous way,
with these techniques. The membrane model proposed
here would make it easy to account for the properties of the plasma membrane, both permeability
properties and immunological characteristics. It
could facilitate understanding how the plasma membrane could be adapted to different specialized
functions, including that of the synaptic membranes. It would leave us possibilities to introduce receptor molecules for transmittor substances
in the membrane and to vary the concentration of
receptor molecules which could lead to variations
in the threshold of stimulation.

The compartmentalization of the cytoplasm. When
searching for a type of cytoplasmic membrane that
could be isolated for a biochemical study, without contamination with other cellular components,
it was found that the apical vesicles in the columnar cells of the intestinal epithelium are particularly suitable (Fig. 21). They represent membranes bounding cytoplasmic compartments in which
during fat absorption triglycerides are accumulating (Fig. 22). It therefore appeared logical to
associate the bounding membrane of these vesicles
with triglyceride synthesis (Sjöstrand, 1963).

A method was worked out by which the intestinal epithelium could be isolated pure with no contamination from the connective tissue of the intestinal villae (Sjöstrand, 1968). Isolation of
the intestinal epithelium after a fatty meal when
the apical vesicles were loaded with fat, followed
by gentle homogenization of the epithelium and step
gradient centrifugation, led to a separation of the
fat filled apical vesicles in a rather pure form
due to their low density.

The content of these vesicles was analyzed by
means of silica gel thin layer chromatography. Although triglycerides were the major component of
the vesicles, a strikingly high concentration of
free fatty acids was also found associated with the
vesicles (Sjöstrand and Borgstrom, 1967). This observation would agree with the concept that triglycerides are synthesized in these vesicles.

The compartmentalization of the cytoplasm in
this case could then be conceived of as allowing
the maintining of a high substrate concentration
at the site of a high enzyme concentration. As a
consequence, the product would accumulate locally
within these compartments.

The compartmentalization of the cytoplasm
therefore might well fulfill the purpose of increasing the rate of chemical reactions by allowing the maintaining of high concentrations of re-
actants locally in the cytoplasm. The enzymes
thus speed up the rate of reactions as catalysts,
the supramolecular structural organization assists
further in increasing the rate of reactions both
by compartmentalization and by securing the proper structural arrangement within the membranes of
the molecules in multienzyme systems.

It seems likely that we are now at the beginning of a new era in electron microscopic analysis
of the molecular structure of cellular components.
With our minds freed from preconceived ideas inherited from the past, and requiring rigorous application of chemical and physical chemical knowledge and controls in electron microscopy as applied to the study of cellular components, we have
opened the door for a great future for the revealing of the molecular structure of these components
and for deducing functional characteristics that
are the consequence of this structural organization and that would never be understood or even
conceived of without this knowledge of the molecular structure of the basic components of living
systems.

REFERENCES

1.   Ambe, K.S. and Venkataraman, A., Biochem.Bio-
     phys. Res. Commun., 1, 133-137, 1959.
2.   Criddle, R.S. and Bock, R.M., Biochem. Biophys.
     Res. Commun., 1, 138-142, 1959.
3.   Criddle, R.S., Bock, R.M., Green, D.E. and
     Tisdale, H., Biochemistry, 1, 827-842, 1962.
4.   Davson, H. and Danielli, J.F., The Permeability
     of Natural Membranes, Cambridge Univ. Press,
     London, 57, 1952.
5.   Dickerson, R.E., Kopka, M.L., Weinzierl, J.,
     Varnum, J., Eisenberg, D. and Margoliash, E.,
     J. Biol. Chem., 242, 3015-3018, 1967.
6.   Fleischer, S. Fleischer, B. and Stoeckenius,
     W., Fed.Proc., 24, 296, 1965.
7.   Goldberger, R., Smith, A., Tisdale, H. and
     Bomstein, R., J. Biol. Chem. 236, 2788, 1961.
8.   Kimura, T. and Singer, T.P., in: Methods of
     Enzymology, Vol. V. (Colowick, S.P. and Kaplan,
     N.O., Eds.), Academic Press, Inc., N.Y., 562,
     1962.
9.   Palade, G.E., Anat. Record, 114, 427, 1952.
10. Palade, G.E., J. Histochem, Cytochem., 1, 188,
     1953.
11. Pease, D.C., J. Ultrastruct. Res., 14, 356-378,
     1966.
12. Quiocho, F.A. and Richards, F.M., Proc.Nat.
     Acad.Sci., U.S., 52, 833-839, 1964.
13. Quiocho, F.A. and Richards, F.M., Biochemistry,
     5, 4062-4076, 1966.
14. Racker, E., Mechanisms in Bioenergetics, Acad.
     Press, N.Y., 1965.
15. Ringler, R.L., J. Biol. Chem., 236, 1192-1198,
     1961.
6.   Sage, H.J. and Singer, S.J., Biochim. Biophys.
     Acta, 29, 663-664, 1958.
17. Schmidt, W. J., Z. Zellforsch. u. mikr. Anat.,
     23, 657, 1936.
18. Schmitt, F.O., Bear, R.S. and Palmer, K.J.,
     J. Cell. Comp. Physiol., 18, 31, 1941.
19. Singer, S.J., Adv. Protein Chem., 17, 1, 1962.
20. Singer, T.P., Kearney, E.B. and Bernath, P.,
     J. Biol. Chem., 223, 599-613, 1956.

21.   Singer, T.P., Massey, V. and Kearney, E.B.,
       Adv. Enzymol., 18, 65-111, 1957.
22.   Sjöstrand, F.S., J. Cell. Comp. Physiol., 33,
       383-403, 1949.
23.   Sjöstrand, F.S., J. Cell. Comp. Physiol., 42,
       15-44, 1953a.
24.   Sjöstrand, F.S., Nature, 171, 30, 1953b.
25.   Sjöstrand, F.S., J. Cell. Comp. Physiol., 42,
       45-70, 1953c.
26.   Sjöstrand, F.S., Experientia, 9, 68-69, 1953d.
27.   Sjöstrand, F.S., Verh. Anat. Gest., Versamml.,
       Stockholm, Gustav Fischer Verlag, Jena, 53,
       1956.
28.   Sjöstrand, F.S., In: Modern Scientific As-
       pects of Neurology (Cumings, J.N., Ed.) Ed-
       ward Arnold, London, 188-231, 1960a.
29.   Sjöstrand, F.S., Radiation Res. Suppl., 2,
       349-361, 1960b.
30.   Sjöstrand, F.S., Nature, 199, 1262-1264, 1,
       1963a.
31.   Sjöstrand, F.S., J. Ultrastruct. Res., 9,
       340-361, 1963b.
32.   Sjöstrand, F.S., J. Ultrastruct. Res., 9,
       561-580, 1963c.
33.   Sjöstrand, F.S., J. Ultrastuct. Res., 8,
       517-541, 1963d.
34.   Sjöstrand, F.S., In: Intracellular Membranous
       Proc. Intern. Symp. Cellular Chemistry, 1st,
       Ohtsu, Japan, Suppl. of Symp. Soc. Cellular
       Chemistry, 14, 103-125, 1963.
35.   Sjöstrand, F.S., In: Biochemistry and Pharm-
       acology of the Basal Ganglia; Proc. 2nd Symp.
       of the Parkinson's Disease Information and
       Research Center, Columbia Univ., N.Y. (Costa,
       E., Cote, L.J. and Yahr, M.D., Eds.), Hewlet,
       N.Y., Raven Press, 17-41, 1965.
36.   Sjöstrand, F.S., Protoplasma, 63, 248-261,
       1967a.
37.   Sjöstrand, F.S., Protides of the Biol. Fluids,
       15, 15-36, 1967b.
38.   Sjöstrand, F.S., J. Ultrastruct. Res., 22,
       424-442, 1968a.
39.   Sjöstrand, F.s., In: Ultrastucture and Func-
       tion of Cellular Membranes (Dalton, A.J. and
       Haguenau, F., Eds.), Academic Press Inc., N.

Y., 151-210, 1968B.

40. Sjöstrand, F.S., In: Regulatory Functions of
     Biological Membranes, (Järnefelt, J., Ed.),
     Amsterdam, Elsevier Publishing Company, 1-
     20, 1968c.
41. Sjöstrand, F.S. and Baker, R.F., J. Ultra-
     struct. Res., 1, 239-246, 1958.
42. Sjöstrand, F.S. and Barajas, L., J. Ultra-
     struct. Res., 25, 121-155, 1968.
43. Sjöstrand, F.S. and Borgström, B., J. Ultra-
     struct. Res., 20, 140-160, 1967.
44. Steinmann, E., Exp. Cell. Res., 3, 367-372,
     1952.
45. Steinmann, E. and Sjöstrand, F.S., Exp.
     Cell Res., 8, 15-23, 1955.
46. Sun, F.F., Prezbindowski, J.S., Crane, F.L.
     and Jacobs, E.E., Biochim. Biophys. Acta,
     153, 804-818, 1968.
47. Tanford, C., Hauenstein, J.D. and Rands, D.
     G., J. Am. Chem. Soc., 17, 6409-6413, 1955.
48. Tanford, C., Swanson, S.A. and Shore, W.S.,
     J. Am. Chem. Soc., 77, 6414-6421, 1955.
49. Tanford, C., Buckley, C.E. III, Paritosh,
     K.D., and Lively, E.P., J. Biol. Chem., 237,
     1168-1171, 1962.

FUNCTIONAL ASPECTS OF SYNAPTOSOMES

R. M. Marchbanks and V. P. Whittaker

SUMMARY

Synaptosomes are pre-synaptic nerve terminals which have been detached by homogenisation of cerebral cortex and isolated by density gradient centrifugation. The synaptosome limiting membrane encloses cytoplasm, synaptic vesicles, and often intra-terminal mitochondria. These may be separated by hypo-osmotic disruption of synaptosomes and further density gradient centrifugation. Synaptosome preparations have been exploited for studies on the content and compartmentation of transmitter substances, and the enzymes which synthesise and degrade them.

Recent studies have suggested that the degree of biochemical organisation and metabolic competence of synaptosomes is greater than was originally envisaged. Permeability studies indicate that the limiting membrane is substantially intact. Synaptosomes have been shown to respire when incubated with glucose, and to produce high energy substances such as adenosine triphosphate and phosphocreatine.

The transport of choline across the synaptosome membrane has many of the properties of the transport of choline in more organised tissues,

159

suggesting that in this respect the synaptosome
membrane is fully functional. There is incorporation of radioactive choline into the acetylcholine
of synaptosomes in vitro. The synthesis and release
of acetylcholine has been studied in synaptosomes
in vitro, ions and pharmacological agents have effects on the acetylcholine metabolism of synaptosomes which are similar to those found in more organised preparations.

These studies suggest that it is possible to
use the synaptosome preparation to investigate in
vitro a wide range of the metabolic functions of
a part of the nerve cell which is of particular
importance for the process of synaptic transmission.

It is a widely held belief that the function
of the brain is determined by the arrangement of
its constituent nerve cells, the characteristics
with which they conduct impulses, and the properties of the process by which impulses are transmitted from one nerve cell to another at the synapse. The extent to which the higher functions
such as learning and memory may be explicable on
this basis is uncertain but at least it provides
a rational framework for investigation into these
problems.

The disciplines of physiological and anatomical enquiry have led to a broad and in some particulars quite detailed knowledge of the arrangement
of nerve cells, and of how they conduct and transmit impulses. The biochemical details of the nerve
cell, and how they relate to brain function are
less extensively known. This is probably because
of the anatomical complexity of brain and of the
greater heterogeneity of the cells of which it is
composed. It is however important to understand
the biochemistry of nervous tissue, as well as the
physiology and anatomy. Physiological and anatomical processes both involve the movement of small
and large molecules, and changes in the activity
of enzymes and other biochemically important sub-

stances. The biochemistry of the action of the
many drugs used in the treatment of nervous disorder must be understood in order for advances in
therapy to be made.

The process of transmission of an impulse from
one nerve cell to the next at the synapse is particularly appropriate for biochemical investigation.
This is not only because of the importance of
transmission in brain function as a whole but also
because the impulse is transmitted at many synapses
by a chemical substance. When a nerve impulse arrives at the presynaptic nerve terminal, at many
synapses it causes the liberation of a chemical
substance which having diffused across the synaptic gap results in the propagation of the impulse
down the next nerve cell (Review: Eccles, 1964).
The presence of small vesicles is characteristic
of the pre-terminal region and they have been termed
synaptic vesicles (de Robertis & Bennett, 1954).
Electro-physiological evidence shows that acetylcholine is liberated from the pre-terminal region
in packets--or "quanta". When an impulse arrives
at the synaptic region a number of quanta of acetylcholine are released (del Castillo & Katz, 1955).
It has been proposed that each of the synaptic
vesicles contains a quantum of acetylcholine. The
chemical identity of other transmitter substances
and the enzymes which synthesize and degrade them
have been the object of much investigation.

The mechanism of the retention of adaptive behaviour and of other higher functions of the brain
may involve changes in the efficiency of transmission at the synapse. For an understanding of these
mechanisms it is important to know the intimate details of the various processes involved in chemical transmission and the factors which affect them.
It has been found possible to isolate in a relatively intact form the synaptic region of the nerve
cells from the brain (Gray & Whittaker, 1962; de
Robertis et al, 1962). The isolation of the synaptic region makes it possible to investigate the
biochemistry of this region its content of transmitter substances and associated enzymes without in-

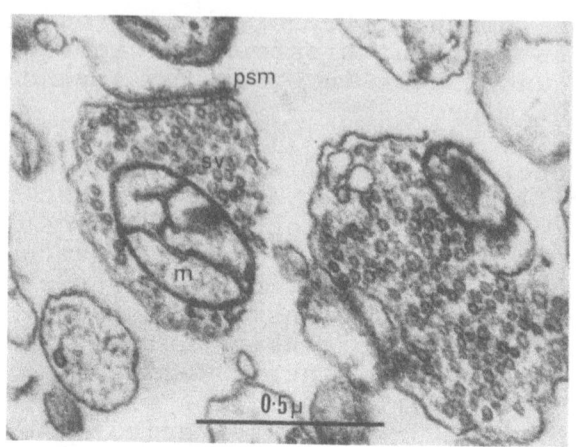

Fig. 1.   Synaptosome fraction. The synaptosome
marked contains a mitochondrion (m) and synaptic
vesicles (sv) surrounded by the extra-synaptosomal
membrane to which a piece of post synaptic membrane
(psm) is adhering (Whittaker & Sheridan, 1965).

terference from other parts of the nerve cell.
This paper will consider the extent to which in
vitro preparations of the synaptic region can be
used to study the biochemistry of the processes
involved in synaptic transmission. The isolation
of pre-synaptic nerve terminals is made possible
because when cerebral cortex is carefully homogen-
ised the nerve terminals nip off but otherwise retain their structure while the rest of the nerve
cell is destroyed. These detached pre-synaptic
nerve terminals have (Fig. 1) been called synaptosomes to indicate their origin. The synaptosome
limiting membrane encloses synaptic vesicles, often mitochondria, and a sample of the intra-terminal cytoplasm.

On centrifugation synaptosomes sediment with
the mitochondria, and they may be further purified
by taking the crude mitochondrial fraction, and
fractionating it on a sucrose density gradient.
The synaptosomes have a density higher than that
of myelin and lower than that of mitochondria so
they come to rest in an intermediate part of the

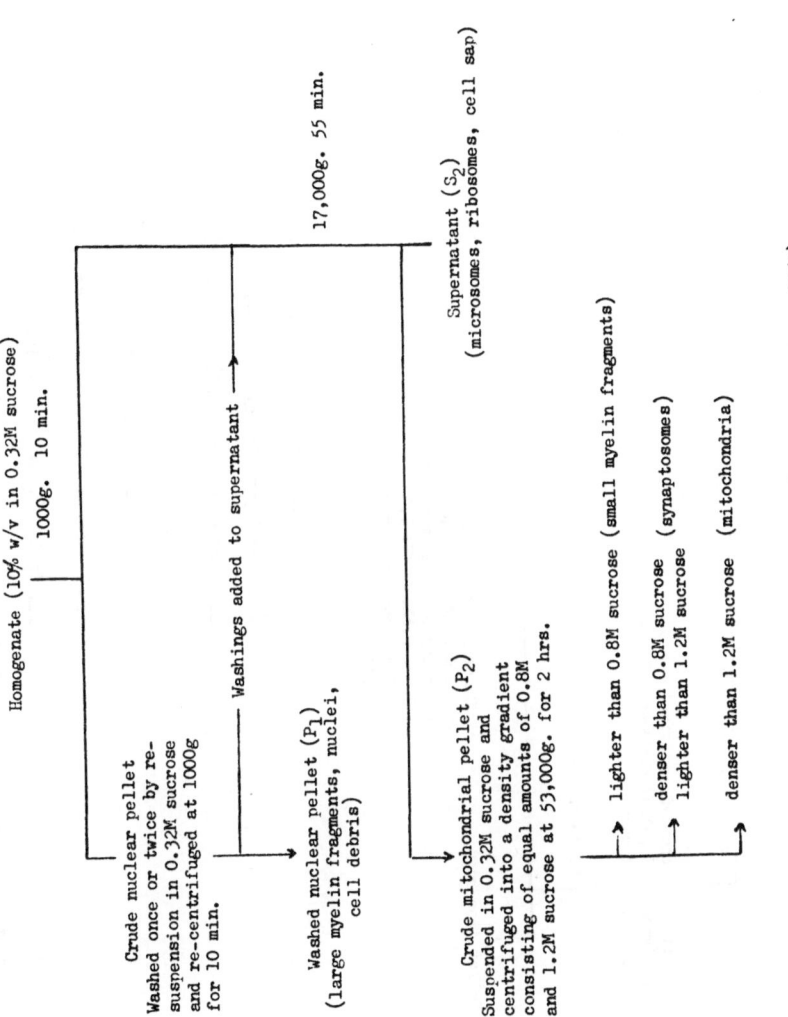

Homogenate (10% w/v in 0.32M sucrose) 1000g. 10 min.

Crude nuclear pellet
Washed once or twice by re-
suspension in 0.32M sucrose
and re-centrifuged at 1000g
for 10 min.

Washings added to supernatant

Washed nuclear pellet (P₁)
(large myelin fragments, nuclei,
cell debris)

17,000g. 55 min.

Supernatant (S₂)
(microsomes, ribosomes, cell sap)

Crude mitochondrial pellet (P₂)
Suspended in 0.32M sucrose and
centrifuged into a density gradient
consisting of equal amounts of 0.8M
and 1.2M sucrose at 53,000g. for 2 hrs.

lighter than 0.8M sucrose (small myelin fragments)

denser than 0.8M sucrose
lighter than 1.2M sucrose (synaptosomes)

denser than 1.2M sucrose (mitochondria)

Table 1.   Scheme for separating synaptosomes. (Gray & Whittaker, 1962).

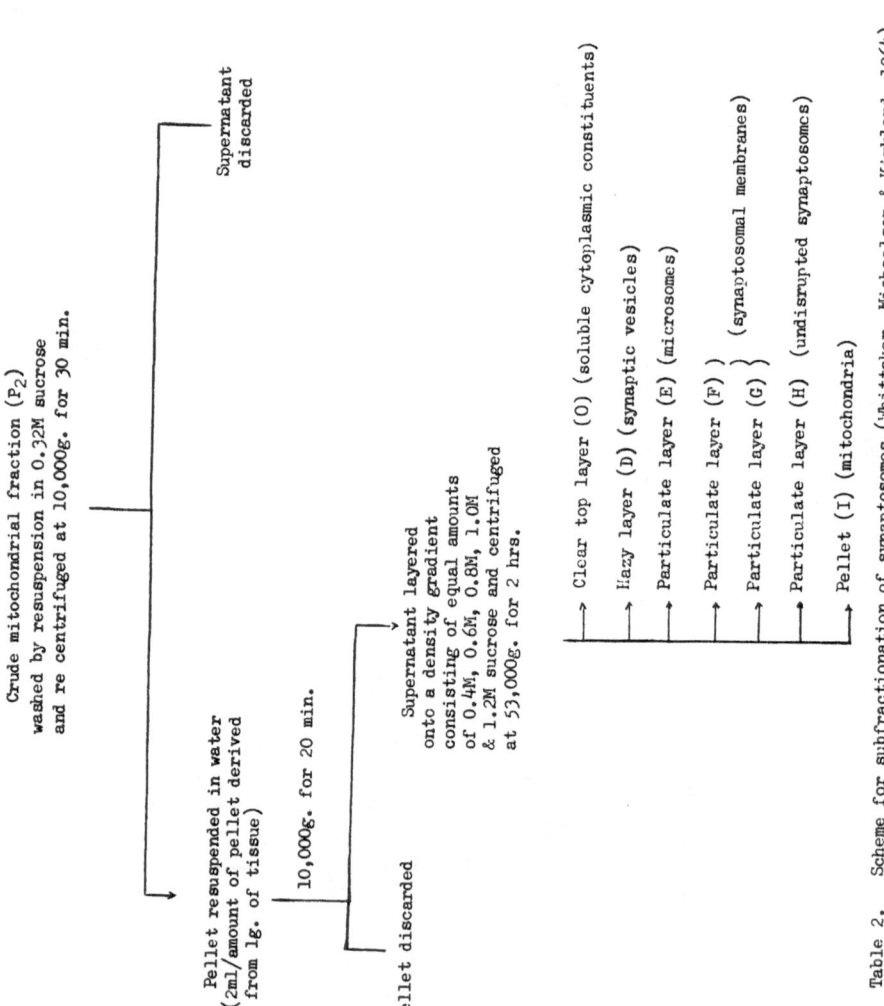

Table 2.   Scheme for subfractionation of synaptosomes (Whittaker, Michaelson & Kirkland, 1964).

gradient.   The method of preparation is shown in Table 1.

Synaptosomes contain the chemical transmitter acetylcholine, and the enzyme which synthesises it-- choline acetyltransferase.   When synaptosomes are hypo-osmotically disrupted only about 50% of the acetylcholine is released, this is called the labile bound acetylcholine (Johnson & Whittaker, 1963), and there is evidence that it is free in the synapto- some cytoplasm (Marchbanks, 1967 b).   By further density gradient fractionation (see Table 2) of ruptured synaptosomes the contents of the synapto- some may be separated and isolated (Whittaker et al, 1964).   Most of the acetylcholine which has survived the hypo-osmotic shock is recovered in the same fraciion of the density gradient in which are found small vesicles (Fig. 2) of diameter about 500 Å.   The acetylcholine associated with the syn- aptic vesicle fraction appears to be retained with- in, rather than bound to, the vesicles because it can be released by osmotic shock (Marchbanks, 1967 b, 1968).

These findings establish the morphological correlate of the quantal unit of chemical trans- mitter substance, at least in the case of acetyl- choline.   The technique of osmotic disruption of synaptosomes and density gradient separation of the soluble cytoplasm, synaptic vesicles, external membrane and intra-terminal mitochondria has been much used to investigate the synaptic localisation of other chemical transmitters, and the enzymes which synthesise and inactivate them (Review: Whittaker, 1965).   There is evidence that the syn- aptosome preparation can be used to study complex biochemical phenomena in the pre-terminal region of the nerve cell.   It is known that synaptosomes contain organelles characteristic of the pre-termi- nal region and soluble cytoplasmic constituents such as the enzyme lactate dehydrogenase, and po- tassium ions.   This suggested that the external limiting membrane had become fully sealed during the initial homogenisation otherwise these con- stituents would be lost during the subsequent iso- lation procedure.

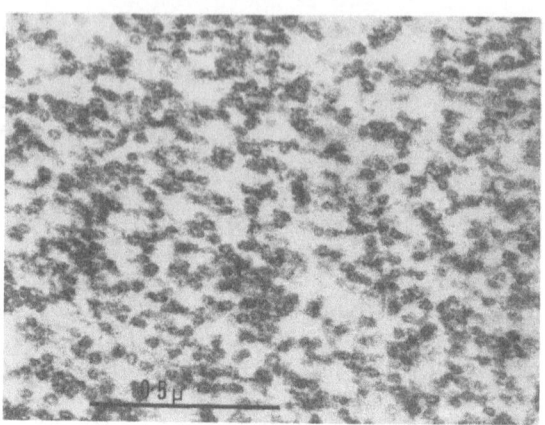

Fig. 2.   Isolated synaptic vesicles (Whittaker &
Sheridan, 1965).

The structural integrity of the outer membrane
was investigated in more detail using a gel filtra-
tion method.   A column made of Sephadex will ab-
sorb small molecular weight substances, but allow
large molecular weight material to pass through.
Any small molecular weight substances which are
bound to, or retained within large molecular weight
substances will also pass straight through the
column.   It is therefore possible with this procedure to separate synaptosomes and their contents
from the suspending medium.   If the synaptosomes
are passed over a Sephadex column that is equilibrated with an iso-osmotic sucrose (0.4M) solution
they retain their internal potassium.   If however
the column is equilibrated with a strongly hypo-
osmotic solution, the osmotic shock causes the
synaptosomes to rupture, and they lose their
internal potassium which is retained on the column.
The procedure is rapid and accurate and it has been
found useful for investigating the amounts of substances that are actually inside the synaptosome
limiting membrane (Marchbanks, 1967a).

If synaptosome preparations are suspended in
media containing small molecular weight substances,
such as potassium, sodium, or radioactive galactose
these substances cross the limiting membrane, and
can be found inside the synaptosome.   Figure 3
shows the uptake of potassium, expressed as a frac-

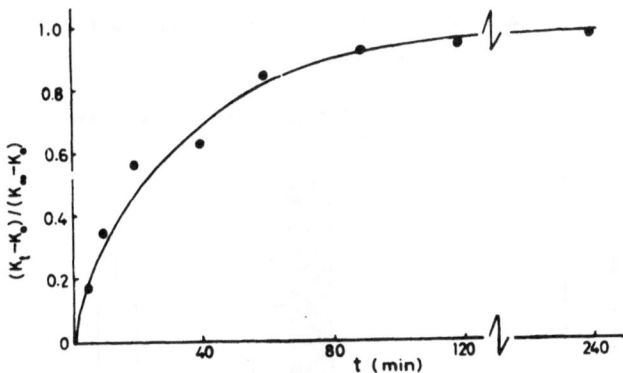

Fig. 3.   Fractional approach to equilibrium of K^+ entering synaptosomes at 5°. K_t is K^+ inside synaptosome at time t, K_o is that at time zero, and K_{oo} is that at equilibrium (more than 3 hours). Concentration of K^+ in medium was 100 mM (Marchbanks, 1967a).

tion of that found at equilibrium as a function of time. The experiments were done at 5°C, and because no concentration of potassium occurred it is believed that under these circumstances only passive diffusion processes operate. It is thus possible to measure the rate of passive diffusion across the synaptosome limiting membrane and to compare the rates with those found in other more intact tissues. The rate of potassium diffusion across the synaptosome membrane was found to be similar to the rate of potassium resting efflux and influx in mammalian non-myelinated nerve fibres (Keynes & Ritchie, 1965). The similarity of the rates suggests that the synaptosome membrane is fully sealed for if it was otherwise and the membrane had large gaps, the rate of permeation would be much greater.

By equilibrating the space inside synaptosomes with a known concentration of a small molecular weight material it is possible to measure the volume of the space enclosed by the synaptosome limiting membrane. This works out at about 40µl. in the amount of synaptosomes derived from 1 gm. of cortex. This estimate may be compared with one of

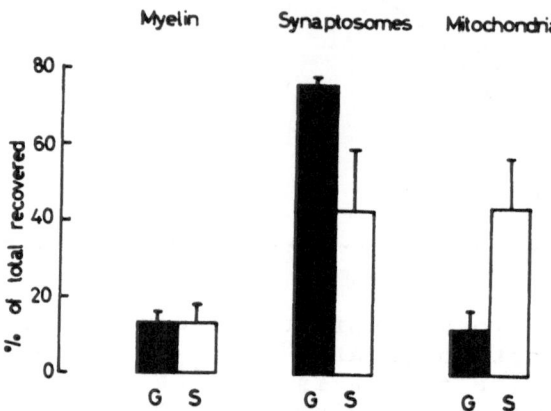

Fig. 4. Ability of the subfractions of a crude brain mitochondrial fraction to use either glucose or succinate as a substrate of respiration.

31 μl. made by measuring the number and size of synaptosomes seen in the electron micrographs. The correlation of the values indicates that by this technique the volume of the cytoplasmic space enclosed by the synaptosome limiting membrane can be measured (Marchbanks, 1966).

In order to investigate more functional properties it was necessary to investigate whether and under what circumstances synaptosomes could provide themselves with high energy phosphate intermediates. Synaptosomes frequently contain mitochondria, and their ability to respire in a physiological medium to which either glucose or succinate or both were added was examined. Figure 4 shows that the ability to respire with glucose alone as substrate is a property of the synaptosome subfraction of the crude mitochondrial fractions. Synaptosomes will also respire with succinate but this is more typically a property of mitochondria, which do not utilise glucose. The ability of synaptosomes

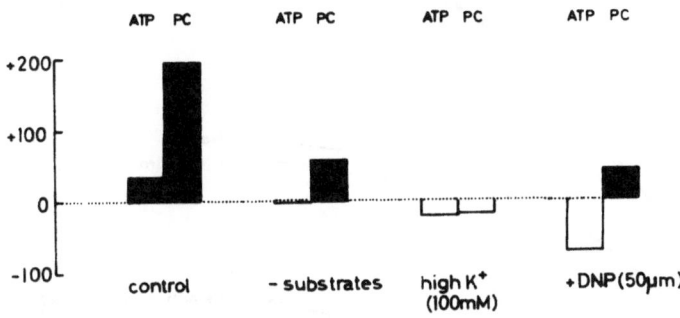

Fig. 5. Synthesis of high energy phosphate derivatives in synaptosomes under various conditions.

to respire with glucose suggests that they contain
a full and functional complement of the glycolytic
enzymes. It was also found that synaptosomes were
capable of synthesizing phosphocreatine and ATP
when incubated in an appropriate medium. The
changes in the amount of phosphocreatine and ATP
inside the synaptosome when they are incubated
under various conditions are shown in Figure 5.
In the presence of high potassium concentrations
the levels of phosphocreatine and adenosine triphosphate were much reduced after incubation. This
is a characteristic feature of the behaviour of
cerebral cortex slices (McIlwain, 1952), and indicates the similarity of the metabolic properties
of synaptosomes to those of more organised tissues.

The transport of radioactive choline across
the synaptosome membrane was studied because of
the importance of choline as a biological precursor of acetylcholine. Also the transport of choline
has been investigated in erythrocytes and cerebral
cortex slices so it was of interest to compare
choline transport in synaptosomes with the trans-

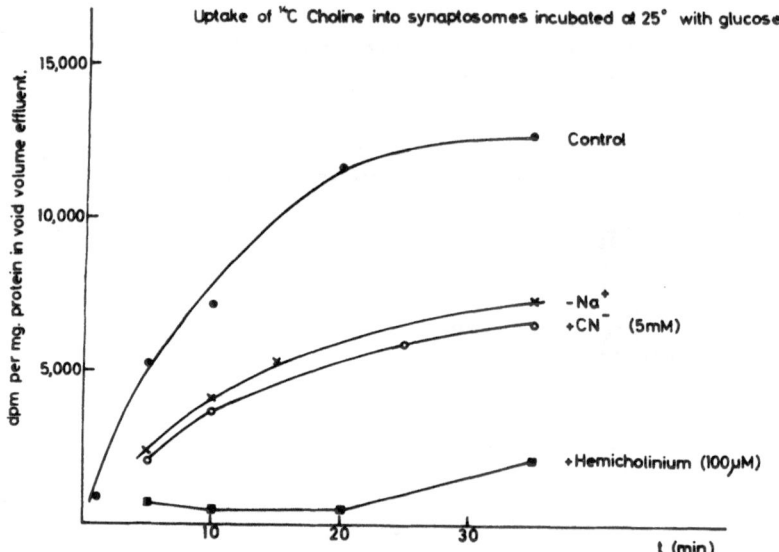

Fig. 6.   Choline transport into synaptosomes sus-
          pended in a physiological medium.

port process in these other more organised tissues.
Figure 6 shows the uptake of the radioactive ^{14}C-
choline into synaptosomes. It was established by
thin layer chromatographic analysis of the Sephadex
column effluent that the radioactivity taken up is
mostly choline, and not a metabolite. It can be
shown that the radioactivity has been taken up into synaptosomes and not some contaminating particle
because the radioactivity can be released by osmotic
shock. Furthermore, if the incubated synaptosomes
are submitted to density gradient centrifugation
the radioactivity is located in the synaptosome
fraction.

     The sodium activation of choline transport
is a characteristic feature of choline transport
in erythrocytes (Martin, 1968) and cerebral cortex
slices (Schuberth et al, 1967). High potassium
concentrations inhibit the choline uptake, again
this is a characteristic of choline transport in
other tissues. The drug hemicholinium-3 is a well
known inhibitor of choline transport.

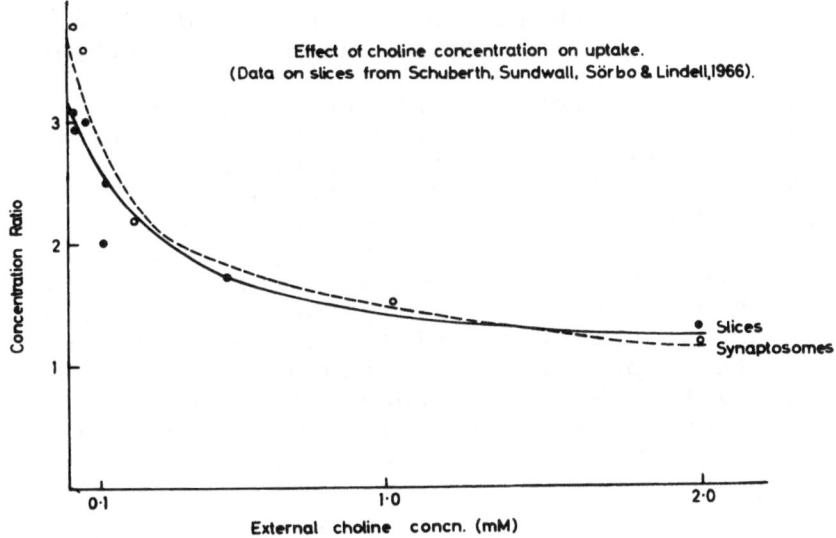

Fig. 7. Increasing choline concentration saturates the carrier so that at high choline concentrations the ratio of choline concentration inside to that outside the membrane falls to 1.

It is a feature of transport processes that are mediated by a carrier mechanism that as the concentration of the transported molecule (in this case choline) is increased the rate of transport should not increase proportionately because at high concentrations the carrier becomes saturated and cannot accomodate any more of the choline to be transported. This means that at high concentrations of choline the concentration ratio of choline between the inside and outside of the membrane that can be maintained by the transport mechanism falls to 1. Figure 7 shows the concentration ratio of choline in synaptosomes that can be maintained by the carrier mechanism at different concentrations of choline. For comparison similar data from experiments on cerebral cortex slices by Schuberth et al (1966) is shown. It can be seen that as the choline concentration increases the concentration ratio falls to one, thus indicating the existence of a functioning carrier mechanism in synaptosomes. Moreover, the quantitative similarities between

the process as seen in synaptosomes, and as seen
in the more organised cortex slices are striking.

It seems therefore that the synaptosome
limiting membrane is not only structurally intact,
but also as far as choline is conerned it is functionally competent. It is not known whether the
uptake of choline is carried out by only a proportion of the synaptosome population, perhaps those
derived from cholinergic neurones, or whether it
is a general property of synaptosomes. It should
be added that competency of the synaptosome membrane with respect to sodium and potassium transport has yet to be established; this important
criteria must be met before it can be confidently
assumed that synaptosomes are fully functional in
vitro.

Nevertheless, the similarity of choline uptake
in isolated synaptosomes and in vivo has encouraged
an investigation of the conversion of choline to
acetylcholine in synaptosomes in vitro. This was
done by separating the radioactivity of the acetylcholine formed from the radioactivity of choline
by thin layer chromatography. By also measuring
the total amount of acetylcholine present within
the synaptosome an estimate of the release of
acetylcholine from the synaptosomes can be made.
The changes in all three components of the system
as synaptosomes are incubated in the usual physiological medium are shown in Figure 8. There is uptake of radioactive choline, and conversion of it
to radioactive acetylcholine, and at the same time
there is release of endogenous acetylcholine.

As mentioned previously acetylcholine exists
in the synaptosome in two compartments, approximately one half associated with the cytoplasm, and
the rest associated with synaptic vesicles. The
compartments into which the acetylcholine was synthesised were investigated by hypo-osmotically dis-
rupting the synaptosomes after the incubation and
then isolating the vesicles by density gradient
centrifugation. The percent incorporation in the
intact synaptosome, compared with that in the iso-

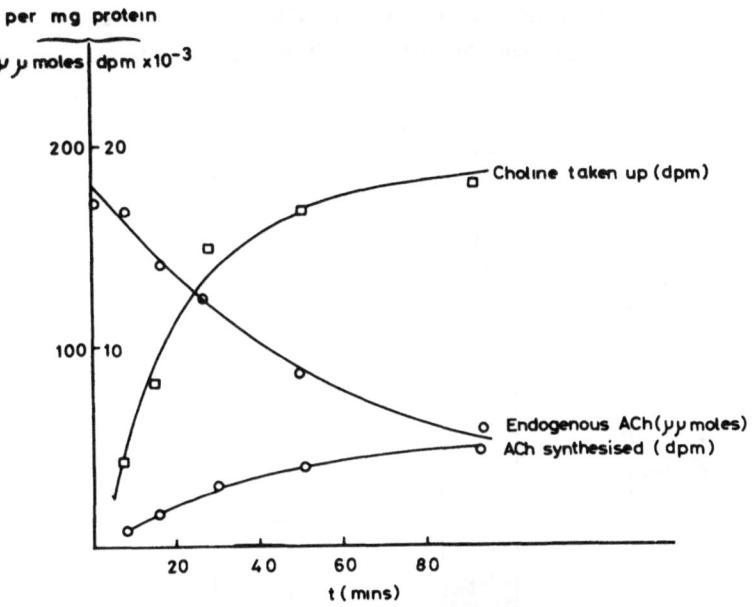

Fig. 8. Uptake of ^{14}C-choline, synthesis of ^{14}C-acetylcholine and release of endogenous acetylcholine from synaptosomes suspended in a physiological medium at 25°C.

lated vesicles is shown in Figure 9. There is little if any incorporation into the acetylcholine of vesicles. This suggests that the incorporation of acetylcholine into vesicles does not take place in the pre-terminal region although a capacity to synthesise acetylcholine exists there. Of course failure to incorporate might be an artefact of the preparation, but Dr. L. W. Chakrin (personal communication) has observed that incorporation in vivo of tritiated choline into the acetylcholine of vesicles is sluggish compared with incorporation into the acetylcholine of the cytoplasmic compartment.

The quantitative aspects of the changes in the levels of choline and acetylcholine in the synaptosome are difficult to evaluate at the moment. However, it is possible to examine the effects of various ions and pharmacological agents on these

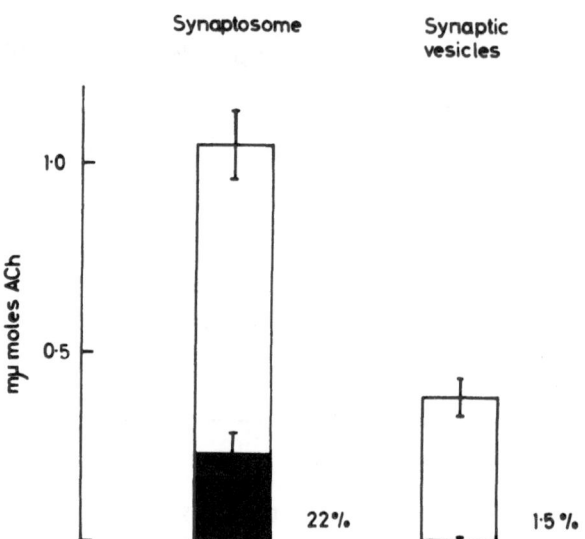

Incorporation of ^{14}C choline into ACh of synaptosomes
incubated for 30 mins. at 25° with glucose etc.

Proportion shaded = fraction incorporated.

Fig. 9. Compartmentation of the ^{14}C-acetylcholine
synthesised from ^{14}C-choline in synaptosomes. After
incubation the radioactivity of acetylcholine and
total acetylcholine were determined in synaptosomes
as a whole (left hand block), synaptic vesicles were
then isolated in the usual manner and the same measurements made (right hand block). Fraction incorporated = specific activity of radioactive acetyl-
choline/specific activity of precursor radioactive
choline.

three processes as a percentage of the control
values. Figure 10 shows some of these. The estimate of the effects of an agent on acetylcholine
synthesis has been corrected linearly for the effect it has on choline uptake. Thus omission of
sodium causes a decrease in choline uptake, and
even when the amount of acetylcholine synthesis
is corrected for lack of choline uptake under

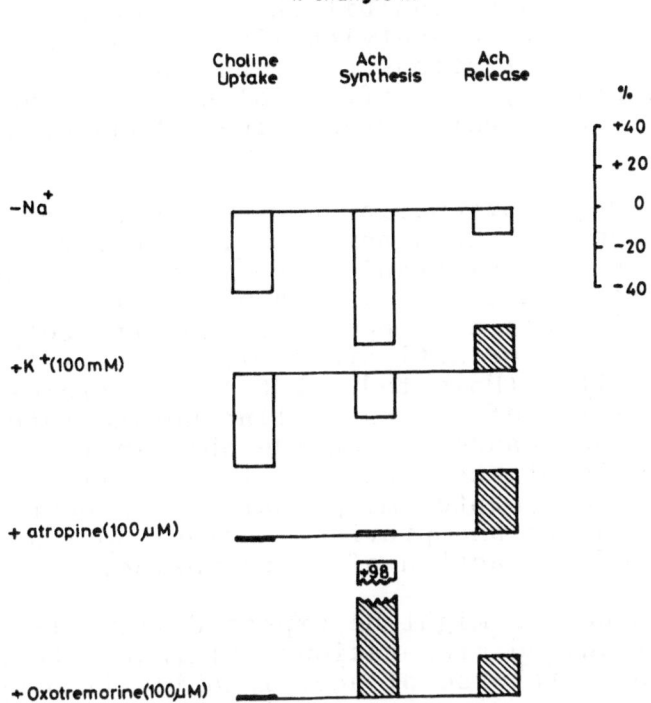

Fig. 10. Effects of ions and drugs (expressed as
% changes from control) on ^{14}C-choline uptake, synthesis of ^{14}C-acetylcholine (Ach) and release of
endogenous acetylcholine from synaptosomes incubated for 30 min. at 25°C in a physiological medium
(see text).

conditions of low sodium there is found to be failure to synthesise acetylcholine at the usual rate.
This finding is of interest since by study of the
superior cervical ganglion preparation Birks (1963)
has found that sodium is essential for acetylcholine synthesis. Sodium has little specific effect
on isolated choline acetylase activity (Morris &
Tucek, 1966). It seems that the effect of sodium
is complex, depending for its expression on the
integrity of parts of the cell structure.

On the other hand high potassium concentrations although they inhibit choline uptake, have much less effect on acetylcholine synthesis. High potassium concentrations cause an increased release of acetylcholine, this effect of potassium has been observed in cerebral cortex slices (Mann et al., 1939).

The drug oxotremorine causes tremor and Parkinsonian symptoms when administered in vivo. It also causes a considerable rise in the concentration of brain acetylcholine when administered in vivo, but it does not appear to inhibit acetylcholinesterase, or activate choline acetyl transferase in vitro (Holmstedt, 1967). Oxotremorine has very little effect on choline uptake into synaptosomes, but causes a considerable increase in the synthesis of acetylcholine from choline. This effect on the synaptosome preparation supports Holmstedt's findings and provides another approach to the study of the action of oxotremorine.

Atropine, as might be expected from its actions on more organised preparations (Giarman & Pepeu, 1964), causes release of acetylcholine from synaptosomes. These studies on acetylcholine synthesis and release in synaptosomes are at a preliminary stage, but they encourage us to hope that there is much to be learnt about the processes of chemical transmitter synthesis and release by using this preparation.

The studies from this laboratory suggest that the synaptosome preparation can be used to investigate complex metabolic functions in the pre-synaptic region. Results from other laboratories support this conclusion, a striking example being the recent demonstration by Morgan & Austin (1968) that synaptosomes in vitro have a capability for protein synthesis, which cannot be entirely due to the intra-terminal mitochondria.

With the aid of the synaptosome preparation it is therefore possible to investigate in vitro, the permeability and metabolic properties of a part of the nerve cell which is of particular importance for the processes of chemical transmission. Because of the ease of manipulation of the preparation it has great experimental flexibility. The possibility of artefacts due to damage must always be borne in mind but it is felt that there are grounds for modest optimism in hoping that useful knowledge about the biochemistry of functional processes at the synapse may come from investigation of the synaptosome preparation.

ACKNOWLEDGEMENTS

We thank Professor F. G. Young for his interest.

REFERENCES

1.   Birks, R. I., Can. J. Biochem. Physiol., 41, 2573, 1963.
2.   Del Castillo, J. and Katz, B., J. Physiol., 128, 396, 1955.
3.   De Robertis, E. D. P. and Bennett, H. S., Fed. Proc., 13, 35, 1954.
4.   De Robertis, E., De Iraldi, A. P., Arnaiz, G. R. de L. and Salganicoff, L., J. Neurochem., 9, 23, 1962.
5.   Eccles, J. C., The Physiology of Synapses, Springer Verlag, Berlin, 316, 1964.
6.   Giarman, N. J. and Pepeu, G., Brit. J. Pharmacol., 23, 123, 1964.
7.   Gray, E. G. and Whittaker, V. P., J. Anat. (Lond.), 96, 79, 1962.
8.   Holmstedt, B., Ann. N. Y. Acad. Sci., 144, 433, 1967.
9.   Johnson, M. K. and Whittaker, V. P., Biochem. J., 88, 404, 1963.
10.   Keynes, R. D. and Ritchie, J. M., J. Physiol., 179, 333, 1965.
11.   Marchbanks, R. M., Biochem. J., 100, 65, 1966.
12.   Marchbanks, R. M., Biochem. J., 104, 148, 1967a.
13.   Marchbanks, R. M., Biochem. Pharmacol., 16, 921, 1967b.
14.   Marchbanks, R. M., Biochem. J., 106, 87, 1968.
15.   Mann, P. J. G., Tennenbaum, M. and Quastel, J. H., Biochem. J., 33, 822, 1939.

16. Martin, K., J. Gen. Physiol, 1968. In press.
17. McIlwain, H., Biochem. J., 52, 289, 1952.
18. Morgan, I. G. and Austin, L., J. Neurochem., 15, 41, 1968.
19. Morris, D. and Tucek, S., J. Neurochem., 13, 333, 1966.
20. Schuberth, J., Sundwall, A., Sörbo, B. and Lindell, J.O, J. Neurochem., 13, 347, 1966.
21. Schuberth, J., Sundwall, A. and Sörbo, B., Life Sci., 6, 293, 1967.
22. Whittaker, V. P., Prog. Biophys. mol. Biol., 15, 39, 1965.
23. Whittaker, V. P., Michaelson, I. A. and Kirkland, R. J. A., Biochem. J., 90, 293, 1964.
24. Whittaker, V. P. and Sheridan, M. N., J. Neurochem., 12, 363, 1965.

THE AMOUNT OF ACETYLCHOLINE IN SYNAPTIC VESICLES

DR. BENSCH: How is this uniform amount of acetylcholine bound into such precise amounts into a quantum unit? Does the amount of acetylcholine in a quantum unit vary from animal species to animal species?

DR. MARCHBANKS: It is not clear how precise the packaging of quanta is because only a limited number of estimations of the amount of acetylcholine in a quantum have been made, and these estimates are very variable. Estimates range between 400 and 10^5 molecules of acetylcholine per quantum.

The process by which acetylcholine is incorporated into vesicles remains very mysterious. We have not been able in our laboratory to get any very significant incorporation of radioactive choline into the acetylcholine of synaptic vesicles in vitro. All the synthesised acetylcholine goes into the cytoplasmic compartment. Of course this may be an artefact of the preparation, but when radioactive choline is injected intracerebrally in vivo and after 1 hour the synaptosomes and synaptic vesicles isolated, there is still greater incorporation into the cytoplasm than into the vesicles. (Dr. L.W. Chakrin, personal communication). It seems that the incorporation of acetylcholine into vesicles is very slow compared with its synthesis in the cytoplasm. This raises the question of the functional importance of the capability to synthesise acetylcholine in the cytoplasm of the pre-terminal region, but we have little knowledge of its role at present.

As to measurements of the quantal amount of acetylcholine in various species, as far as I know comparative measurements have not been made in the same tissues from different species. This is because the size of the acetylcholine quantum can only be measured in favourable preparations.

DYNAMICS OF PROTEIN METABOLISM IN THE NERVOUS SYSTEM

Abel Lajtha and Neville Marks

In recent years a great deal of knowledge
has been accumulated on processes of protein
metabolism in the nervous system. This is a
very large topic on which there has already been
some discussion in the symposium, but which clearly concerns all aspects of brain function and mal-
function, and therefore has more aspects than
even a symposium completely devoted to it could
adequately cover. Because of its importance, and
the many recent and exciting advances, it would be
worth while to briefly, if somewhat superficially,
discuss some aspects of cerebral protein metabolism here.

The contents of the brain do not fluctuate
widely, but they are not in a closed system, but
rather in dynamic equilibrium, as we showed some
years ago for the free amino acid pool. If labelled amino acids are administered to living animals in tracer doses, which do not change plasma
levels, the label rapidly enters the brain from
the blood. Such experiments, since physiological

equilibrium is not disturbed, demonstrate that
under normal conditions there is a rapid movement
of metabolites between plasma and brain. Although
there is a rapid flow of metabolites from the blood,
the contents of the brain do not increase under normal conditions; therefore this movement must be bi-
directional-inflow and must be balanced with outflow.
The rate of this exchange can be measured as shown
in Table I. The half-life of free amino acids, that
is, the time that it takes half of the content to be
exchanged between plasma and brain by this bi-dir-
ectional flow is in minutes. More recently we studied this exchange further, and we could show that
although it is stable it undergoes alterations in a
number of conditions such as stimulation, extreme
changes in diet, or drugs; thus, this dynamic exchange, although stable under normal conditions,
does respond to the requirements and influences of
cerebral environment.

Table I. The flux of amino acid from blood to brain
(counter balanced by an equal flux from brain to
blood) is sufficiently high to replace half of the
brain amino acids in minutes.

μmoles amino acid per 100g brain

Amino Acid	Concentration in the brain	Flow per minute from blood	Half-life in brain, minutes
Leucine	3.6	1.14	2.2
Phenylalanine	5.3	0.32	11
Lysine	19.3	0.69	19
Histidine	8.1	0.23	24
Glutamic acid	990	1.56	440

TABLE II

COMPOSITION OF CEREBRAL FREE AMINO ACID POOL IN VARIOUS SPECIES

Amino Acid	μmoles amino acid per gm fresh brain				
	Mouse	Rat	Pig	Hen	Frog
Glutamic acid	9.9	11.6	9.5	12.2	5.4
Taurine	8.3	6.6	2.4	3.8	2.7
Aspartic acid	2.2	2.6	2.0	2.8	0.86
Glycine	0.91	0.68	0.71	0.42	0.83
Alanine	0.55	0.65	0.60	0.70	0.20
Lysine	0.23	0.21	0.09	0.07	0.09
Arginine	0.13	0.11	0.13	0.13	0.22
Valine	0.11	0.07	0.09	0.07	0.09
Leucine	0.08	0.05	0.05	0.05	0.09
Tyrosine	0.08	0.07	0.04	0.05	0.08
Methionine	0.05	0.04	0.03	0.02	0.07

Some of the amino acids of the cerebral acid soluble (free) amino acid pool are
shown in order of decreasing concentration.
The sequence of concentrations are closely similar in most brains investigated.

The free amino acid pools which undergo this
exchange and which have been already discussed by
Dr. Himwich in the present symposium, for different
species, are shown in Table II. The composition of
this pool is fairly characteristic for each species,
for each brain area, and for each developmental
stage. Indeed it has been called a finger-print,
since from the composition of the pool one could
recognize the particular species, brain area, or
developmental stage from which this composition
derived. In spite of this characteristic individu-
ality, there also is a basic similarity in composition. In Table II the components are put in sequence of their concentration, starting with the
amino acid present at the highest concentration in
the brain. If this order is compared among the
various species, it can be seen that the highest

TABLE III

TURNOVER RATES OF PROTEINS IN VARIOUS AREAS OF THE MONKEY BRAIN

	Half-life time in days	
	Short term (5 min) experiment	Longer term (45 min) experiment
Cerebral cortex	5.5	12.7
Midbrain	6.1	16.8
Pons-Medulla	6.6	13.7

There is regional variation in turnover rates, and longer experiments give
lower metabolic rates indicating that each area is composed of proteins
with various rates of turnover.

TABLE IV

TURNOVER OF PROTEINS IN MOUSE BRAIN

		Half life time in days	
		5 min experiment	60 min experiment
Adult	Lysine	4	15
	Leucine	8	27
Young	Lysine	2	6
	Leucine	1.2	4

The rate of metabolism of proteins is higher in young brains and again
higher in short term experiments. Young brain seems to have more of the
rapidly metabolized proteins in comparison to adult brain.

amino acid is highest in each brain, the second
highest similarly is at the second spot in each
brain, and so on, the lowest being the lowest in
each species. These similarities of composition
indicate not only that the basic biochemical properties and mechanisms are similar in the brains
from the various species, but also that the metabolite composition must have an important functional significance. This is further underlined by
the fact that those amino acids that are used for
metabolic purposes as well are at higher levels in
all brains than those amino acids that are utilized
for protein formation only.

The dynamic state of the constituents as well
as the apparently consistent and characteristic
composition requires a carefully balanced control
mechanism, which maintains cerebral homeostasis
and which also can respond to the requirements of
the organ by supplying a specific metabolite, depending on the particular need. This must be a rather intricate control mechanism, since the amino
acids are not homogeneously distributed in the
brain. Amino acid distribution varies in different
brain areas, and is not uniform within a single
brain cell.

The inhomogeneous distribution which is the
logical consequence of the structurally complex
nervous system shows also that the utilization of
the same compound may differ from one structure
of the brain to another. It is interesting to
speculate that the transport processes, which
must have a controlling influence on the distribution and the level of metabolites, influence,
thereby, the utilization of such metabolites
throughout the nervous system.

From the free amino acid pool the amino acids
are incorporated into brain proteins. This incorporation is also surprisingly rapid as shown in
Table III, although much slower than the flux of
the free amino acids from plasma to brain. The
half-life (that is, the time to replace half) of
the proteins can be measured in days. The values

given in the table, which show a significant regional variation, are average figures, indicating that
the metabolic rate is not the same for all proteins.
This is shown by the lower turnover in longer experiments. Shorter experiments measure components
of higher than average turnover rates, whereas in
the longer experiments the slower components pre-
dominate. Some proteins therefore may have a replacement rate of a few minutes, others, of weeks,
with the 15 days half-life only an approximate average for proteins.

Since protein formation occurs in the non-
growing adult brain, it must be counterbalanced by
an equivalent breakdown, showing that the proteins,
like the amino acids, are in dynamic equilibrium,
that is, that although their absolute levels do
not change under most conditions, they undergo
continuous and rapid formation and breakdown.

The rate of metabolism of the proteins is altered during development. Table IV shows a comparison of half-lives of cerebral proteins of young
and adult mice. In mice also there is an increase
of turnover with longer experimental times, which,
as discussed before, is an indication that there
are proteins of higher and proteins of lower metabolic rates in the brain. The average half-life
values are in young, showing that young brain has
a higher content of proteins with a higher metabolic rate. It is difficult at present to decide
whether the composition is similar during development, with the metabolic rates of the same components being higher during early stages of development, or whether the early stages of development
are characterized by abundance of those proteins
that are rapidly metabolized both in young and
adult, except that they are present in the young
at higher levels.

The half-life of about 15 days is a surprisingly high rate of turnover, since it obviously
would mean that the components of the brain are
replaced several times throughout the lifetime of
the animal. For this reason, the question arose

whether such an active metabolic state exists for
all brain components or only for a small portion.
Obviously experiments of a few hours test only a
small portion of the brain components if these
have half-lives of about 15 days. The experiment
that we did, of testing stability of cerebral proteins, consisted of feeding animals with food containing lysine of constant specific activity.
The mothers were fed before conception till the
specific activity of the amino acid in their plasma was the same as in the food, then they were
mated and fed through pregnancy, and the new-born
were fed till adulthood with the same radioactive
food. This way the experimental animal throughout its development had access to only labeled
lysine, and therefore all the proteins throughout
the organism had the same specific activity as
that of the food. This could also be verified
experimentally. When in adult animals the radioactive food was replaced with food containing unlabeled lysine, the radioactivity in the brain
rapidly decreased, as shown in Table V, and no
evidence of stable components was found; since
more than 95% of the label disappeared, more than
95% of the cerebral proteins must be in the metabolically active state. The average half-life
was close to 15 days throughout these experiments,
again showing that a fairly large majority of constituents have half-lives close to average and
that the very rapidly and very slowly metabolized
fractions comprise only a small fraction of the

TABLE V

Dynamic State of Most of the Cerebral Proteins

Days following the replacement of the radioactive diet	Per cent of radioactive proteins in the brain	Average half life of proteins in days
0	100	
30	29	15
60	7.0	16
150	2.4	41

brain. The dynamic state of the brain that is the
depository of long-term information is rather unexpected, and it does require a dynamic concept
of memory storage since the structure in which
memory is stored is not a stable one in itself.

The dynamic state of brain proteins requires
a continuous and active breakdown of proteins that
balances the continuous, active formation of neurons. We have much less knowledge about breakdown than about synthesis of proteins, but it
seems fairly certain that the site of formation
is primarily on the ribosomal structures from
which the newly formed protein is released into
the cell and that breakdown occurs at another site.
This difference in sites requires an information
transfer between sites of formation and breakdown
to keep the two mechanisms in balance.

Breakdown, like synthesis, may play an important role in the function of the nervous system;
for example, the growth and development of the
nervous system may be affected by increased rate
of formation, or by a partial inhibition of breakdown, since during active turnover, if the breakdown part is slowed down, the result is a net formation of proteins. The enzyme families responsible for cerebral protein breakdown, the acidic
and neutral proteinases and the various peptidases,
do undergo changes during development as shown in
Figure 1; it can also be seen that the pattern is
rather complex, with the temporal changes in the
various enzymes being different in the various
structures of the nervous system.

Since protein synthesis is heterogeneous,
some proteins being formed at high and others at
low rates, for the composition to stay constant,
the breakdown must follow a similar pattern, with
a much higher rate of breakdown of those components
whose rate of synthesis is high. There is indication that this may be so in the changes of specificity in proteinases in development, as can be
shown in Figure 2; the peptide fragments resulting
from the interaction of insulin with cerebral pro-

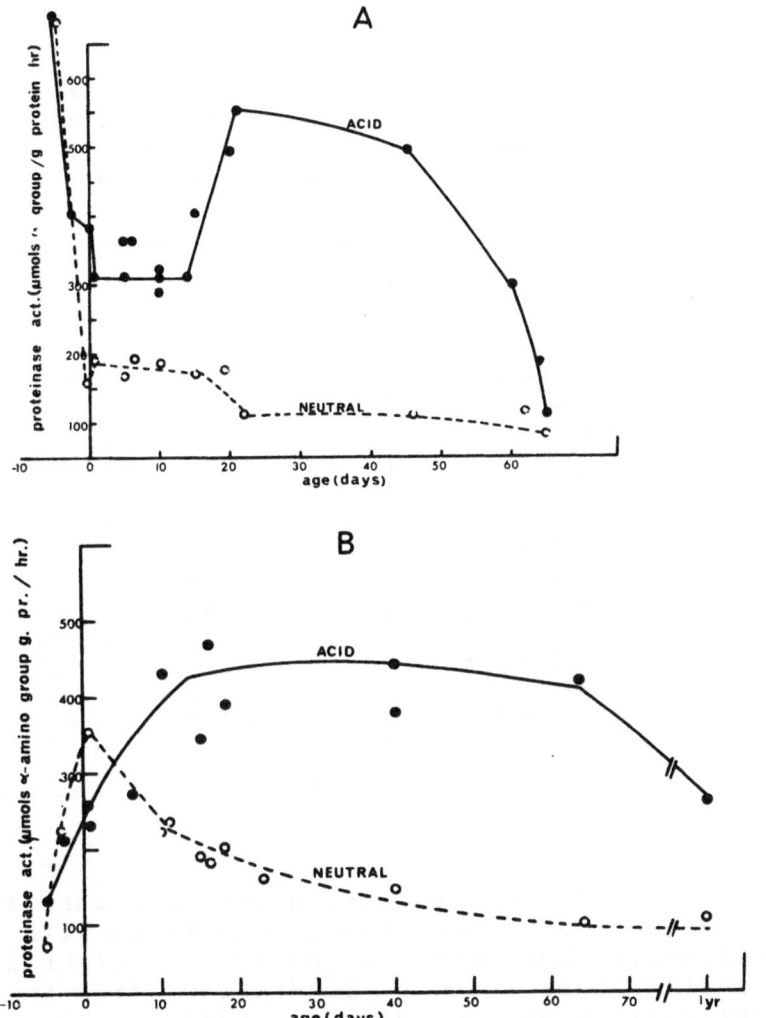

Fig. 1. A) Change in proteolytic enzyme content of crude mitochondrial fractions (containing myelin, synaptosomes and lysosomes) with age. Acid and neutral proteinase fell prior to parturition; acid proteinase remained at a low level prior to myelinization. B) Change in nuclear enzyme content. Note the increase in the level of enzyme prior to parturition followed by a rise in acid but a fall in neutral proteinase with development in the young animal.

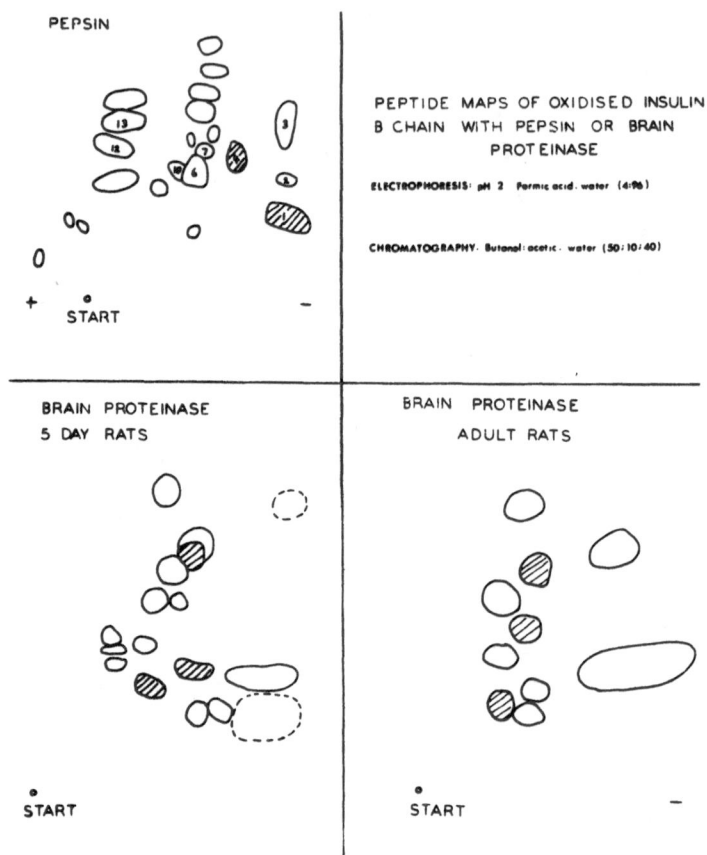

Fig. 2. The different peptide patterns resulting
after incubation of insulin as substrate with various enzymes show that the specificity of brain
intracellular proteinases is distinct from an extracellular enzyme active at an acid pH. Also
the differences between the young and adult brain
enzyme could indicate a change in enzyme speifi-
city (or protein) with age.

teinases show a different pattern with enzymes
from adult brain in contrast to enzymes from young
brain, showing that the substrate specificity of
these enzymes (as shown by studying the peptide

bonds split in insulin), undergoes changes during development.

Protein breakdown may have important roles in addition to participation in protein turnover. Partial breakdown may result in modifying proteins, thereby modifying their biological activity. It is well known, for example, that digestive enzymes can be formed from their inactive precursors by splitting off small peptide fragments from such precursors, such as the trypsinogen-trypsin conversion, and somewhat similar conversion occurs in a fibrinogen-fibrin conversion in blood clotting. Enzymes (or other biological mechanisms) could be either activated or inactivated by proteinases, and in the nervous system, biologically active peptide products could be formed or broken down.

Of particular interest are enzymes that are on membrane surfaces, and as shown in Table VI, a number of enzymes active in peptide bond splitting can be identified as bound to membranes. Such membrane bound proteinases may have important roles

TABLE VI

DISTRIBUTION OF HYDROLASES IN MITOCHONDRIAL MEMBRANE FRACTIONS

Fraction	Acid Proteinase	Neutral Proteinase	Arylamidase Arg-βNA	Leu-βNA	Aminotripeptidase Leu-Gly-Glu
	Specific Activity (units/mg. protein)				
Mt-synaptosomes	5.6	0.6	8.2	7.1	5.2
Purified Mt-	18.1	1.1	6.0	3.9	5.6
Inner membrane	0.6	--	0.7	0.6	1.2
Outer membrane	1.7	--	0.5	0.5	5.9
Heavy (inner) membrane	0.1	0.6	1.3	0.6	0.9
Light (outer) membrane	8.0	0	0.3	1.0	1.0

Inner and outer membranes were prepared by treatment of purified mitochondria with phospholipase A and the heavier and lighter membrane fractions separated by centrifugation after large amplitude swelling. Outer membrane has more aminotripeptidase and acid proteinase.

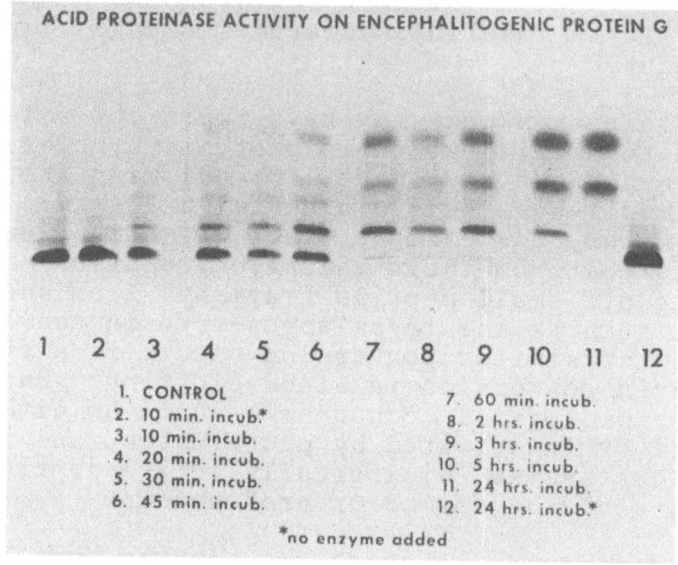

Fig. 3.   After 5 hours incubation with acid proteinase the original encephalitogen split to yield
three products one of which is identical to encephalitogen G previously described by E.R. Einstein
(E.R. Einstein, J. Csejtey, W. Davis and N. Marks,
Int. Arch. Allergy, 37, 1969, in press).

in pathological alterations in the brain. As a
consequence of damage to membrane, a normally inactive enzyme could be liberated, resulting in
pathological changes. This can be illustrated
with experiments designed to test whether in allergic encephalitis damage to membranes liberates
proteinases that split myelin proteins to a sufficiently small size to be able to get out of the
brain and induce immune reaction against brain
constituents. This could be a mechanism in some
auto-immune diseases. We could show in cooperation
with Dr. E. Einstein that activity of a protein
causing experimental allergic encephalitis can be
increased by breaking it down with cerebral acid
proteinase into a smaller molecular weight compound,
which in electrophoretic pattern is different from
the original compound (Figure 3) and has a smaller

molecular weight. Further breakdown of this component results in biologically and pathologically
inactive compounds, showing that another possible
role of proteinases is the inactivation of biologically active compounds.

Until now we have mostly discussed protein
metabolism in the central nervous system. This
metabolism is equally important in the peripheral
nervous system, where significant regional differences are found, especially in the proximal-distal
direction, as shown in Figure 4, with gradients
in amino acids and peptide splitting enzymes
along the nerve. The flow of proteins from the
nerve cell body along the axon, the local formation of proteins, and finally the breakdown of
the proteins resulting in activation or inactivation are important aspects of functional significance of protein metabolism in the nerves.

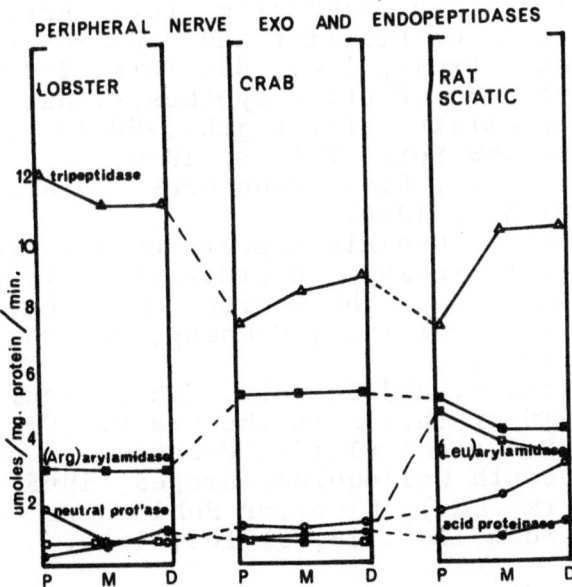

Fig. 4. The proximo-distal (P-M-D) gradient of
aminotripeptidase (Leu-Gly-Glu) fell in the lob-
ster but rose in the crab and the rat sciatic
nerve. Acid proteinase rose in all species but
other enzymes were variable.

It is clear from this very brief summary that
the proteins of the nervous system undergo constant metabolic turnover, and although the composition is not significantly changed, the high
rate of turnover allows the organ to respond to
physiological stimuli. These responses are highly
specific, as far as brain area, brain structure,
or even specific protein is concerned, either in
formation or in breakdown. Clearly, protein
metabolism in its multiple aspects mirrors the
complex function of the nervous system.

REFERENCES

Reviews

1. Lajtha, A., Protein metabolism of the ner-
    vous system, Intl. Rev. Neurobiol., (C.C.
    Pfeiffer and J.R. Smythies, Eds.), Academic
    Press, Inc., New York, 6, 1-98, 1964.
2. Lajtha, A., Alteration and pathology of cer-
    ebral protein metabolism, Intl. Rev. Neuro-
    biol., (C.C. Pfeiffer and J.R. Smythies, Eds.),
    Academic Press, Inc., New York, 1-40, 1964.
3. Roberts, S., Protein Synthesis. Handbook of
    Neurochemistry, (A. Lajtha, Ed.), Plenum
    Press, New York, Vol. 4, 1969.
4. Datta, R.K., Brain ribosomes. Brain Research,
    2, 301-322, 1966.
5. Hydén, H., Dynamic aspects on the neuron-
    glia relationship--a study with micro-chem-
    ical methods. The Neuron (H. Hydén, Ed.),
    Elsevier Publishing Company, New York, 179-
    219, 1967.
6. Roberts, S. and Zomzely, C.E., Regulation of
    protein synthesis in the brain. Protides of
    the Biological Fluids, Proceedings of the
    Thirteenth Colloquium, Bruges, 1965, (H.
    Peeters, Ed.), Elsevier Publishing Company,
    Amsterdam, Vol. 13, 91-102, 1966.

Typical Studies

7. Murthy, M.R.V. and Rappoport, D.A., Bio-
    chemistry of the developing rat brain. VI.
    Preparation and properties of ribosomes,

Biochim. et Biophys. Acta, 95, 132-145, 1965.

8. Murthy, M.R.V, Protein synthesis in growing-rat tissues. II. Polyribosome concentration of brain and liver as a function of age. Biochim. et Biophys. Acta, 119, 599-613, 1966.

9. Appel, S.H., Davis, W. and Scott, S., Brain polysomes: Response to environmental stimulation, Science, 157, 836-838, 1967.

10. Furst, S., Lajtha, A. and Waelsch, H., Amino acid and protein metabolism of the brain--III. Incorporation of lysine into the proteins of various brain areas and their cellular fractions, J. Neurochem., 2, 216-225, 1958.

11. Roberts, S., Regulation of cerebral metabolism of amino acids--II. Influence of phenylalanine deficiency on free and protein-bound amino acids in rat cerebral cortex: Relationship to plasma levels, J. Neurochem., 10, 931-940, 1963.

12. Koenig, E., Synthetic mechanisms in the axon-II. RNA in myelin-free axons of the cat, J. Neurochem., 12, 357-361, 1965.

13. Austin, L., Bray, J.J. and Young, R.J., Transport of proteins and ribonucleic acid along nerve axons, J. Neurochem., 13, 755-759, 1966.

14. Toschi, G., Dore, E., Angeletti, P.U., Levi-Montalcini, R. and de Haën, Ch., Characteristics of labelled RNA from spinal ganglia of chick embryo and the action of a specific growth factor (NGF), J. Neurochem., 13, 539-544, 1966.

15. Oja, S.S., Studies on protein metabolism in developing rat brain, Ann. Acad. Sci. Fennicae, 131, 1-81, 1967.

16. Bondy, S.C. and Roberts, S., Messenger ribonucleic acid of cerebral nuclei, Biochem. J., 105, 1111-1118, 1967.

17. Hydén, H. and Egyházi, E., Glial RNA changes during a learning experiment in rats, Proc. Nat. Acad. Sci. U.S., 49, 618-624, 1963.

18. Lajtha, A., Protein metabolism in nerve, Chemical Pathology of the Nervous System (J. Folch-Pi, Ed.), Pergamon Press, New

York, 268-275, 1961.
19. Appel, S.H., Inhibition of brain protein syn-
     thesis: An approach to the biochemical basis
     of neurological dysfunction in the amino-acid-
     urias, Trans. N.Y. Acad. Sci., 29, 63-70, 1966.
20. Talwar, G.P., Chopra, S.P., Goel, B.K. and
     D'Monte, B., Correlation of the functional
     activity of the brain with metabolic para-
     meters--III. Protein metabolism of the occi-
     pital cortex in relation to light stimulus.,
     J. Neurochem., 13, 109-116, 1966.
21. Moore, B.W. and Perez, V.J., Specific acidic
     proteins of the nervous system. Physiological
     and Biochemical Aspects of Nervous Integration
     (F.D. Carlson, Ed.), Prentice-Hall, Englewood
     Cliffs, 343-359, 1968.

PROBLEMS OF GENETIC EXPRESSION IN THE NERVOUS SYSTEM

Paul Mandel

INTRODUCTION

Our views on the future in neuroscience are a result of an association of our knowledge, hypotheses and sometimes of utopias. Our interest in neuroscience is in some aspects a result of our curiosity: to know how the most sophisticated human organ is functioning and our desire to influence central nervous system activity and to improve the life of people developing mental diseases. The questions which arise are extremely complicated. When we attempt to answer these questions, the direct way, which seems the shortest, is not always the best. During the last twenty years we have learnt that in biology, success depends on the choice of models and projects as simple as possible. The mechanisms of neural activity, of our behavior and memory and of mental diseases are a result of high specialization of nerve cells, their functional pathways and their connections. One of the main ways to answer the questions we are interested in is to look at the differentiation and ontogenesis of the nervous system in parallel to the functional activity.

On the other hand, the basic mechanisms depend upon the structural molecules, mainly proteins

and lipids, and upon the enzymatic activities.
These molecules, their synthesis and their control
mechanisms are in the center of our interest. Thus,
the main approaches to solve problems set by neuroscience are those of transcription, that is of RNA
synthesis, and of translation, that is protein
synthesis in which RNAs are involved.

We will briefly discuss some problems concerning nervous cell differentiation, RNA synthesis and
the investigations on effects of nervous cell stimulations on this synthesis.

I. NERVOUS CELL DIFFERENTIATION

There are only very few data concerning biochemical phenomena or transcription in early cell
differentiation (1,2). Nevertheless, it is well
known that the appearance of nucleoli is among the
most interesting morphological phenomena during
the formation of neurons. Since the nucleolus is
the site of ribosomal RNA formation (3), the production of Nissl bodies which characterizes the
neuronal morphology depends on the activity of this
subnuclear fraction. The knowledge of these elementary mechanisms during differentiation and ontogenesis and their control may be of a great help
for our understanding of the function of nerve
cells. Investigations should be devoted to this
problem in the future.

Recently, it was found in our laboratory (4)
that in neuroblasts and spongioblasts of chicken embryo brain, RNA synthesis is mainly limited to the
nucleus until about 15 days of embryonic life. Only
later, in parallel to the differentiation of nucle-
oles, does a fast increase of RNA appear in the cytoplasm. The problem arises as to how this important phenomenon is controlled. Several hypotheses
can be formulated. One of the possibilities is
that in the early differentiation, the amount of
mitochondria producing nucleoside triphosphates,
precursors of RNA, is very low. Thus, RNA synthesis is limited and the polyribonucleotides produced
remain mainly in the nuclei. When in later periods

of the ontogenesis the brain metabolism becomes ae-
robic, the production of more mitochondria is induced, nucleoside triphosphate synthesis increases,
as does the synthesis of RNA. Starting from this
moment, RNA is transferred in high amount into the
cytoplasm, and the translation phenomena, that is
protein and enzyme synthesis which contribute to
the establishment of the cell morphology, may
actively begin.

II. RNA SYNTHESIS IN ADULT BRAIN

DNA is the carrier of the genetic information
and the template for the biosynthesis of ribonucleic acid (RNA) from nucleoside tri-phosphates, cata-
lyzed by RNA polymerase. This biochemical reaction
presented in Figure 1 is called the transcription
phenomenon since polyribonucleotides receive the
genetic information from DNA. Three main types of
RNA can be produced: transfer, ribosomal and messenger. All actively participate in protein synthesis translating genetic information. The polyribonucleotides produced are complementary to one
DNA strand. Many control mechanisms are involved
in the regulation of transcription phenomena, as
was demonstrated first in bacteria (5). In eukariot
cells and mainly in animal cells, supplementary
mechanisms and regulations appeared during evolution, insuring a specialization of some cells in
order to realize several peculiar functions. Moreover, high specialization leads to a distribution of
the different cellular activities among subcellular
structures and compartments. The nucleus which
contains the major part of DNA is coding for the
major fraction of RNA, and from it of proteins. In
spection of the biochemical reaction in Figure I
gives rise to several questions:

1) how is the supply of the precursors of
    nucleoside triphosphates controlled?

2) how does the enzyme producing the re-
    action act?

3) how does DNA participate in this reaction,

and what are the mechanisms and extent by which the copy of DNA is produced?

We will try to discuss these questions briefly.

A.   THE REGULATION OF THE PRECURSOR POOL

The first problem arises how is the supply of free nucleotides controlled. There are three possibilities:

1)   synthesis of NTP in the nucleus.

2)   transfer of NTP from the cytoplasm.

3)   liberation of nucleotides from storage molecules.

a) Synthesis of NTP in the nuclei.

The enzymatic equipment of nuclei is mainly glcolytic (6) and does not present the optimal conditions for a production of ATP and via transphosphorylation of other nucleoside triphosphates. Only a low amount of free nucleotides is present in the nuclei, whereas the rate of RNA synthesis is very high.

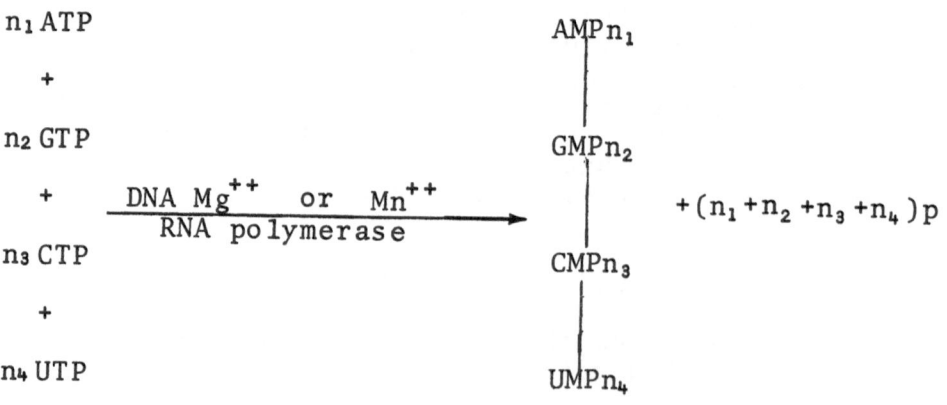

Figure 1

It is difficult to imagine that the nucleus can realize the fourteen chemical reactions which are necessary for the synthesis of one NTP when the signal arrives for RNA synthesis. The second and the third eventuality we propose seem more likely.

b) Transfer of nucleotides from the
cytoplasm to the nucleus.

The available data concerning a transfer of nucleotides from the cytoplasm to the nucleus are contradictory (7,8,9). The prevailing opinion has been that only nucleosides are able to be transferred through the nuclear membrane. We were able to demonstrate that AMP as a whole can be transferred from the cytoplasm to the nuclei following NAD synthesis which occurs in the nuclei (10).

Nuclei were incubated in a medium containing double labelled ATP, that is C^{14}, αP^{32}-ATP. In these conditions, if the ratio of the radioactivity of αP^{32} to C^{14} remains the same in NAD as it was in AMP in the presence either of cold phosphates or of cold nucleosides, it will mean that the nucleotide was transferred as a whole. The transfer of nucleosides after dephosphorylation of nucleotides will lead in the presence of cold phosphates to a decrease of this ratio and in the presence of cold nucleosides to an increase of this ratio. The answer obtained was that the ratio of C^{14} of the adenine to αP^{32} of the nucleotide was the same in NAD as in AMP. This permitted the conclusion that AMP is transferred as a whole (Table I).

c) The eventual role of homopolymers.

The third possibility evoked was that there are storage forms of nucleotides. There is some evidence in favor of the presence of storage forms:

1) The existence in the nuclei of enzymes able to synthesize homopolymers.

2) The existence of enzymes able to produce nucleoside triphosphates from homopolymers.

Table I

Synthesis of NAD by rat brain nuclei incubated
in vitro in the presence of nicotinamide mono-
nucleotide and double labelled ATP adenine ^{14}C α
^{32}P.

Exp.	Addition	Ratio of Radioactivity ^{14}C/^{32}P	
		ATP	NAD
1	none	1.19	1.15
2	none	1.19	1.2
3	phosphate	1.19	1.21
4	phosphate	1.19	1.2
5	phosphate	1.0	1.04
6	phosphate	1.0	1.17
7	phosphate	1.0	1.025
8	adenosine	1.61	1.51

Table II

Synthesis of Homopolymers by Particulate Enzymatic
Extracts of Brain Nuclei

pH	NTP	Incorporation $\mu\mu$ moles / mg protein
9	UTP	13.5
7.5	CTP	24.0
9	ATP	22.2
9	ATP	24.0

The reaction mixture contains one labelled NTP,
Mn^{++} as cation, and the enzymatic extract. For
exact concentrations see Table IV.

3)   The presence of homopolymers.

1)   Homopolymer synthesis and degradation.

As it appears from Table II, it is possible to produce from enzymatic nuclear extracts polyadenylate, polycytidylate, polyguanylate and polyuridylate (11,12,13).

After incubation of polyadenylate P^{32} with a nuclear extract, we obtained ATP, ADP, AMP (14). Among these three nucleotides, ATP represents 8.8% and ADP 25.5%. The distribution of these nucleotides suggests the possibility of a rapid production of a nucleoside triphosphate from the homopolymer. The reaction is probably going through two steps, the first liberating ADP, the second involving adenylate kinase activity present in the nuclei giving ATP and AMP. The characterization of the enzyme hydrolysing polyadenylate is underway.

2)   Presence of homopolyribonucleotides in the nuclei.

Edmonds and Abrams have demonstrated the presence of polyadenylate in thymus nuclei (15). In collaboration with Doly, we were able to demonstrate the presence of a peculiar polyadenylic compound: polyadenosine diphosphate ribose (polyADPR) in nuclei stimulated by nicotinamide injections (16). This homopolymer is produced from NAD by an enzymatic extract of the nuclei (17). It may be also considered to be a storage form of a precursor of NAD. In addition, it should be noted that the presence of polyA on microsomes also has been reported (18). The evaluation of the quantity of homopolymer present in cerebral cell nuclei we are studying is not easy because the degradation by the nuclear enzymes of this homopolymer is extremely fast.

B.   REGULATION OF THE NTP POOL AND THE PROBLEM OF LIMITING FACTORS

Since the major part of the nucleotides for RNA synthesis is provided by the cytoplasm, it

would be interesting to know how the pool of NTP
is controlled. We have to take in mind that NTP
participates in several metabolic reactions as it
appears from Figure 2. Stimulation of one of these
reactions: phosphatide synthesis, polysaccharide
synthesis or NAD synthesis, may influence the pool
of NTP and thereby the synthesis of RNA. With
Revel (19), we demonstrated that a stimulation of
NAD synthesis in kidney nuclei leads to a decrease
of nuclear RNA synthesis and even to an inhibition
of compensatory renal hypertrophy, inhibiting RNA
synthesis (20).

One nucleoside triphosphate, CTP, is present
in adult tissues in very low amount (Table III).
In addition, we could observe that the biosynthesis
of CTP is also the lowest amount among the four NTP.
This suggests that CTP may be a limiting factor in
RNA synthesis (14). During growth or during cell
differentiation, the amount of CTP is much higher,
suggesting that there is a kind of adaptation of
CTP synthesis to the high level of RNA biosynthesis.

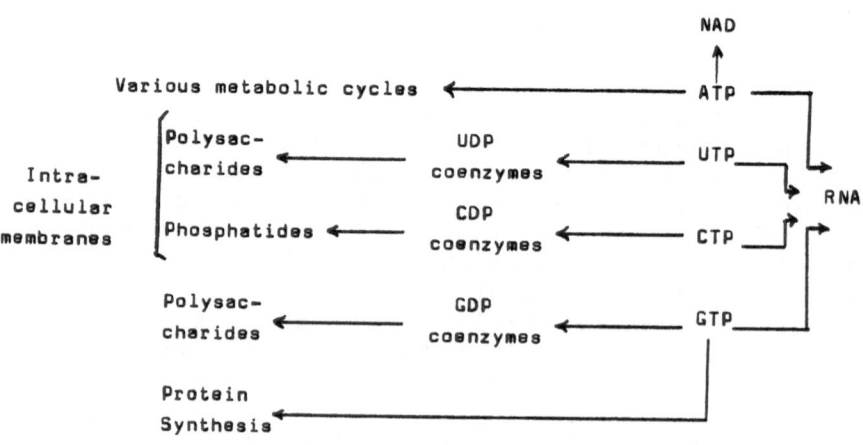

Figure 2
Summary of the intervention of the nucleoside di-
and triphosphates in different metabolic cycles.

Table III

Distribution of Nucleoside Triphosphates in Different Rat Tissues.

μmoles/100 g wet weight

	ATP	GTP	UTP	CTP
Liver Adult	277.0	49.83	42.57	26.07
Lens Adult	132	25.4	18.6	trace
Brain Adult	240	27.9	20.6	trace
Brain 1 day old	188	36.8	53.5	13.0

Hypoxia leads to a decrease of brain ATP more
or less pronounced according to duration (21,22).
Functional hyperactivity noted after injection of
imino-dipropionitrile (23) or pervitine (24) produces an increase of brain ATP. Convulsions produce a decrease of ATP content of the brain. This
may be due to the high consumption under reduction
of oxidative phosphorylation due to hypoxia. Hypo-
thermia also decreases the level of ATP and the
ratio of ATP to ADP even if oxigenation is maintained by artificial respiration. The high psychomotor
activity produced by amphetamine does not affect
significantly the level of NTP in the brain. The
amount of ATP is not usually changed. However,
it was observed that the turnover of ATP measured
by the incorporation of P^{32} as α-phosphate of ATP
increases during this functional hyperactivity
produced by amphetamine (25). Since α-phosphate
is incorporated in RNA, even if the level of RNA
biosynthesis does not change, the specific activity
becomes higher. This experiment demonstrates the
possibility of errors in interpreting a higher in-

corporation of P^{32} into RNA as an increase of RNA synthesis in functional hyperactivity. Thus, it appears that there are several possibilities of changes in the NTP pool. It must, however, be noted that the modification in the amount of NTP does not necessarily affect the synthesis of RNA which is certainly regulated by large number of other factors.

In summary, the synthesis of RNA in nuclei which contains a small pool of NTP raises the question of the regulation of the supply of these precursors in order to insure RNA synthesis. Since this nucleotide can be provided by the cytoplasm or by homopolymers, and since CTP seems to be a limiting factor, several control mechanisms are necessary in order to maintain the harmony of one of the main cellular functions: transcription. Our knowledge in this field is very limited and calls for some new investigations. It is important to point out that the phenomenon designated as "blood brain barrier" may play a role in the supply of the brain by precursors or fractions of nucleotide molecules. Thus it may be that the optimal level of RNA synthesis is not always achieved.

III. RNA POLYMERASE ACTIVITY

The activity of an RNA polymerase similar to that described by Weiss (26) in the liver was found in brain nuclei simultaneously by Barondes (27), Bondy and Waelsch (28) and by us (29). This polymerase activity requires the presence of the four NTP, as it appears from Figure 1, as well as of DNA and of a divalent cation, Mg^{++} or Mn^{++}. In addition to the Weiss enzyme, we have demonstrated the presence of a soluble RNA polymerase in brain nuclei in which DNA dependence could be established as shown by Ramuz et al., (30) (Table IV).

We found, as did Bondy and Waelsch (28), that when the content of DNA is used as a reference, the RNA polymerase activity attached to the chromatin, that is the aggregate Weiss enzyme, is higher in the brain than in liver (31). However, this difference

Table IV

RNA Polymerase Activity of Brain Nuclei

Reaction mixture	Incorporation GMPα^{3e}P
	$\mu\mu$mol/mg protein
Soluble enzyme	
Complete	49.1
Omit UTP, α CTP	3.0
Omit DNA	1.0
Particulate enzyme	
Complete	110
Omit UTP, α CTP	12.0
+ DNASE	13.5
+ Actinomycin, 50 μg	10.0

The complete reaction mixture (0.25 ml) contains
160 mM Tris buffer pH 7.2, 4 mM $MnCl_2$ or 30 mM $MgCl_2$
8 mM 2-mercaptoethanol, 1mM each of CTP, ATP, UTP,
GTP one labelled in α with ^{32}P and approximately
0.2 mg protein from the enzyme extract. In the
case of soluble enzyme 60 μg of DNA were added.
Incubation for 10 min at 37°.

may be due to a higher RNA polymerase activity as
well as to a lower ribonuclease activity, that is
to a higher rate of biosynthesis or to a lower degradation rate by endogenous nucleases. This hypothesis was tested and it appears that the nuclease
activity is really lower in brain nuclei; but the
difference cannot explain the higher level of synthesis with brain aggregate enzyme. The nuclease
activity in brain is only about 15% lower than in
liver whereas the RNA polymerase activity is at
lease twice as high (31). In addition, higher
values were obtained in the brain by the measurement of RNA polymerase activity at a high ionic
strength (27,31) which permitted the evaluation of
the relative amount of RNA polymerase molecules
fixed on the chromatin (32) (Table V). Thus, RNA

Table V

RNA polymerase activity of the aggregate RNA polymerase of Rat Liver and Brain Nuclei.

	mμ moles ^{32}P GMP incorporated / mg DNA	
	- $(NH_4)_2$ SO_4	+ $(NH_4)_2$ SO_4
Brain	1.2	7.0
Liver	0.75	2.5

The incubation medium (0.25 ml) contained: 100 mM
Tris-HCl buffer (pH 8), 4 mM MnCl2, 1mM ATP, 1mM
CTP, 1mM UTP, 1mM GTP (GTP labeled α^{32}P: specific
activity 3-5 μC/μmole), 12 mM 2-mercaptoethanol
and 0.1-0.2 ml of the aggregate enzyme. 0.4 M
ammonium sulphate (saturated at 4° and adjusted to
pH 8 with NH4 OH) was added when the incubations
were conducted at high ionic strength.

polymerase activity in brain appears to be higher
than in an organ like liver synthesizing a great
amount of proteins. The real meaning of this phenomenon has to be investigated. It suggested an intense synthesis of a great variety of proteins in
brain cells. This work was done on whole brain.
It would be useful to investigate different types
of brain cells, not only neurons and glia, but
also the different types of neurons.

IV. INVESTIGATIONS ON IN VIVO RNA BIOSYNTHESIS

The discovery of rapidly labelled RNA involved in protein synthesis greatly enriched our
knowledge on the role of RNA in cell metabolism.
Among these rapidly labelled RNAs two types could
be distinguished. One messenger RNA, containing a
DNA-like base distribution, which carries information from DNA to the ribosomes where the translation
occurs. The other is a precursor form of ribosomal
RNA, as shown by Scherrer and Dannell (33) and by
Perry (34). Investigations on the regulation of
the biosynthesis of these two types of RNA might
permit the establishment of correlations between
the functional activity of nerve cells and a differentiated protein synthesis.

It was shown by several authors and in our
own laboratory that at a short time after the injection of P^{32} labelled uridine or orotic acid,
highly labelled RNA appeared in brain. In our own
work (35), only 30 minutes after intracisternal
injection of P^{32}, a high molecular weight RNA
appears in the central nervous system. This RNA
is eluted from a methylated albumin Kieselguhr
column after the ribosomal RNA, and appears to be
very heterogeneous. The base distribution, which
is similar to that in DNA, suggests that this kind
of RNA might be messenger in nature. By this type
of experiment, it might be possible to follow in
the future the production of different messengers
in relation to nervous activity. However, methods
should be elaborated which permit the fractionation
of different kinds of messenger RNA in order to

discover the qualitative changes which occur. This
kind of work is only now beginning. In addition, it
should be pointed out that the fundamental proof
of a messenger property of an RNA is really only
accomplished when a specific protein is synthesized
or is directed by this messenger. Until now, except in the case of reticulocyte RNA and hemoglobin
synthesis (36), it has not been possible to obtain
any protein synthesis in an acellular system with
purified RNA from mammalian cells. Controls permit
only the demonstration that in some experiments
reported concerning such a synthesis, the observed
biosynthesis was due to an artifact and represented
only a Mg^{++} effect (37).

All types of RNA, messenger, transfer and
ribosomal, are copied from the DNA. The comple-
mentarity of base sequences may be demonstrated by
DNA-RNA hybridization, a technique introduced by
Hall and Spiegelmann (38). By this technique it
can be shown that RNA produced in a particular cell
type hybridizes with DNA of this cell but not with
heterologous DNA. By determining the amount of hybrids formed when the DNA is incubated with an
excess of DNA-like RNA, or of rRNA, one can estimate the extent of the region of DNA which is
transcribed. In brain, rRNA seems to account for
about 0.15 per cent of the genome, and messenger
type RNA for about 1.4 per cent (39). However,
some of the nuclear RNAs are not found in cytoplasm.
The saturation of DNA with cytoplasmic RNA shows
that only 0.6 to 0.8 per cent of the DNA is coding
for cytoplasmic RNA. The remaining should correspond to specific nuclear RNA (39). Similar results
have been found in HeLa cells by Schearer and Mc-
Carthy (40). On the other hand, Harris (41) has
demonstrated that a high proportion of nuclear RNA
is destroyed in the nuclei and never transported
to the cytoplasm.

If we assume that only one of the two strands
is used for transcription, it would appear that
about 2.7% of the genome is expressed in brain
cells. From these data the number of cistrons
which are transcribed can be evaluated. It appears

that about 6,000 cistrons were coding for the two
types of ribosomal RNA and 50,000 to 500,000 for
dRNA (39). The significance of these data has to
be investigated deeply. It is important to determine if the number of cistrons which can be transcribed corresponds to the amount of specific proteins which can be produced by different cell
types and by different neurons. These preliminary
data suggest that an extremely large variety of
proteins can be produced by nervous system cells.
However, these figures which result from hybridization experiments should be treated with caution.
Thus, we do not know the exact requirements for
ribonuclease resistance which is used to distinguish
true hybrids. In addition, an important redundancy
of cistrons is very probable. It may be interesting
to note that investigation of the extent of gene
expression is easier in brain than in other tissues,
because of the relative stability of the precursor
pool in brain compared to other tissues.

Very few data are available concerning the
regulation of tRNA. Nevertheless, this type of
RNA may play a role in the regulation of protein
biosynthesis in differentiated cells. Thus, we
were able to demonstrate that in differentiated
lens cells which produce mainly three types of
crystalline proteins there is a strong correlation
between the level of different types of tRNA and
the distribution of the amino acids in these proteins (42).

V. EFFECT OF STIMULATION ON RNA METABOLISM

In order to establish the correlation between
RNA synthesis and brain function, it is useful to
investigate the effects of different kinds of stimulation on RNA synthesis in specific cells. During
ontogenesis, visual stimuli are necessary for an
induction of a normal RNA biosynthesis. A retardation in the evolution of this biosynthesis was
observed when the animals were brought up in total
darkness (43). In the first days after unilateral
section of the optic nerve in monkeys, a rapid increase of the RNA content in the lateral geniculate

nucleus could be demonstrated. This may be related
to an increase of the activity in this area. Similarly, in adult animals, a vestibular stimulation
provokes an increase of the RNA in Purkinje cells
(44), as well as in Deiters cells (45). Olfactory
stimulation seems also to increase in fish brain
the RNA content, in addition to a change in base
composition of nuclear RNA (46).

Many studies have been devoted to effect of
learning on the quantitative and qualitative RNA
synthesis. This problem will be reported in this
Symposium by other authors (47). An inhibition of
RNA synthesis produces different kinds of impairment
in learning (48). All these data suggest that RNA,
which is a vector of information for brain protein
synthesis, is involved in nervous cell activity.
However, all these data should be treated and interpreted with caution. This is especially the case
when the conclusions are drawn from experiments
which produce stress, and when labelling with radioactive precursors is used for detection of an increase in metabolism or changes in base distribution.
Learning which involves a stress may produce an
overall metabolic increase without specificity for
the phenomenon studied. In labelling experiments,
an increase of the synthesis of free nucleotide
precursors of RNA will also produce an increase of
incorporation of the radioactive precursors and of
RNA. In such conditions, the higher specific activity of RNA may express not a higher rate of synthesis of this compound but the presence of a more
labelled compound. Thus, it is necessary in the
future to relate data concerning RNA synthesis to
the immediate precursors. Moreover, changes in the
distribution of RNA bases in nuclei may be a result
of a variation in the ratio of biosynthesis of
ribosomal RNA to messenger RNA. In addition, we
should keep in mind that at least 1,000 different
messengers are produced for the normal common metabolism of different cells. The addition to these
messengers of a new one will be difficult to detect.
It means that we need elaboration of extremely
sensitive methods in order to be able to correlate
a specific nervous activity with the biosynthesis

of specific messenger. From extremely interesting experiments suggesting a synthesis of new messengers as an answer to the basis of different kinds of learning, a hypothesis was formulated that a synthesis of new kinds of RNA may be involved in memory. The interpretation of these experiments needs great criticism. It is obvious that only such an RNA can be synthesized for which there exists a complementary structure in the DNA genome. If every engram of memory is related to a synthesis of a new specific RNA molecule, we should admit that all that we will memorize in the future is already written in our genes and this seems unlikely. Therefore, more experiments with more sophisticated methods are necessary in order to establish a strong correlation between behaviour phenomena and specific RNA synthesis.

REFERENCES

1. Himwich, W.A. and Himwich, H.E., The developing brain, Progress in Brain Research, Elsevier Publishing Co., Amsterdam, Vol. 9, 267, 1964.
2. Purpura, D.P. and Schade, J.P., Growth and maturation of the brain, Progress in Brain Research, Elsevier Publishing Co., Amsterdam, Vol. 4, 289, 1964.
3. Perry, R.P., In: Progress in Nucleic Acid Research and Molecular Biology, J.N. Davidson and W.E. Cohn (Eds.), Academic Press, New York and London, Vol. 6, 220, 1967.
4. Sensenbrenner, M. and Mandel, P., Z. Zellforsch 82, 65, 1967.
5. Jacob, F. and Monod, J., J. Mol. Biol., 3, 318, 1961.
6. Siebert, G., Advances in Enzymology, 27, 239, 1965.
7. MacEwen, B.S., Allfrey, V.G. and Mirsky, A.E., J. Biol. Chem., 238, 758, 1963.
8. Pennial, R., Liu, Sh.M., and Saunders, J.P., Biochim. Biophys. Acta, 76, 170, 1966.
9. Karjalainen, E., Acta Chem. Scand., 20, 586, 1966.
10. Simler, S., Popovic, D. and Mandel, P., Bull. Soc. Chim. Biol., 49, 1509, 1967.
11. Jacob, M. and Mandel, P., In: Protides of

the Biological Fluids, H. Peeters (Ed.),
Elsevier Publishing Co., Amsterdam, 63, 1965.

12. Pete, N., Chambon, P. and Mandel, P., In:
Variations in Chemical Composition of the
Nervous System, G.B. Ansell (Ed.), Pergamon
Press, London, 88, 1965.

13. Mandel, P., Pete, N. and Dravid, A., J.
Physiol., 59, 261, 1967.

14. Mandel, P., Bull. Soc. Chim. Biol., 49,
1491, 1967.

15. Edmonds, M. and Abrams, R., J. Biol. Chem.,
238, 1186, 1963.

16. Doly, J. and Mandel, P., Comptes Rendus
(Paris), 264, Serie D, 2687, 1967.

17. Chambon, P., Weill, J.D., Doly, J., Strosser,
M.T. and Mandel, P., Biochem.Biophys. Res.
Commun., 25, 638, 1966.

18. Eisenstadt, J.M. and Brawerman, G., Proc.
Natl. Acad. Sci. U.S., 58, 1560, 1968.

19. Revel, M. and Mandel, P., Cancer Res., 22,
456, 1962.

20. Revel, M., Mandel, L., Wintzerith, M., Klein,
N. and Mandel, P., Bull. Soc. Chim. Biol.,
43, 91, 1961.

21. Thorn, W., Pfleiderer, G., Frowein, R.A.,
and Ross, I.R., Arch. Ges. Physiol., 261,
334, 1958.

22. Gerlach, E., Doring, H.J. and Fleckenstein,
A., Arch. Ges. Physiol., 266, 266, 1958.

23. Mandel, P., Harth, S. and Rebel, G., In:
Chemical Pathology of the Nervous System,
J. Folch-Pi, (Ed.), Pergamon Press, London,
551, 1958.

24. Palladine, A.V., Wiener Klin. Wochenschr.,
473, 1954.

25. Edel, S., J. Physiol., 59, 234, 1967.

26. Weiss, B.S., Proc. Natl. Acad. Sci., U.S.,
46, 1020, 1960.

27. Barondes, S.H., J. Neurochem., 11, 663, 1964.

28. Bondy, S.C. and Waelsch, H., Life Sciences,
3, 633, 1964.

29. Pete, N., Wintzerith, M., Mandel, L. and
Mandel, P., Comptes Rendus (Paris, 258, 5283,
1964.

30. Ramuz, M., Doly, J., Mandel, P. and Chambon, P., Biochem. Biophys. Res. Commun., 19, 114, 1965.
31. Munoz, D. and Mandel, P., C.R. Soc. Biol., 1968, in press.
32. Chambon, P., Ramuz, M., Mandel, P. and Doly, J., Biochim. Biophys. Acta, 157, 504, 1968.
33. Scherrer, K., Latham, H. and Darnell, J.E., Proc. Natl. Acad. Sci., U.S., 49, 240, 1963.
34. Perry, R.P., Proc. Natl. Acad. Sci., U.S., 48, 2179, 1962.
35. Jacob, M., Stevenin, J., Jund, R., Judes,C. and Mandel, P., J. Neurochem., 13, 619, 1966.
36. Schapira, G., Dreyfus, J.C. and Maleknia,N., Biochem. Biophys. Res. Commun., 32, 558, 1968.
37. Kempf, J. and Mandel, P., Bull. Soc. Chim. Biol., 50, 773, 1968.
38. Hall, B.D. and Spiegelman, S., Proc. Natl. Acad. Sci., U.S., 47, 137, 1961.
39. Stenenin, J., Samec, J., Jacob, M. and Mandel, P., J. Mol. Biol., 33, 777, 1968.
40. Schearer, R.W. and McCarthy, B.J., Biochem., 6, 283, 1967.
41. Harris, H., Nature, 213, 809, 1967.
42. Virmaux, N., Garel, J.P. and Mandel, P., unpublished data.
43. Goswamy, S., Mandel, P. and Karli, P., In: Biochemistry of the Eye, U. Dardenne and J. Nordmann (Eds.), S. Karber, Basel, 514, 1966.
44. Jarlstedt, J., In: Macromolecules and the function of the neurons, Z. Lodin and S.P.R. Rose (Eds.) Excerpta Medica Foundation, Amsterdam, 1968.
45. Hyden, H. and Pigon, A., J. Neurochem., 6, 57, 1960.
46. Rappoport, D.A. and Daginawala, H.F., Abstr. 1st. Intern. Meeting of the Intern. Soc. Neurochem., Strasbourg, 175, 1967.
47. Hyden, H., this Symposium.
48. Agranoff, B.W., this Symposium.

ODD ELECTRON MOLECULAR EFFECTS IN EYE, NERVE, AND BRAIN

B. David Polis

In the fatty regions of the cell where ionization is more difficult than in aqueous regions, we may expect the predominance of excited state reactions. In conformity with this concept, high lipid concentrations are found in biological organelles like chloroplasts and retina that make direct use of excitation energy.

Our in vitro studies on bioenergetic control mechanisms were predicated on the hypotheses that a biomolecular system in an excited state, such as is obtained with the unpairing of electrons to generate free radicals, might be especially susceptible to controlling factors such as hormones or drugs, and that physiological as well as metabolic events could be initiated or controlled via these excited state structures. It is, of course, recognized that an active intermediate or excited state does not necessarily imply an unpaired electron and that indeed only a few type reactions generate free radicals. Yet, the unequivocal identification of a free radical by electron spin resonance (ESR), coupled with the hyperreactivity of a free radical, make the search for free radical generation and regulation in bioenergetic

systems worthwhile. Free radical excited states
already have been unequivocally demonstrated in a
number of enzymatic reactions, and the low temperature photoparamagnetism observed in the quantasome has been interpreted as the primary step in
the quantum conversion in photosynthesis. The
approach taken was, therefore, not without prece-
dent or rationale.

The spinning electron has an associated magnetic moment which is randomly oriented in field
free space. In a magnetic field, only two orien-
tations of the electronic magnetic moment are
allowed - paralleled and antiparalleled to the
applied field. These two states correspond to
different energies, and transitions between the
states are accompanied by absorption or emission
of radiation. In a field of 3,000 G, the transition energy is 0.8 cal. per mole, which corresponds to light in the microwave region of 3.6 cm
wavelength. Detection and quantitation of free
radicals were accomplished with a Varian ESR
spectrometer.

Frequently, the initial experimental valida-
tion of a biochemical hypotheses depends on the
fortuitous choice of the right species of life
and even the right organ. Subsequent definition
of the critical factors then make the initial
observations more general. We chose the eye for
our inaugural free radical search; first because
the eye is an extension of brain into the environment, and we were primarily concerned with regulatory factors in cerebral energetics during
stress. Secondly, it was believed that the molecular structures of the eye were of the type
that would be excited to free radicals by visible
light. Anatomical structures of the eye like
retina, pigment layer and choroid were physically
separated from each other by dissection. These
were further resolved into isolated sub-cellular
components, such as melanin granules, rods, mitochondria and nuclei by differential centrifugation procedures after homogenization in isotonic
sucrose. Suspensions of the melanin granules under

near physiological conditions of ionic environment
and temperature, when injected in the quartz flow
cell in the ESR cavity and illuminated with vis-
bile light, gave the typical free radical spectrum
shown in Figure 1. The curve represents essentially a plot of the derivative of the microwave
power absorbed (dA/dH) against the magnet field
strength (H). By adjusting the magnet field
strength to correspond to the peak of the spectrum, it is possible to follow the kinetics of
free radical generation in the light and decay in
the dark. In these experiments the signal to
noise ratio was so poor that it was impossible to
obtain reasonable reproducible and relatively
smooth curves. To offset this difficulty, the
signal from the ESR apparatus was fed into a computer for averaging transients and a light on -
light off cycle was repeated enough times to pick
the signal out of noise. In the experiments shown
in Figure 2, the light was turned on for 4 seconds
and off for 28 seconds to allow for practically
complete decay of the radical. The curve shown
represents the summation of 20 cycles. With the
filtering in the circuit, the rise time may still
be too fast for any precise evaluation, but the
peak height, which is a function of the number of
free radical sites, and the decay time are amen-
able to study.

Teleological considerations of eye melanin
granules lead us to believe that the melanin free
radical may have a function in the eye other than
that of a mere energy sink or stray light scrubber.
If these structures were active in the visual
process, it was possible that some of the neurohormones commonly found in nerve tissue might interact with the free radical generating sites.
Butanol extracts of melanin granules from cow eyes
exhibit an intense fluorescence at 350 and 450mµ
characteristic of catecholamine or indol alkyl
amine neurohormones and their precursors. Because
of structural analogies between melanin and serotonin, this neurohormone was investigated first
for its effect on the light generated free radical
kinetics. The lower curve of Figure 2 shows this

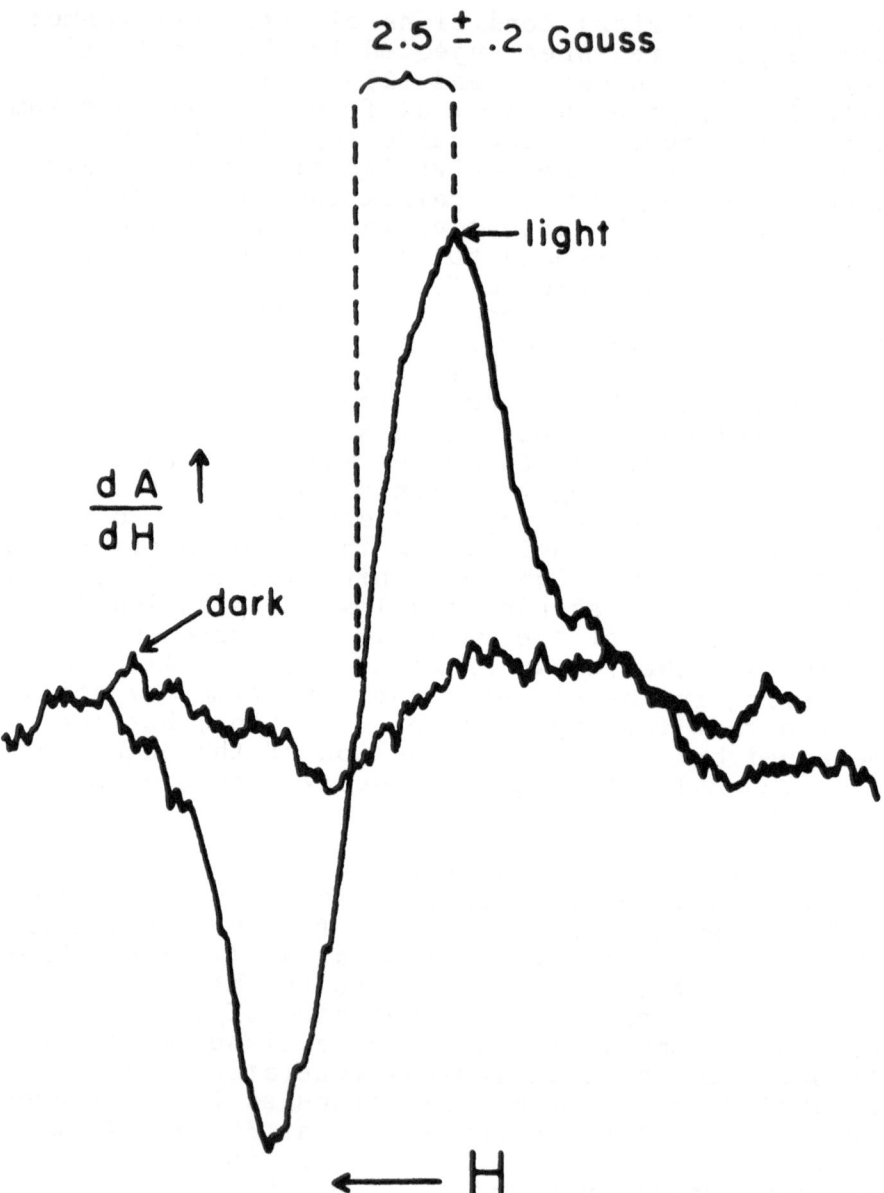

Fig. 1.   Light generated free radical spectrum of
isolated melanin granules.

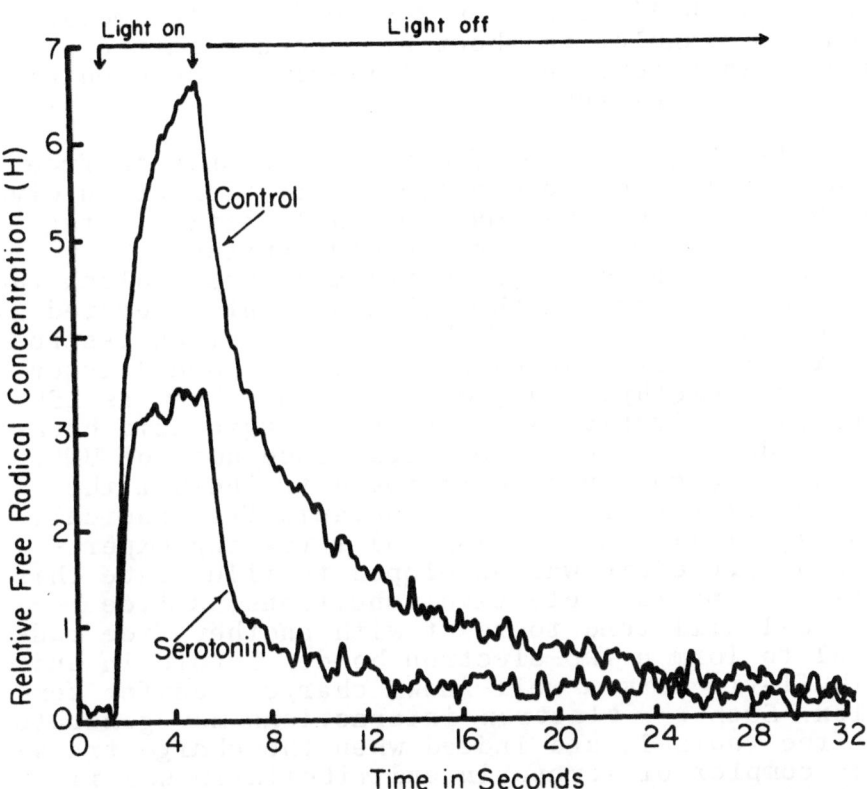

Fig. 2. Kinetic curves of melanin free radical generation in light and decay in dark. The lower curve shows the effect of added serotonin on the peak height and decay rate of the light generated free radical.

effect. It is evident that with serotonin the
free radicals generated approach an equilibrium
value faster, that the peak height or relative
free radical concentration in a unit of time is
lower, and that the decay time is appreciably
faster. Quantitatively, serotonin caused a decrease in half life from 2 seconds in the melanin
control granules to about 0.8 seconds in the granules with serotonin. Similar effects were obtained with norepinephrine.

If the _in vitro_ effect of serotonin on free
radical kinetics were a true reflection of in-vivo
mechanisms, then pharmacologically active antagonists to serotonin also should manifest their
action in the isolated system and should block the
serotonin effect on the melanin light generated
free radical. Some of the most potent anti-sero-
tonin drugs are the psychomimetic compound lyser-
gic acid diethylamide, more commonly know as LSD,
and the derivative 1-methyl-D-lysergic acid butanolamide with the abbreviated trade name of UML.
Both these compounds were found to inhibit the
action of serotonin on the melanin free radical.
An especially interesting and revealing experimental procedure was developed to illustrate this.
Barring special selective conditions, a free
radical will tend to react with another free radical to form a two-electron bond. Serotonin and
riboflavin form a well-known charge transfer complex. Such π π electron interactions may generate
a free radical, and indeed when the charge transfer complex of serotonin and riboflavin was illuminated with visible light, electron spin resonance evidence was obtained for free radical generation. The addition of LSD or UML to this complex blocked free radical formation by light. It
was now possible to study the effects of free radical generating systems reacting with one another,
and to evaluate the action of neurohormone antagonists. The topmost curve in Figure 3 is the dark
free radical pattern of light generation and dark
decay obtained with isolated melanin granules.
When this reaction takes place in the presence of
free radicals formed by the interaction of sero-

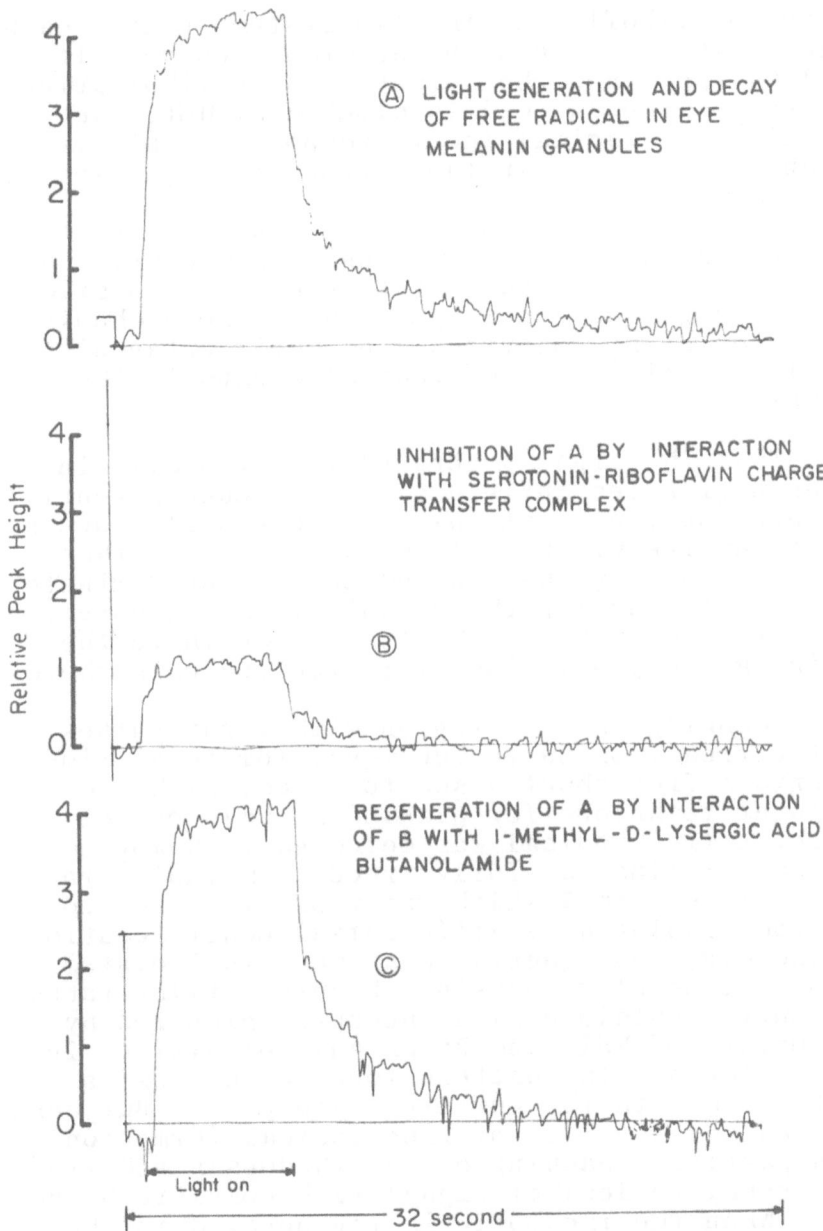

Fig. 3.   Inhibition of light generated free radical in melanin granules by serotonin-riboflavin and the reversal by 1-methyl-D-lysergic acid butanolamide.

tonin and riboflavin, the two systems cancel each
other. This is shown in the middle curve. It
follows that when the formation of a riboflavin-
serotonin free radical is blocked by UML, the
melanin free radical should reappear. This is
shown by the curve at the bottom of the series.

The demonstration of a free radical event in
isolated structures of eye, and its mediation by
neurohormones and counteraction by antagonistic
drugs, suggested a new approach to the mechanism
of action of neurohormones and their antagonists.
It also implicated free radical events in the
visual process.

That the light generated free radicals in
melanin granules influence rod and cone potentials
and play an important role in the visual process
is not unlikely. But it is apparent also that
the reactions in the rod and cone layer dominate
the events inducing the visual signal in nerve.
We therefore turned our attention again to the
retina and began to look for free radicals there.

Scanning the G-2 region with light pulses
that varied from 60 to 200 msec. for total time
intervals from about ¼ sec to 1 sec. with the
computer technique for picking signals out of
noise, a free radical was detected in homogenates
of whole retina carefully freed of melanin granules. This signal which was rapidly traced to
the rods isolated by differential centrifugation
in sucrose, subsequently was found in isolated
preparations of rhodopsin. Figure 4 illustrates
the curves obtained with rhodopsin prepared by
the method of Wald and Brown and solubilized in
2% digitonin. The pattern found with rods is
reiterated with the purified rhodopsin. But here
one can see the loss of free radical formation
with partial bleaching of the rhodopsin and finally complete loss of signal with complete bleaching. When the rhodopsin is regenerated in the
dark, the free radical signal returns again in a
form very similar to the original.

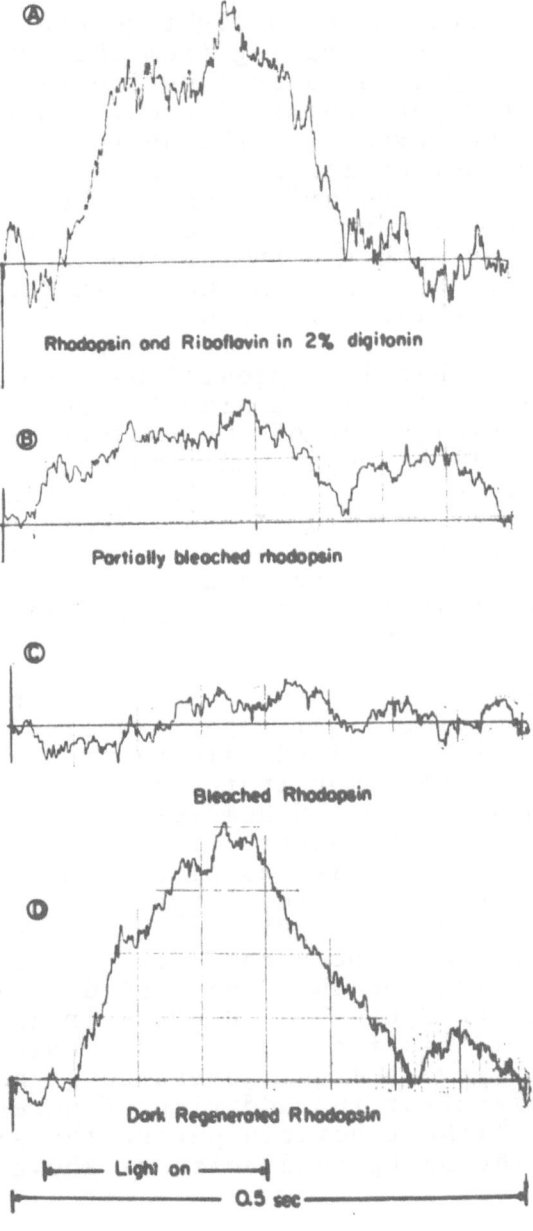

Fig. 4. Free radical kinetic curves of generation in light and decay in dark for isolated rhodopsin solutions.

The indications of a light generated free radical in rods originating from rhodopsin offers at least an approach to an experimental mechanism connecting the photochemical events in the retina with electrical events in the nerve. So far, the gap between photoexcitation of pigment molecule and nerve fiber depolarization has been bridged by structures with very little experimental under-pinning. Even with the indication of a free radical generated by light in rhodopsin there still remains the question of whether a free radical event can initiate a nerve response. That this is indeed possible is suggested by reports of Lyudskovskaya from the Russian Academy of Sciences on the photoexcitation of the isolated giant axon of the cuttlefish sensitized with dyes like neutral red. Her data shows the formation of spike potentials obtained upon illumination of the squid nerve stained with neutral red. These ef-fects could be explained by a free radical event which might be involved in the generation of the action potential.

Support for this hypothesis would be the demonstration of a light generated free radical in nerve or brain stained with neutral red. The data in Figure 5A demonstrates free radical form-ation in whole rat brain homogenate in sucrose with .003% (10^{-4} M) neutral red. The curve repre-sents about 250 repetitions of a light on - light off period summated on the computer for averaging transients. The free radical generating activity of brain could be concentrated in an insoluble particulate fraction that sedimented in isotonic sucrose between 2,000 and 10,000 G in 15 minutes. Suspension of this fraction in phosphate-saline buffer again gave a light generated signal with dye, but only after the addition of norepinephrine, Figure 5B. Without norepinephrine the signal faded into the background noise as shown in Figure 5C.

The nerve and brain fractions which were norepinephrine dependent for free radical gener-ation were soluble in high salt concentrations.

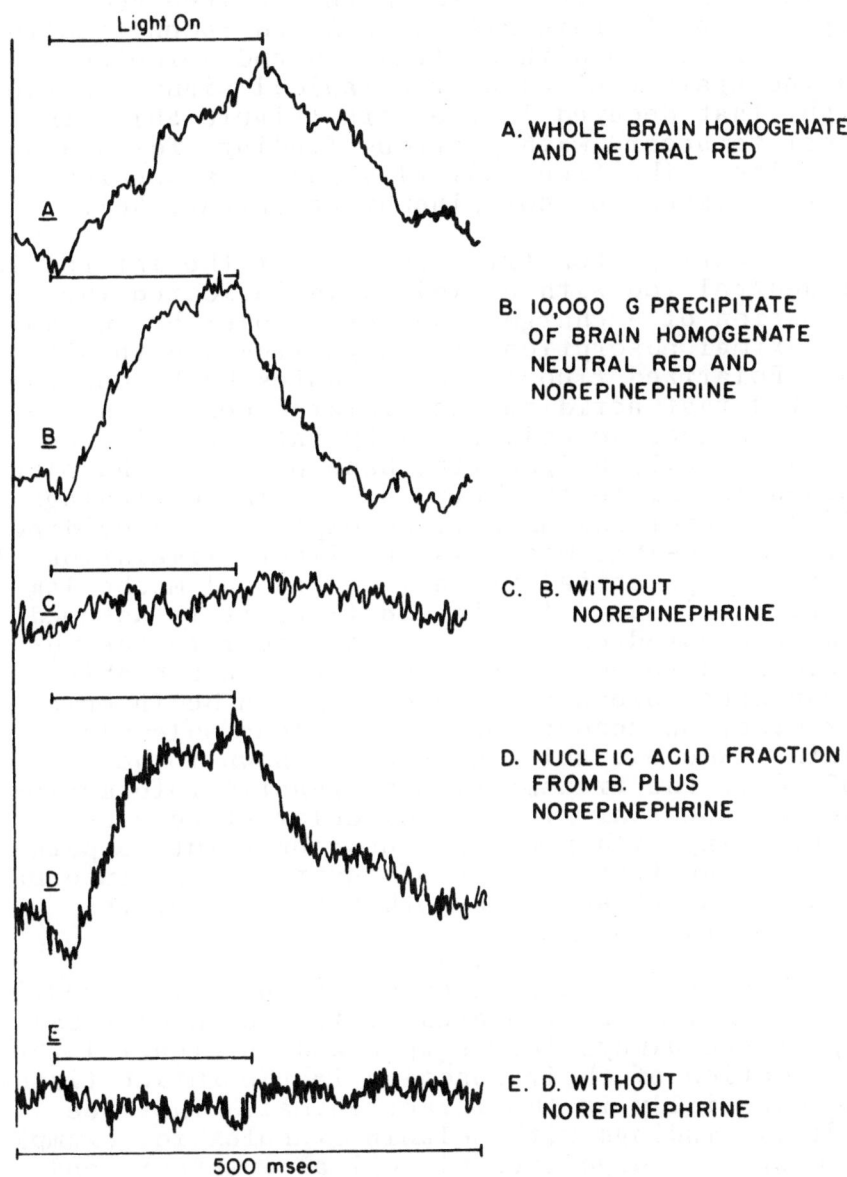

Fig. 5. Light generated free radical formation of brain neutral red charge transfer complex.

This and some other optical properties led us to
believe that nucleic acids might be involved.
Separation of brain RNA and DNA and reconstitution
of the reaction with neutral red and norepinephrine again yielded a free radical signal in the
light that decayed in the dark (Figure 5D).   In
complete accord with previous findings for brain
fractions, the free radical signal disappeared
in the absence of norepinephrine (Figure 5E).

Spectrophotometric analysis of the interaction
of neutral red with nucleic acid indicated the
formation of a charge transfer complex by the shift
in maximal absorption of thedye from 450 to 555
mµ.   Polarized fluorescence studies by Lerman indicated that acridines and related compounds are
bound to nucleic acid (DNA) by intercalation between normally neighboring base pairs in the plane
perpendicular to the helix axis.   These findings
offer theoretical as well as experimental evidence
for the possible mechanism of light stimulation of
an action potential via a free radical mechanism
in nerve tissue.   If, in addition, it is assumed
that the noradrenalin attaches itself to the nucleic acid to act as an oscillator in a highly
conjugated molecule, a mechanism can be inferred
for electron conduction along a long molecular
chain.   Nucleic acids have been shown to be involved in the transmission of genetic information.
The concept that similar molecular structures
interacting with neurohormones constitute a pathway for the transmission of environmental information is an attractive hypothesis that invites
experimentation.

The experimental demonstration of free radicals in molecular isolated systems of biological
importance always leaves open and inferential the
implication of their function in the intact tissue
and especially in the intact animal.   The free
radical findings with melanin granules for example,
although of suggestive biological interest, and
probably of importance in offering a mechanism for
neurohormonal function, are not too unexpected
from a chemical viewpoint.   It is well known that

defects produced in solids by radiation fields can
enter into chemical reactions on the surface and
that absorbed species can be made more or less
active for interaction with molecules in the adjacent liquid or gas phase.

If the observed free radicals in the isolated
systems had any biological significance, it should
be possible to demonstrate an ESR spectrum in intact nerve. To accomplish this, we used the giant
photoactive ganglion and attached pedal nerves
dissected out of the sea snail, Aplysia caniforn-
icus. Electrophysiologists have found that these
ganglia, when illuminated with blue or red light,
trigger spike potentials that can be detected in
the nerve fiber. These nerves have dimensions of
length and thickness that are just suitable to
permit the nerve fiber to be drawn into the quartz
flow cell used for free radical detection, while
the photoactive ganglion is caught in the upper
constriction of the cell and remains out of the
region of ESR detection. The problem then became
one of revealing a fast ESR signal buried in noise.
Using signal averaging techniques coupled with
judicious selection of filters and reaction times,
the summation of a thousand repetitions of a light
pulse revealed the generation and decay of a free
radical signal that coincided with the on and off
period of the light. Figure 6 shows the variation
of the ESR signal with variation in the time of
the light pulse. These kinetic indications of the
ESR spectrum were complicated by the generation of
a slow dark signal of variable frequency that
followed the light-induced signal. These may be
correlated with the slow spontaneous dark spike
potentials following illumination of the pigmented
ganglion reported by neurophysiologists. It is
evident that even under the best conditions the
signal observed is uncomfortably close to the
noise level. If the signals observed truly represented a kinetic reflection of free radical generation and decay, then by changing the magnet field
strength to coincide with the upper and lower peaks
of the ESR derivative spectrum, the light-generated
signal also should change its direction. That this

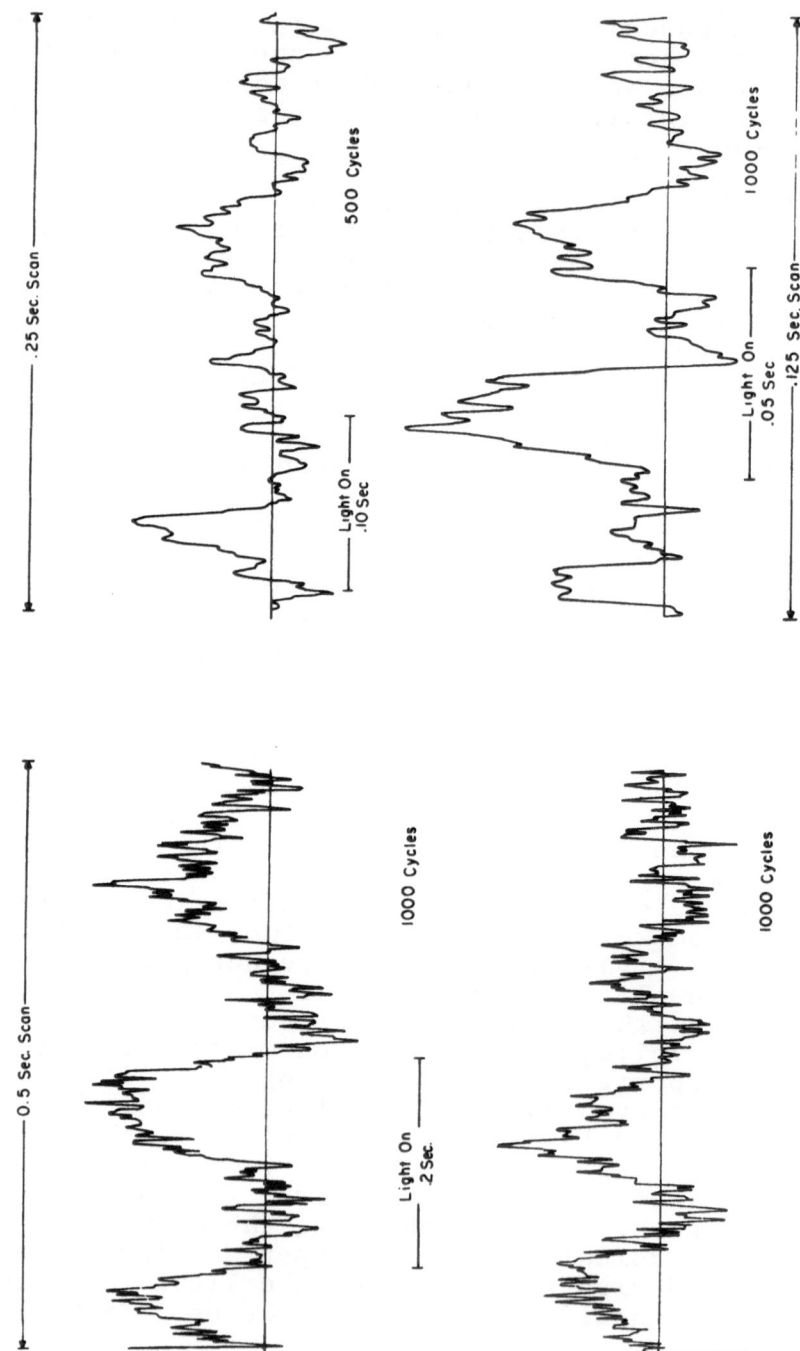

Fig. 6.   Light generated free radical signals in Aplysia nerve.

indeed does occur is shown in Figure 7. It is believed then that these series of experiments effectively argue for free radical events in brain
and nerve.

We then sought to demonstrate a free radical
effect in whole animals. When a protein, like
crystalline serum albumin, containing the amino
acid residues of tyrosine and tryptophan is partially oxidized with nitrosyl disulfonate at 5° C
under controlled conditions, then these amino acid
residues are converted to free radicals stable in
aqueous solution for long periods of time (two to
three weeks at 5° C). These stable free radical
proteins have an absorption spectrum in the visible (490 m maximum) and show a characteristic ESR
spectrum in solution at room temperature (Figure
8). The free radical protein is of considerable
interest and possible practical importance since
intravenous injection of the free radical albumin
produced an unusual state of cerebral excitement.
Measurement of the rabbit EEG by Dr. L. Goldstein
of the New Jersey Psychiatric Research Institute
showed that protein free radicals in small concentrations of a few mg/kg reversed the mild sed-
ation induced with pentobarbital. The result was
a sudden EEG arousal which was unique in that it
alternated with returns to the sedated state.
With increasing amounts of the free radical, the
EEG experiments revealed hyper-arousal which was
correlated with behavioral changes in the animal.
These electroencephalographic and behavioral
changes of the rabbits injected with protein free
radical did not resemble the effects found by Dr.
Goldstein with amphetamine but were more like the
reactions obtained with LSD.

The free radical events that we have found
in nervous tissue are representative of the known
mechanisms for the generation of free radicals.
In melanin granules, light stimulated free radicals probably are formed by the migration of electrons and holes in some quasi-ordered lattice.
Interactions with serotonin and norepinephrine
offer a pathway for the transfer of this excited

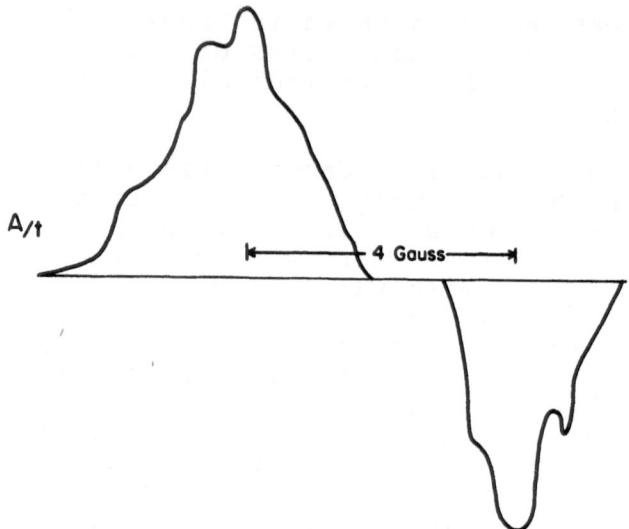

Fig. 7.   Light generated free radicals in Aplysia
nerve.   Correlation of the kinetic signal with the
derivative peaks of ESR spectrum.

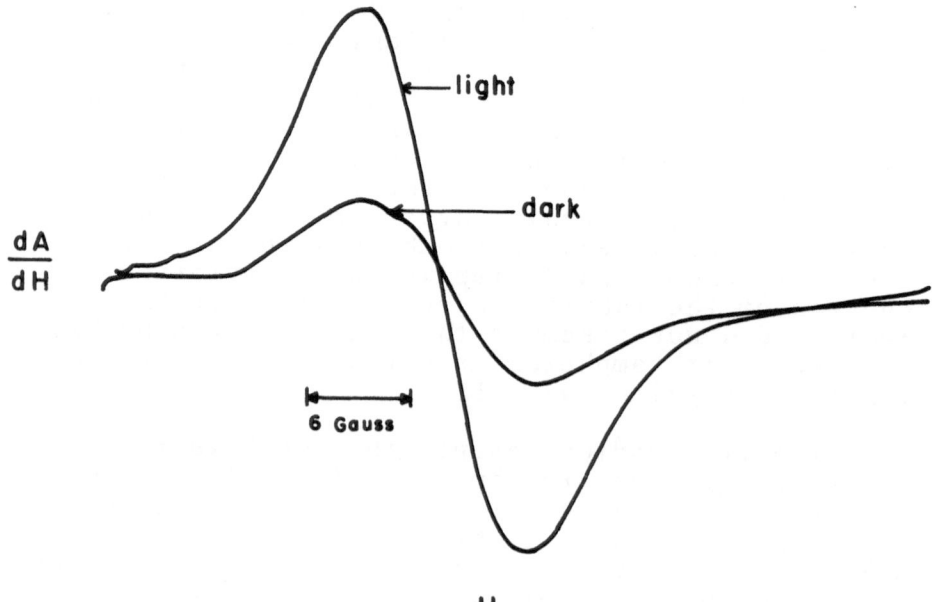

Fig. 8.   ESR spectrum of serum albumin free radical in light and dark.

state energy and define a system for governing
the energy level of the excited structure. Light
generated free radicals occur in vitro in brain
homogenates through the formation of a charge
transfer complex between nucleic acid and neutral
red in the presence of norepinephrine. The
radical formation in rhodopsin may be considered
as a representative of the photodissociation of
a bond involving migration of fragments and re-
combination in the dark. Finally the stable protein free radical probably represents oxidation
to semi-quinone and nitrogen-oxo derivatives of
tyrosine and tryptophan in the peptide chain.

Admittedly there is still considerable
speculative associations between the in vitro
chemical observations and the in vivo biological
results. Nonetheless, it is believed that the
study of excited state intermediates in isolated
systems coupled with the in-vivo effects of excited molecules offers a unique and powerful
approach to the bioenergetic mechanisms of nerve
tissue.

FREE RADICAL CHARACTERISTICS: THEIR POSSIBLE
RELATION TO TRANSDUCTION IN THE VISUAL SYSTEM.

DR. FREED: This has been a very interesting
paper. What is the lifetime, roughly of these protein free radicals which you illuminated at room
temperature?

DR. POLIS: In vitro they will hang around
for weeks. In vivo they will disappear within
24 hours. We have the peak level which occurs
one hour after injection and then it starts to degrade.

DR. RAPPORT: I would like to know, are there
classes of substances, let us say, amongst the
lipids that interfere with free radical formation.

DR. POLIS: In lipids one would be more
likely to find free radical formation than the
interference of free radical formation.

When you say interference with free radical
formation, the question is interfere how, by combination, by utilization of the energy?

The whole point is that we are so accustomed
to thinking in terms of ionization and gross bond
contracting that we fail to realize that there are
a lot of excited state intermediates that take
place in the biological system.

I think we cannot dismiss the mechanism of
formation or generation and control of excited
state intermediates as merely a bizarre phenomenon
because we are not familiar with them or because
they are poorly understood. I think Calvin's work
on the quantasome shows that these mechanisms are
primary to the living system. Certainly if they
take place in an important mechanism like photosynthesis, one can expect that they will continually
take place in other biological phenomena.

DR. FREED: It is well known that free radicals
occur in the peroxidation of lipids.

DR. BOGOCH: I found the data of Dr. Pollis on the visual system of particular interest. If free radical formation occurs here, it might well be important in the process of transduction. In other words, if the free radical reaction is indeed the critical reaction which is set up in relation to bleaching and resynthesis of visual pigment compounds under the influence of light, and this free radical is transferred to nerve substrates at this junction, it may well be the means of setting up an electrical impulse which is then propagated in the nervous system through the visual system.

DR. POLIS: I think you have stated the problem correctly. This is exactly our purpose in doing these studies and the reason why we began to look to see whether there was any mechanism for the propagation of a free radical along a nerve. We have been able to actually isolate free radical proteins that are native to the nerve and to the brain. It is possible that there are free radical, or other excited state mechanisms important in the transmission of information in the nervous system, and of the transmission of energy from one compartment to another compartment within the cell.

THE EFFECT OF ENVIRONMENT UPON THE DEVELOPING

BRAIN

Williamina A. Himwich

Events in neurochemistry have moved rapidly
in the last thirty years, but perhaps no single
development is of more interest and importance
than the demonstration by several laboratories
that environment can play a definite role in the
development of the brain, both chemically and
functionally. Among the earliest work was that of
Levine and Alpert (1959), who demonstrated that
the handling of young rats influenced the rate at
which phospholipids accumulated in the brain.
Monumental studies have also appeared from the
laboratories of Krech, Rosenzweig and Bennet
(e.g., 1956, 1960, 1963), who have been interested for many years in the effects of various environmental treatments on protein, acetylcholinesterase and, in the last analysis, on the intelligence of rats. Nutrition both of the mother
and the young must also be considered in studying
the effects of environment on the brain. The
simplest dietary change treatment is a reduction
of total calories keeping the proportions of dietary constituents the same or undernutrition (Dobbing, 1969). Another way is to rearrange the constituents of the diet so that one, e.g., protein.
is low and the animal receives an unbalanced diet.
This is called malnutrition. The third form, of
course, is to completely remove some element, or

to add it in excess, producing again a type of
malnutrition which is more severe.

Unfortunately, we have much more data on the
effects of the environment on brain composition
than on function. Studies of behavior which
are directly and undeniably linked to changes in
the composition of the brain are far and few between. This link is, of course, the one of
greatest interest. If we can determine how to
influence by diet or other means, at just the
right time, the brains of those who live in over-
populated and chronically undernourished countries, we can make an enormous contribution to
the well being of these populations. Again, and
even more important, if we can learn how to influence the development of the brains of our own
children so that they will be, by a combination
of environmental factors, including diet, more
able and more intelligent, we have again made an
enormous step toward improving mankind. These are
problems of great importance but the factors affecting them are so inter-locked in complexity
that to separate them and study them individually
is very difficult. The problems represent not
only the question of the environmental treatment
of the animal, which can be varied in an infinite
number of ways, but also the so-called "vulnerable" periods, during which function or behavior
can most easily be affected (Dobbing, 1969; Fox,
1966). Frequently, an effect can be produced only
during a "vulnerable" period, and not at any other
time. Since the "vulnerable" periods vary for
practically every facet of development and behavior and since their chronological arrangement may
differ from species to species, the problem is of
enormous complexity. In this field, I believe
that certain planning conferences would be of
great value because no one laboratory or group of
laboratories can completely cover the problems.
It would be beneficial to have a number of laboratories attacking simultaneously the same or
similar "vulnerable" periods in various species of
animals under conditions as nearly identical as
possible.

Since our meeting is on the future of the
brain sciences. I would like to pose several
questions, partial answers to which are available
but which need more research. The illustrations
will use results from our laboratory which have
a bearing on the questions. Space will not permit me to discuss related work from other laboratories.

I.   Is there a non-specific effect of stress
     on the brain as there is in the rest of
     the body? As corrolaries to this ques-
     tion, how long does this effect last?
     What are the parameters affected?

Let us start first by considering data which Dr.
Michael Fox accumulated in our laboratory in collaboration with Dr. Harish Agrawal and myself on
the effects of isolation and handling on the
young dogs (Agrawal, Fox and Himwich, 1967). In
this study the animals were subjected to partial
isolation and partial sensory deprivation for one
week between four and five weeks of age. The puppies were sacrificed immediately after testing for
one hour at the end of the isolation period. They
showed behavioral abnormalities, were hyperactive,
and their responses to stimuli were more diffuse
than we would expect, and were, in general, inap-
propriate. These reactions were considered by
Dr. Fox to be the effects of abnormal development
of normal fear reactions. The biochemical studies
of the various parts of the brain for the free
amino acids, glutamic, glutamine, GABA and aspartic, showed a tendency for glutamic acid to increase as did also GABA (Fig. 1). The responses,
however, were different in the various parts of
the brain. In other words, biochemical responses
which we saw here were about as diffuse as the behavioral reactions and served to tell us nothing
beyond the fact that changes were occurring. This
study was the first we undertook on changes in the
brain due to environment, and we were not happy
with the results. Puppies as experimental animals
are very costly so we turned to problems which
could be done on larger groups of animals, such as

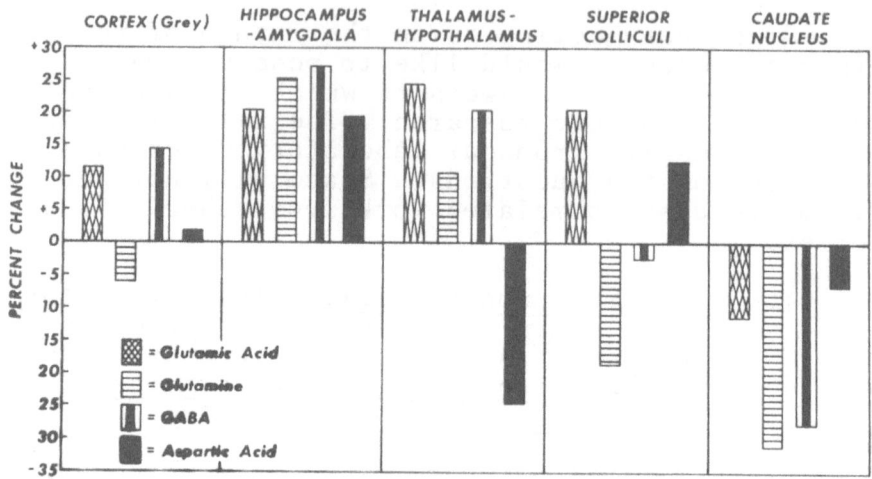

Fig. 1. Effect of isolation on the amino acids of the brain in 5 week old dogs (For experimental details see Agrawal, Fox and Himwich, 1967.).

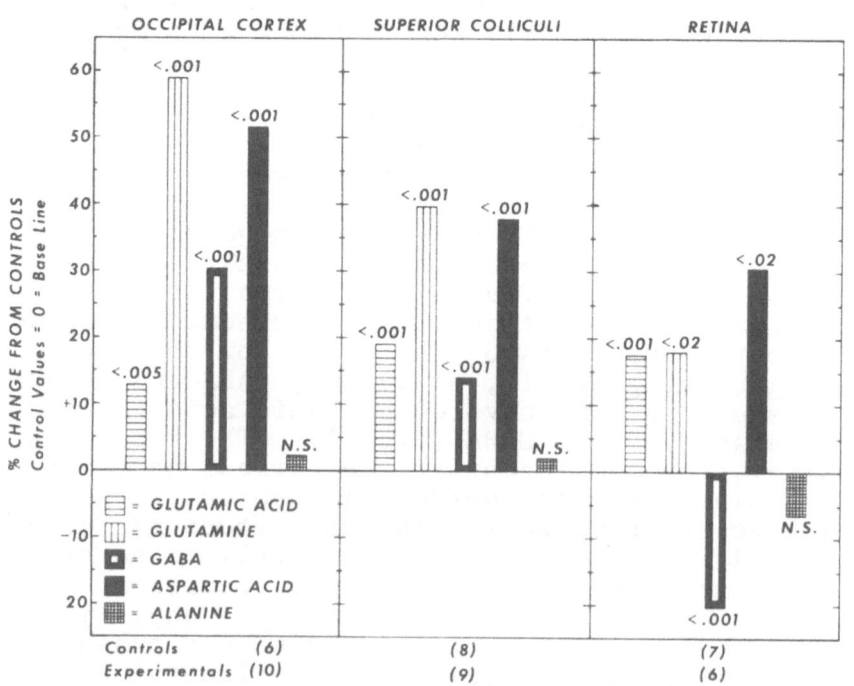

Fig. 2. Effect of visual deprivation on amino acids in discrete areas of the visual system of rabbits (For experimental details see Himwich, 1968.).

rats or rabbits. However, for studies of spontaneous behavior, we still prefer the dog.

Our next study was concerned with young rabbits. We followed effects of visual deprivation
between 8 and 11 and 8 and 15 days of age on the
development of the visual tract (Himwich, 1968).
I wish to report here on only three parts of the
visual system: the occipital cortex, the superior
colliculi and the retina. The visual deprivation
resulted in an initial increase in the amino acids
in all the parts studied. If visual deprivation
was continued for only three days (Fig. 2) there
was a marked and significant tendency for the free
amino acids studied to increase. The only decrease occurred in the GABA of the retina. If,
however, visual deprivation was continued for
another four days until the animals were 15 days
of age, a different picture appeared. The increases of amino acids in the occipital cortex
have all decreased and have become relatively insignificant (Fig. 3). The same is also true of
the superior colliculi. The decrease of GABA in
the retina has remained fairly large, but is less
significant than it was before. Aspartic acid and
alanine also have shown considerable increase.
Our rather fragmentary information about the role
of aspartic acid and of alanine is nonetheless
great enough to suggest to us that when there is
an increase in these substances, there may very
well be unusual metabolic activity. In these
animals, we see again, as we saw in the isolated
animals (Fig. 1) a tendency for a diffuse reaction
to occur in the concentrations of several of the
amino acids. With my colleagues, Dr. Agrawal and
Mr. Davis, I have also done some studies in which
we followed the effect of early weaning on young
rats (Himwich, Davis and Agrawal, 1968). These
animals were weaned at 15 days or at 19 days and
were all killed at 21 days. If we compare the
two groups, there is in the 15 day weanlings a
significant increase in most of the amino acids
in the cortex, as well as in acetylcholinesterase,
as compared to the controls (Fig. 4). In the
cortex of the animals weaned at 19 days, however,

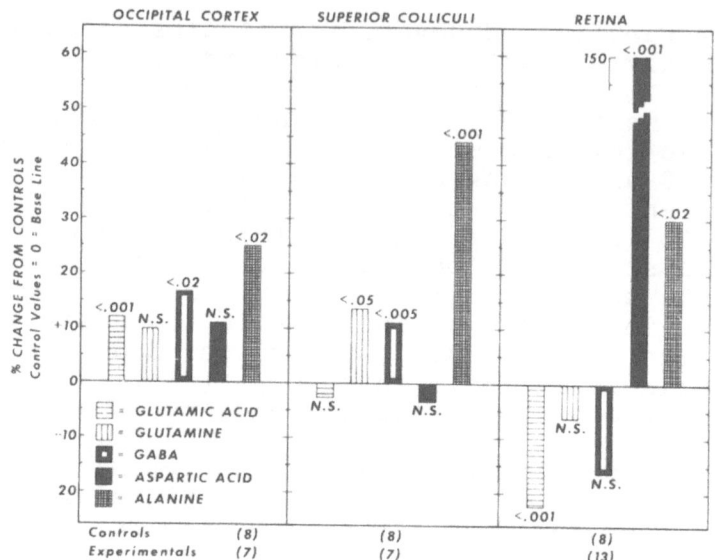

Fig. 3. Effect of visual deprivation for seven days
on amino acids in discrete areas of the visual system of rabbits (For experimental details see Himwich, 1968.).

most of the increases have become less, and in
some cases, they have actually fallen to non-sig-
nificant decreases. On the other hand, if we look
in the midbrain, under these same circumstances,
we find that even at the 15 day weaned animals,
the changes in amino acids are either insignificant or in a few cases show a significant decrease (Fig. 5).

These data appear to be in many ways rather
a "mish-mash" of results, yet I believe that we
are beginning to identify a patterned response in
all these studies. I would like therefore to suggest that any kind of environmental change may
serve as a stress. The first effect of this
stress may be an immediate <u>non-specific</u> response
in the brain, resulting in an increase in some of
the important members of the free amino acid pool
of the brain and of cerebral acetylcholinesterase.
This response in varying degrees follows undernu-

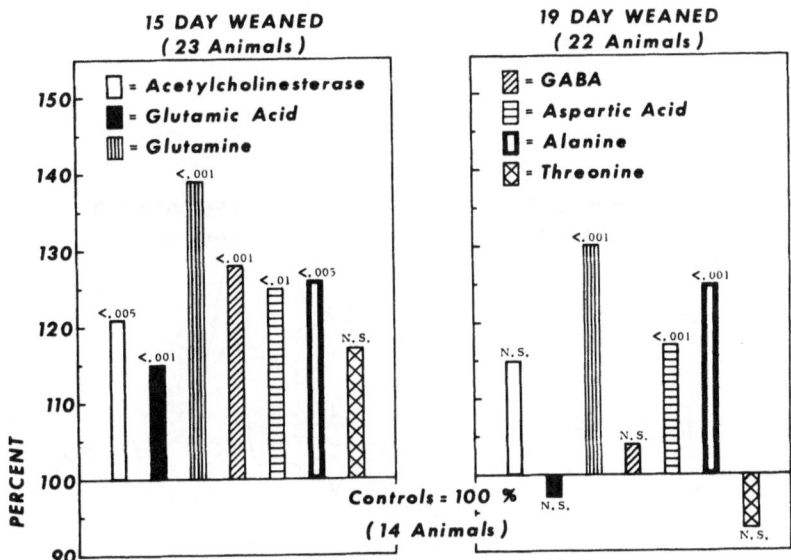

Fig. 4. Effect of weaning on acetylcholinesterase
and amino acids in rat brain cortex (For experimental details see Himwich, Davis and Agrawal, 1968.).

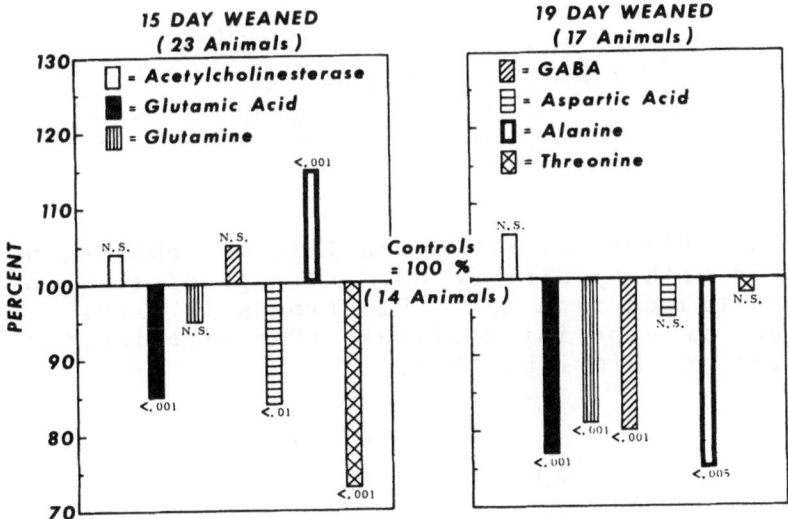

Fig. 5. Effect of weaning on acetylcholinesterase
and amino acids in rat midbrain (For experimental
details see Himwich, Davis and Agrawal, 1968.).

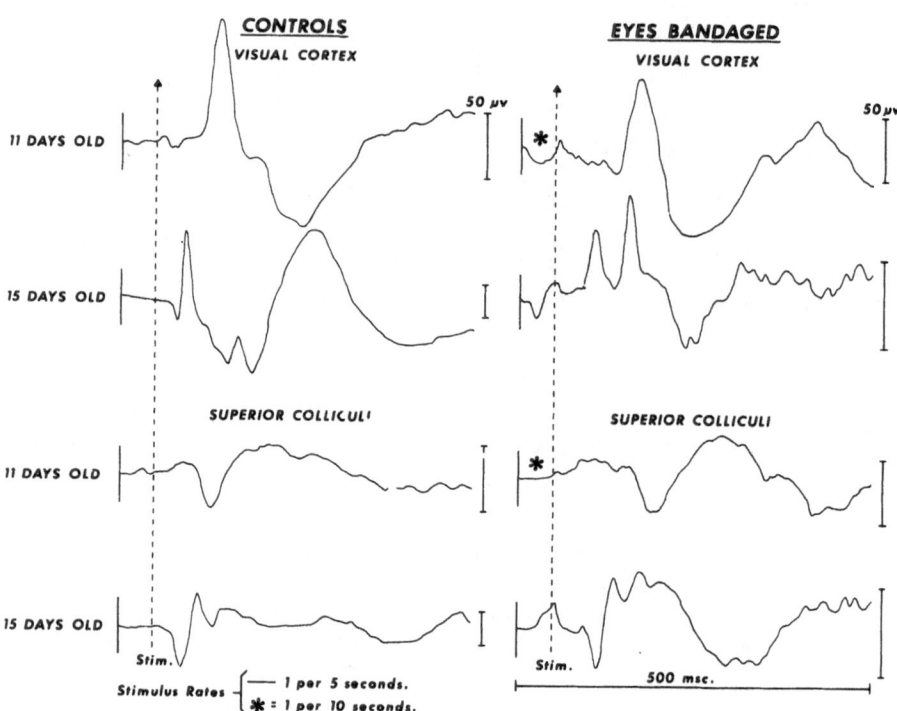

Fig. 6.   Visual evoked potentials of rabbits, normal and with eyes bandaged.   Average of ten responses traced from CAT.   Electrodes in visual cortex and in superior colliculi (For experimental
details see Himwich, 1968).

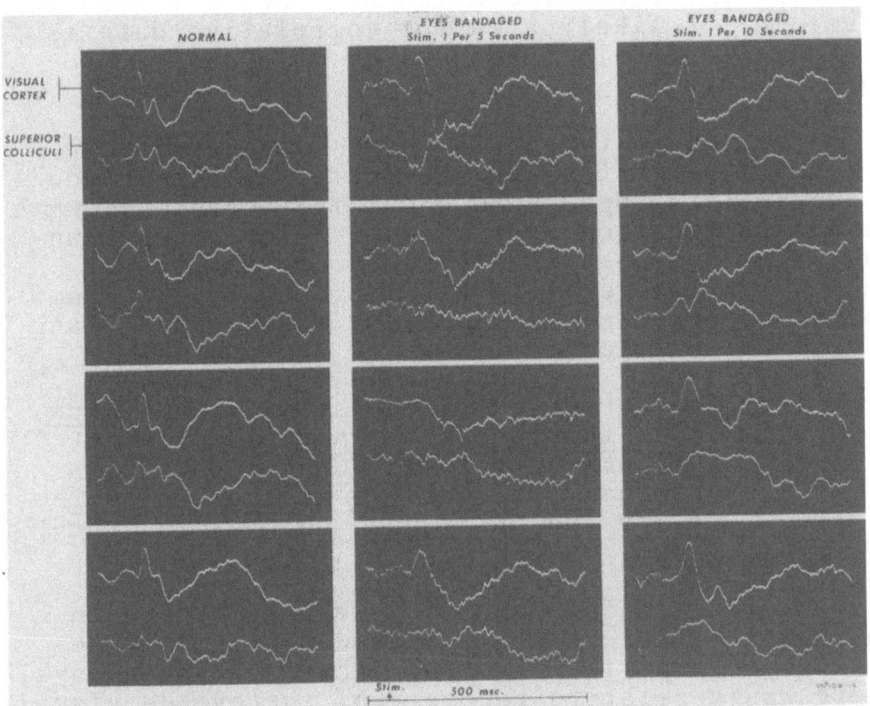

Fig. 7. Visual evoked potentials of rabbits, at eleven days of age, normal and with eyes bandaged. Electrodes in visual cortex and superior colliculi. Single responses photographed from CRT (For experimental details see Himwich, 1968).

trition, visual deprivation, early weaning, and
isolation. Although we do not yet have enough
animals for the data to be statistically significant, we have some indication of this same response occurring in puppies which have received
additional handling. The exact reasons for this
increase in free amino acids have not yet been determined. A likely area for correlative data is
a study of protein synthesis or of types of protein formed.

What does it mean if these "non-specific"
effects level off as the period of stress is continued? This appears to occur in the animals with
visual deprivation, for example. Does this mean
that the "non-specific" effects are fading out,
and that if there is a real long-lasting biochemical difference we must look for more subtle changes? I am inclined to think so.

   II.   The second question is <u>what do we know
         about these effects in terms of func-
         tion of the animal? Can we have bio-
         chemical, or close to biochemical nor-
         mality and still have changes in func-
         tion?</u>

We believe that this may very well be true. In
the rabbits, for example, that were visually deprived for a period of either three or seven days,
there was a relative recovery of the evoked visual
potential in deprived animals tested at 15 days,
considered as recorded from a CAT (computer of
average transients) which averaged ten responses.
Even in the averaged evoked potentials, the animals
mals with the eyes bandaged, had at 15 days a
greater immaturity of form for the response
(Fig. 6).

At 11 days of age, the normal animal responded well to stimuli if repeated either every 5 secs
or every 10 secs. The animals, whose eyes had
been covered, however, showed poor responses after the second stimulation when the stimulation
time was every 5 secs (Fig. 7). If the time was

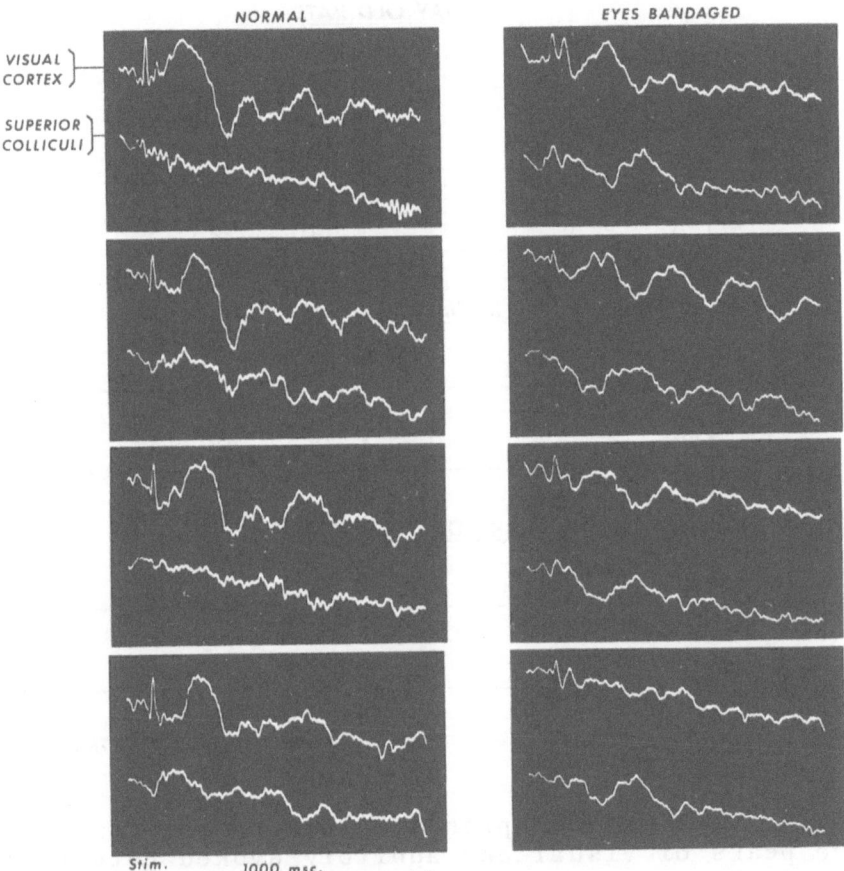

Fib. 8. Visual evoked potentials of rabbits, at fifteen days of age, normal and with eyes bandaged. Electrodes in visual cortex and superior colliculi. Single response photographed from CRT (For experimental details see Himwich, 1968).

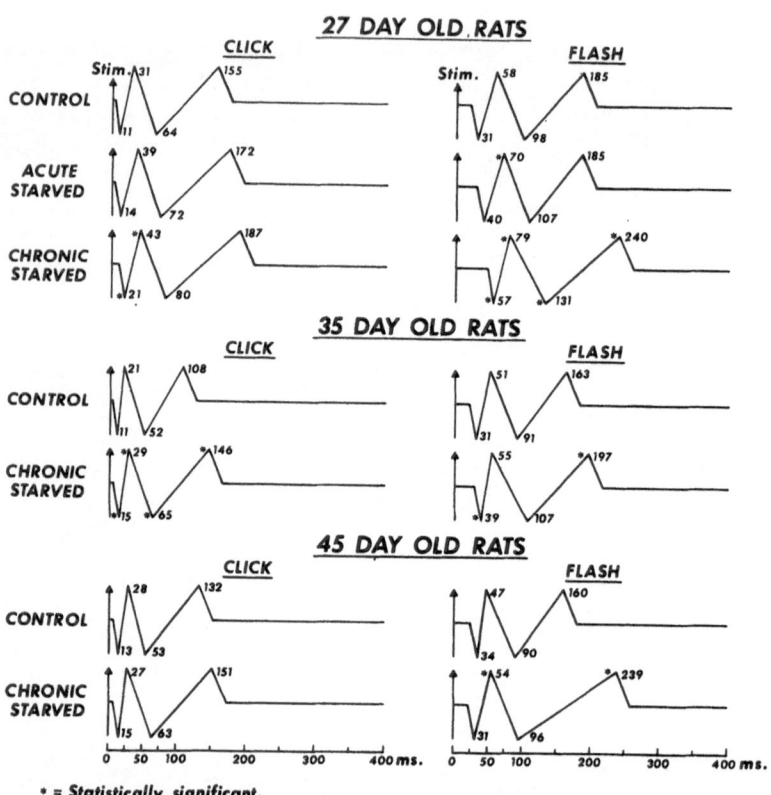

Fig. 9. Schematic representation and latencies of wave peaks of visual and auditory evoked potentials of animals undernourished for 10-12 hrs. a day from 5-10 days of postnatal life. Acutely starved animals were separated for one 24 hr. period (For experimental details see Mourek, Himwich, Myslivecek and Callison, 1967).

doubled (1/10 sec), the response was somewhat better. The 15 day old animal was not much better
when we considered each evoked potential (Fig. 8).
After one stimulus at 1/5 secs, he showed deterioration of the evoked potential in both the visual cortex and in the superior colliculus. This
poor response occurred in spite of the fact that
the chemistry of the colliculi and of the cortex
had been returning to normal. Also, from the
work of Mourek (Mourek, Himwich, Myslivecek and
Callison, 1967) done in our laboratory on the
chronic and acutely starved animals, differences
in the latency time of the various peaks of the
evoked potential to either visual or auditory
stimuli can be seen as late as 45 days of age
(Fig. 9). This difference may not be apparent,
unless the latency of the last negative wave is
also considered.

III.   The third question is, is it possible
        that some amount of stress may actually
        serve as a stimulus?
In the study described above on early weaning, the
brains of animals weaned at 15 days were closer
to those of the controls both in actual weight
and in percentage (Fig. 10) than brains of animals
left with their mother for 4 more days (weaned at
19 days). These differences could be considered,
however, a short term effect similar to the "non-
specific" stress response described earlier.

As evidence of a long-lasting effect, data
obtained from two groups of animals are pertinent:
one group (5 days undernourished) was taken away
from the mother for eight hours every day for
five days between the ages of five and ten days;
the other group was treated in exactly the same
fashion but taken from the mother for an additional five days, that is, from 5 to 15 days. In
these experiments. the effects of temperature and
himidity were carefuly controlled so that we hoped they might have no effect upon the development of the young animals, and the results would
be only the effects of undernourishment. There

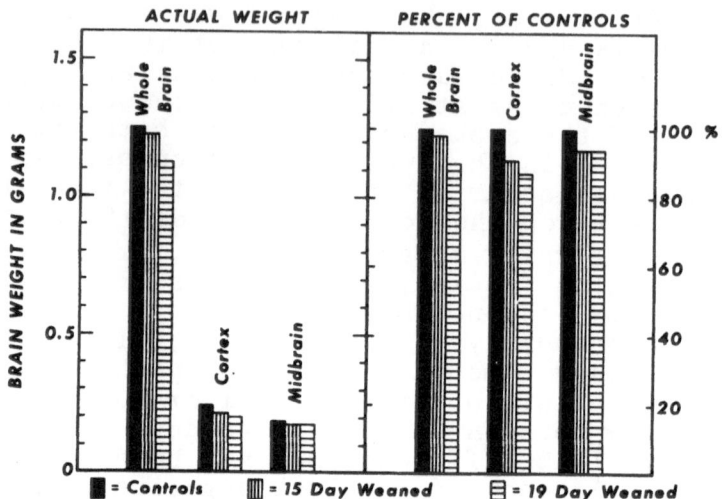

Fig. 10.   Brain weights of control and rats weaned at 15 or at 19 days.   Weights taken at 21 days are expressed as actual weights, and as percentages of control weights (For experimental details see Himwich, Davis and Agrawal, 1968).

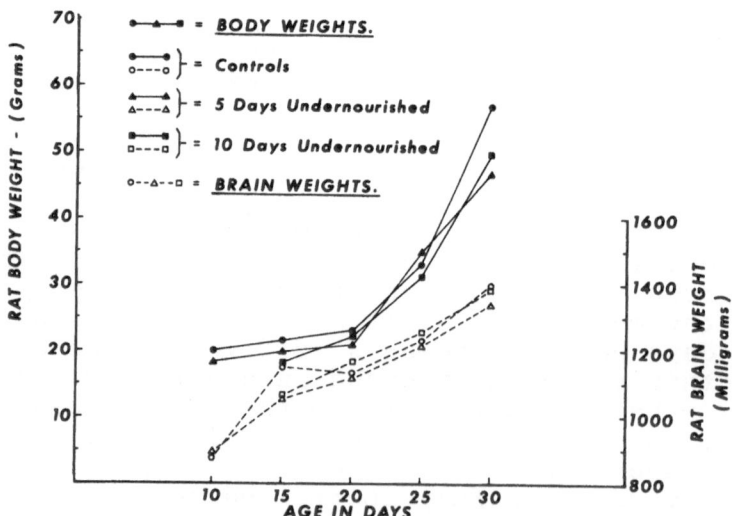

Fig. 11.   Body and brain weights of rats undernourished from 5-10 (5 day undernourished) or 5-15 (10 day undernourished) (For experimental details see Himwich, Sargent and Agrawal, 1969).

was a definite effect of the extra five days
of removal from the mother (Fig. 11). These animals responded in body weight and brain weight as
if they had been stimulated. At 30 days of age,
the brain weights were close to those of the
controls, and the body weights were more than the
five day undernourished group. It seems possible
that a stimulus or stress applied at the proper
time may have a stimulating effect.

SUMMARY

The data presented here give only partial
answers to questions which must be answered before definitive experiments on the effects of environment can be planned. If there is a "non-
specific" effect of stress on the brain, then investigations must be arranged so that the specific and "non-specific" responses can be reliably
separated. In order to correlate changes in function with cellular chemistry, more sensitive
tests of function as well as of single cell (both
neuron and glia) composition are needed. This
area of investigation will be more difficult to
explore than the relatively gross studies of composition and overall response described here.
The possibility of a stimulating effect of stress
applied at the right time in the right amount
must be differentiated from the non-specific response to stress and probably from other effects
of the stress. When these and all the other problems that will undoubtedly appear have been solved, we will be ready to manipulate the environment to improve the brains of mankind.

REFERENCES

1.   Agrawal, H.C., Fox, M.W., and Himwich, W.A.,
     Neurochemical and behavioral effects of iso-
     lation-rearing in the dog, Life Sciences, 6,
     71-78, 1967.
2.   Dobbing, J., Historical introduction, in W.
     A. Himwich (Ed.), Developmental Neurobio-
     logy, C. C. Thomas, Springfield, 1969.
3.   Fox, M.W., Neuro-behavioral ontogeny. A syn-

thesis of ethological and neurophysiological concepts, Brain Research, 2, 3-20, 1966.

4. Himwich, W. A., Multi-disciplined studies of the visual system in developing rabbits, in L. Jilek and S. Trojan (Eds.), Ontogenesis of the Brain, Proceedings of the International Symposium Neuroontogeneticum, Praga, 1967, Charles University, Prague, 1968, in press.

5. Himwich, W. A., Davis, J.M., and Agrawal, H. C., Effects of early weaning on some free amino acids and acetylcholinesterase activity of rat brain, in J. Wortis (Ed.), Recent Advances in Biological Psychiatry, Vol. X, Plenum Press, New York, 266-270, 1968.

6. Himwich, W. A., Sargeant, S. M., Agrawal, H. C., Effect of undernutrition during neonatal life on the biochemical maturation of the rat brain, in J. Wortis (Ed.), Recent Advances in Biological Psychiatry, Vol. XI, Plenum Press, New York, 1969.

7. Krech, D., Rosenzweig, M.R., and Bennett, E. L., Dimensions of discrimination and level of cholinesterase activity in the cerebral cortex of the rat, J. Comp. Physiol. Psychol., 49, 261-268, 1956.

8. Krech, D., Rosenzweig, M. R., and Bennett, E. L., Effects of environmental complexity and training on brain chemistry, J. Comp. Physiol. Physiol. Psychol., 53, 509-519, 1960.

9. Krech, D., Rosenzweig, M.R., and Bennett, E. L., Effects of complex environment and blindness on rat brain, Archives of Neurology, 8, 403-412, 1963.

10. Levine, S., and Alpert, M., Differential maturation of the central nervous system as a function of early experience, A.M.A. Arch. Gen. Psychiat., 1, 403-405, 1959.

11. Mourek, J., Himwich, W.A., Myslivecek, J., and Callison, D.A., The role of nutrition in the development of evoked cortical responses in rat, Brain Research, 6, 241-251, 1967.

ON BRAIN FREE AMINO ACID POOLS

DR. FREED: I wonder in some cases where the biochemistry did not seem to go along with behavior, whether it is possible that changes in pool size within the brain may be important.

DR. HIMWICH: I agree with you; this is an important possibility and one that must be looked into.

DR. LAJTHA: I think these metabolites have an important role, and marked changes in their concentration may have some functional significance.

I think it is not unimportant that the concentrations of free amino acids used more for general metabolic purposes are at higher levels than the ones used for protein formation, and I think changes in levels would almost necessarily mean changes in the rates of utilization.

DR. HIMWICH: What we are seeing is a change in utilization and with this in mind, we are attempting to run it down in terms of protein formation in the various parts. I appreciate your asking me the first question. I think you know one of my pet speculations. I am referring to the fact that if you look at the glutamic acid and GABA content of the cortex of the adult animal, you find that the level of glutamic acid is greater than of GABA, that is the glutamic acid has increased during development whereas in the cortex at least, the change in GABA with development has not been so striking.

Then if you go to a structure like the superior colliculi the ratio between GABA and glutamic acid is not quite one but it approaches it. It is in the superior colliculi that GABA shows the most marked increase with development whereas glutamic acid does not.

I have been trying very hard to pin some of
my anatomically minded friends down to say there
are more inhibitory neurones in the superior colliculi than in the cortex but I haven't gotten one
to really come out and do anything more than speculate about something like that.

We have been quite excited about Dr. De Robertis' material showing vesicles containing GABA in
the nerve endings and we would like very much to
see if we could find more of these vesicles in the
superior colliculi.

DR. MANDEL: Could you comment on the methods
of sacrifice and conditions of extraction?

Dr. HIMWICH: We are aware of the problems
that arise from killing the animal and the circumstances under which the tissues are extracted.
For this reason we have a routine method in which
all animals are killed in a cold room at -4°C.
They are brought in, guillotined, the brain removed
and placed on blocks of dry ice within less than
four minutes.

We have used this method in preference to
using liquid air because I am not happy about dissecting out nuclei or small brain parts in a sample
that has been completely frozen. We realize that
this is a compromise as many things are but it is
the routine method we use for all our studies.

DR. SACKS: I should like to make a few remarks in regard to the question raised by Dr. Himwich concerning a possible function for the free
amino acids found in brain. A few years ago we
published a paper in which we examined 13 years of
data obtained with our arterio-venous technique for
determining cerebral metabolism in humans in vivo
(Sacks, W., J. Appl. Physiol., 20, 117-130, 1965).
It was concluded that there was a significant dilution (about 40 fold) of the ^{14}C of glucose-^{14}C in
traversing the Krebs cycle in human brain. A scheme
for the cerebral metabolism of glucose in humans
in vivo was proposed based on our studies and those

of other investigators. Its primary feature was
the inclusion of glutamate, GABA and succinic
semialdehyde in the Krebs citric acid cycle. A
small metabolically active glutamate pool in the
Krebs cycle was considered to be in slow equilibrium with a large metabolically inactive glutamate pool in the brain. An N-acetylaspartate pool
was thought to be a source of substrate in emer-
gency conditions. Therefore, it may be that the
free amino acids, especially glutamate, may function as metabolic buffer pools to insure uninter-
rupted, constant brain metabolism.

DR. HIMWICH: I would agree with you.

DEVELOPMENTAL APPROACH TO PSYCHOBIOLOGY:

A RESEARCH PROGRAM

Joseph Altman

In altricial mammals profound changes take place after birth in the organization of brain and behavior. The drastic alterations that occur in the morphology, biochemistry and physiology of the central nervous system during postnatal development, which is concurrent with the acquisition of new behavioral capacities and propensities, make this period ideal for the study of brain-behavior relationships. Recent studies, particularly those employing thymidine-H^3 autoradiography, showed that the structural organization of the brain is far from completed at birth. The acquisition of entire neuronal systems (such as the basket, stellate and granule cells of the cerebellar cortex, the granule cells of the hippocampal dentate gyrus and of the olfactory bulb), and of the vast majority of glia cells in many brain regions, begins after birth in rats and cats [1-4]. Indeed, the recruitment of microneurons and of glia cells continues in some brain regions into adulthood. The proliferation of primitive cells in the brains of infant and adolescent rats and cats represents the first event in the complex process which ends with their subsequent interdigitation into the changing architecture of the brain. These include such processes as migration, chemical and morphological differentiation, and the establishment of varied synaptic connections and intercellular relationships. These processes require considerable time (matters of weeks), further extending the period during which changes occur in the basic morphology of the brain (5-6).

The first objective of our research program is to establish the time table of cell proliferation and migration in selected brain regions in rats and cats by means of thymidine-H^3 autoradiography, and to study aspects of their subsequent differentiation and maturation with histological (including Golgi), histochemical, autoradiographic and electron microscopic techniques. An example of this effort is our recent study (6) in which we succeeded in constructing a time table of the postnatal development of the cerebellar vermis on basis of autoradiographic evaluation of the onset of differential of granule cells in different lobules of the vermis (Fig. 1). The results of these studies will provide us with qualitative and quantitative standards of the development of the brain under "normal" conditions and allow us to assess deviations from it as a consequence of experimental manipulations. Such studies will hopefully also produce data for the formulation of some of the principles of neural development. Finally, attempts will be made to relate the chronology of development of selected brain regions to concurrent development of behavioral functions that are dependent on them.

The second objective of this program is to study the effects of pathological interferences with the early development of the brain on its subsequent maturation and on the development of behavior. We are investigating the effect of three pathological manipulations during infancy: irradiation of selected brain regions with low-level x-ray; nutritional deprivation throughout or during some period of infancy; and early surgical intervention with the normal development of the brain. X-irradiation was selected because proliferating and migrating cells are extremely radiosensitive. Radiation doses that leave apparently

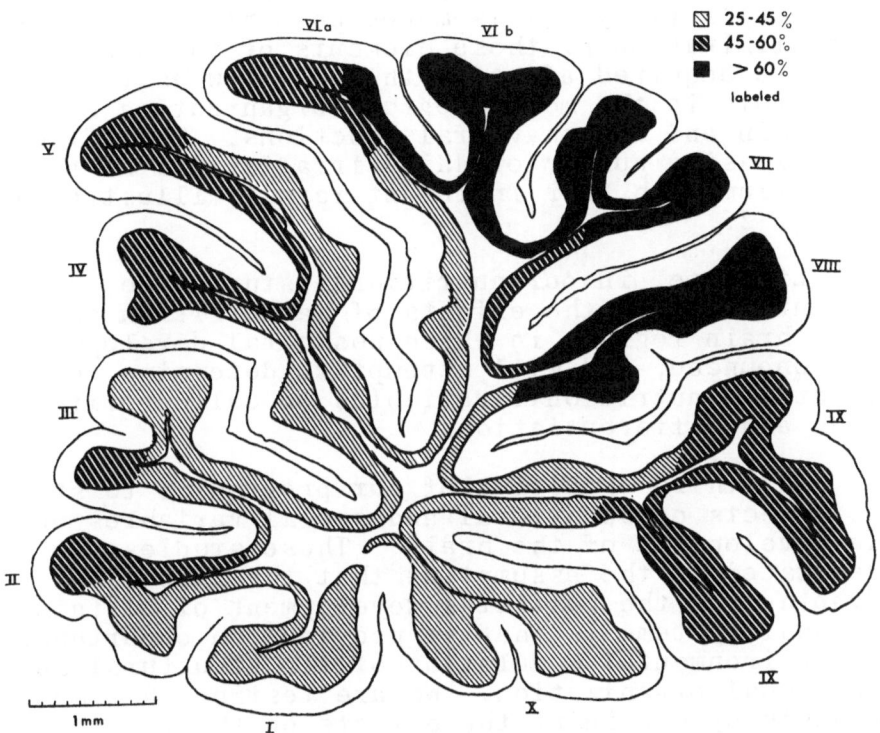

Figure   1

Ontogenetic development of the cerebellar vermis
based on determination of the proportion of granule
cells formed after 11 days of age.   Numerical data
that were obtained in coronal sections were transferred to a sagittal tracing of the vermis and an
arbitrary tripartite classification was made.   (a)
Light diagonal lines: early maturing regions, with
only 25-45% of granule cells formed after the 11th
day.   (b) Heavy diagonal lines: areas intermediate
in maturation.   (c) Black: later maturing regions;
more than 60% of cells formed after the 11th day.
(From Altman, 1969).

unharmed differentiating and mature cells that are in their final locations, will kill out a large proportion of the multiplying and migrating primitive cells (7-9). This makes possible the selective elimination of those elements of the brain that are acquired after birth and allow us to study their role in the morphological organization of the brain and in behavioral functions. The effects of increasing number of daily irradiations of the cerebellum with 200r in infant rats is illustrated in Fig. 2.

Likewise, in our nutritional studies we are concentrating on the effects of malnutrition on those brain regions in which postnatal development is pronounced, and will attempt to determine the effects of nutritional variables on cell proliferation and differentiation.

The third objective of our program is to study the effects of behavioral and social variables on the development of the brain. These studies are predicated on the assumption that the functional advantage of the postnatal development of certain neuronal systems is that it provides an opportunity for environmental modulation of its structural and functional organization. We are testing this hypothesis by examining the effects on the growth of the brain of such variables as "isolation" from and exposure to a complex environment during infancy (non-handling vs. handling); rearing animals after weaning in restricted or enriched environments; the effects of prolonged and intensive sensory and motor exercise. Some preliminary results on the effects of these variables on the growth of the cerebrum are presented in Fig. 3.

The fourth objective of this program is to study the conditions under which neural recovery may occur. Our recent studies have shown that after its subtotal destruction with low-dose x-ray, the proliferative matrix of the cerebellar cortex reconstitutes itself (12-13). This is illustrated in Fig. 4. The nature and conditions of early neural repair will be examined in the cerebellum

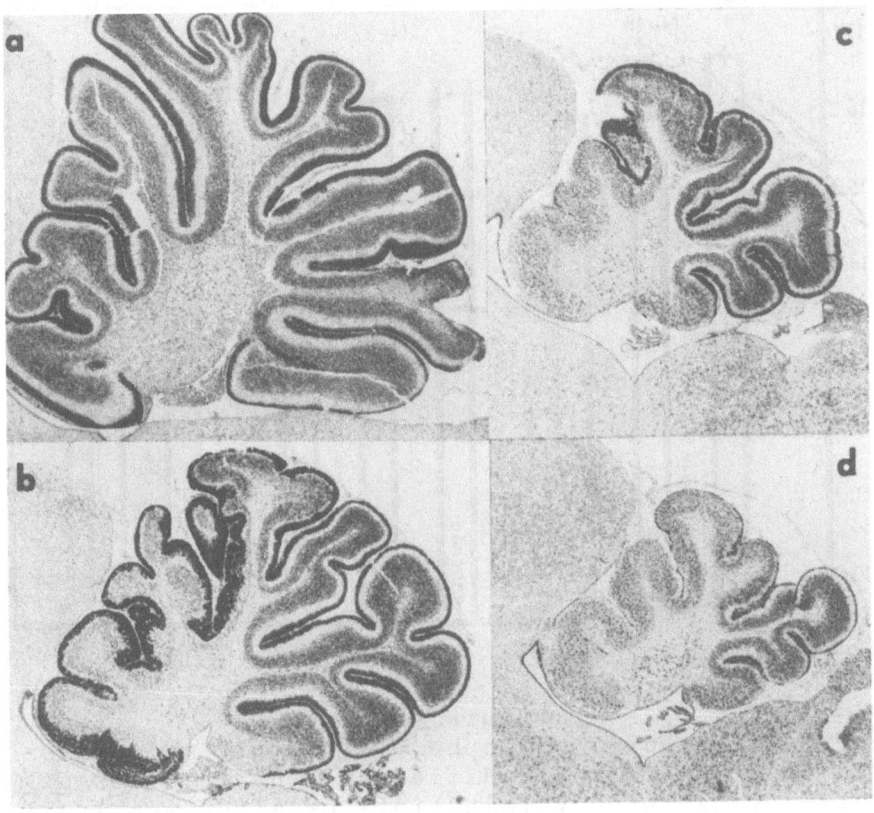

Figure   2
Low-power photomicrographs of the cerebellar vermis
from 10-day old rats with the following antecedent
irradiation history: (a) 1 x 200r;   (b) 5 x 200r;
(c) 8 x 200r;   (d) 10 x 200r.   (From Altman, Anderson and Das, 1969).

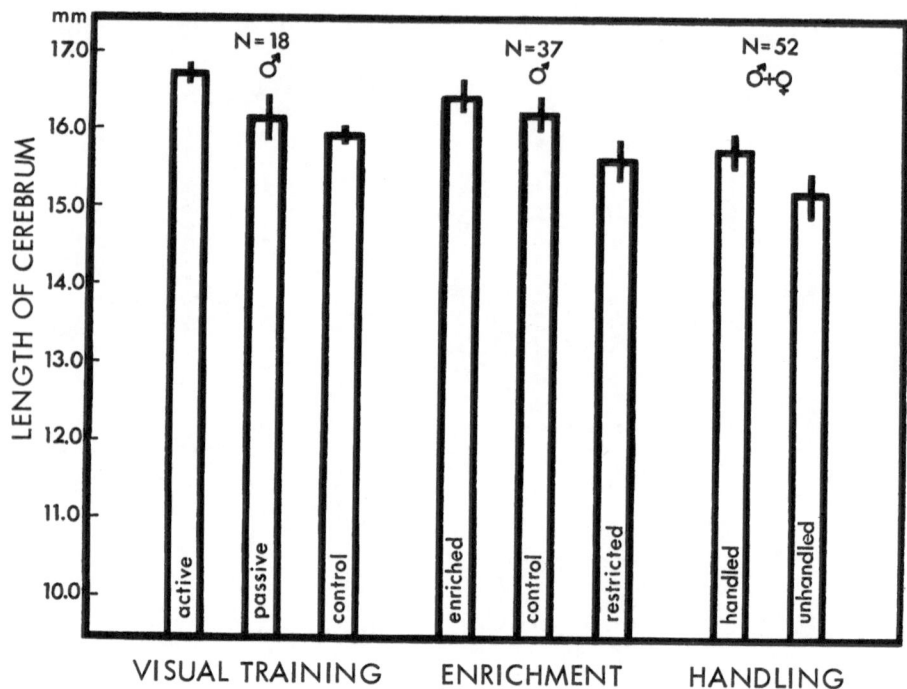

Figure 3
Length of cerebral hemispheres in adult male rats
as a function of three different behavioral treatments during infancy (handling) or after weaning
(visual training or enrichment). Note that the
results suggest that visual training and enrichment enhance cerebral growth in the anteroposter-
ior dimension, whereas rearing in restricted environment and absence of infantile handling retard
it.   (From Altman, Wallace, Anderson and Das, 1968).

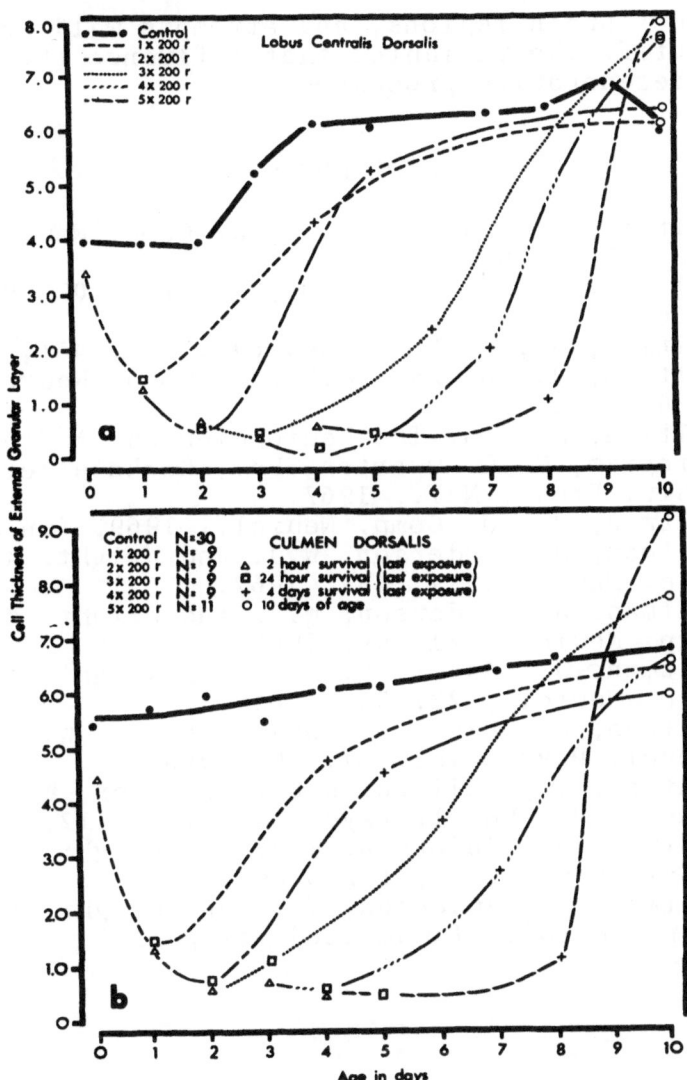

Figure 4
Cell-thickness of external granular layer in lobulus
centralis (dorsalis) and culmen dorsalis as a function of number of daily exposures, age, and survival
period after last irradiation. Note regeneration of
the external granular layer in all instances in the
10-day old animals. (From Altman, Anderson and Das,
1969).

and other brain regions, and attempts will be made
to identify the variables that influence these
early regenerative processes.

REFERENCES

1. Altman, J. and Das, G.D., J. Comp. Neurol.,
    124, 319, 1965
2. Altman, J. and Das, G.D., Nature, 207, 953,
    1965.
3. Altman, J., J. Comp. Neurol., 128, 431, 1966.
4. Altman, J. and Das, G.D., J. Comp. Neurol.,
    126, 337, 1966.
5. Altman, J., In: The Neurosciences, A Study
    Program, G. C. Quarton, Ed., Rockefeller
    Univ. Press, N.Y., 1967.
6. Altman, J., J. Comp. Neurol., 1969, (in press).
7. Altman, J., Anderson, W.J. and Wright, K.A.,
    Exp. Neurol., 17, 481, 1967.
8. Altman, J., Anderson, W.J. and Wright, K.A.,
    Exp. Neurol., 21, 69, 1968.
9. Altman, J., Anderson, W.J. and Wright, K.A.,
    Exp. Neurol., 22, 52, 1968.
10. Altman, J., Das, G.D. and Anderson, W.J.,
    Devel. Psychobiol., 1, 10, 1968.
11. Altman, J., Wallace, R.B., Anderson, W.J. and
    Das, G.D., Devel. Psychobiol., 1, 112, 1968.
12. Altman, J., Anderson, W.J. and Wright, K.A.,
    Anat. Rec., 1969, (in press).
13. Altman, J., Anderson, W.J. and Wright, K.A.,
    Exp. Neurol., (submitted for publ.).

B. NUCLEIC ACIDS IN MEMORY AND LEARNING

TRENDS IN BRAIN RESEARCH ON LEARNING AND MEMORY

Holger Hydén

The future reality will no doubt be formed by the intensive activity within basic and applied research. A most important task will be to determine the effect of priority to a certain field. Therefore, an analysis of the goal will be important, and the indicators showing that this goal is approaching. The research of interest has a broad basis in natural sciences, medicine and psychology. The future society will be more and more dependent on scientists, and the scientists dependent on the state. An increasing amount of the budget will go to research. For the man living in such a society it is important that his genetically given potentialities will be realized, and that aggression among his fellows will be channelized so as to become positive factors.

Since brain research which deals with learning and memory also means analysis of the highest form of learning, problem solving, this is a good occasion to scrutinize where we stand now and if some meaningful guesses may be hazarded on two main problems: Learning and memory mechanisms and ways to realize the genetical potentialities of the brain. The implications of the problems which this conference will take up are certainly

far-reaching, not only as scientific and technical questions but for the future society and its
inhabitants. Therefore, it may be appropriate
to sketch a frame, to relate questions in brain
research to problems which have been formulated
for the society to come.

An old theory says that when something is
learned and stored, there occurs in millions of
nerve cells a facilitation of the messages at
the synapses. This is an attractive view, because the activity flow can be made very specific and one big neuron may have 10,000 synaptic
knobs on its surface. The theory only scrapes
the surface of the problem, however.

Learning means the capacity of a system to
react in a new or modified way as a result of experience. Memory is the capacity to store information which can later be retrieved with high
distinction to steer the function which is correlated with the new information.

In learning, a short-term memory is first
established. This lasts for seconds to hours and
is labile and susceptible to interference of various kinds. Short-term memory differs in nature
from long-term memory. This storage or fixation
process takes place during training to learn and
immediately after. Long-term memory may last for
the better part of a life-cycle. It is remarkab-
ly resistent to poisoning and shocks, and can be
retrieved in a fraction of a second.

It is important to consider that there are
three principle components of the brain. Neurons,
which differ bio-chemically e.g. with respect to
transmittor substances (1) and RNA composition (2).
They do not divide and only a small part of their
genome can be assumed to be active. Glia, the
second type of brain cells also seldom divide and
are electrically inexcitable. They are rich in
lipoproteins and in rapidly turning over RNA(3).
The third compartment of the brain tissue consists of an extra cellular space which seems to
amount to around 20% (4). It presumably contains

mucopolysaccharides and mucoproteins.

1. The genetic roots of learning mechanisms:

If the outer part of the neuron is the site for the pulse-coded nerve impulse, the inner part handles the energy demands, but also the synthesis of specific substances, and regulates the expression of the gene activities. Synaptic facilitation, therefore, can be expected to be regulated by genetically controlled mechanisms.

Any substance which can serve a memory mechanism for both innate behavior and experiential learning for a life time, can be assumed to have an easy access to the genes.

It may be argued that learning involves an additional activation of the genome of the neuron which, therefore, becomes richer patterned biochemically and structurally, and can respond within a wider range with more selectivity. Such a reasoning could obviously lead to a deterministic view. Many people would find such an idea hard to accept. The limitation of the capacity of the brain would be set. It would mean that only the genetically preprogrammed of our brain cells could be selected for a certain purpose and utilized. This means a selection theory which goes back to the ideas of Plato (5). On the other hand, one can ask whether there exists an instructional mechanism in the brain for learning at the highest level, the problem solving, and which operates at the system level in the brain hierarchies, and in the individual brain cells. This goes back to John Locke (6) who maintained that the brain at birth is a tabula rasa on which experience chisels the writing. Phrased in modern language and concepts, both a selectional and instructional mechanism is supposed to operate.

Under the influence of the achievements within molecular biology, genetics, and ethology, interest is now focussed on molecular events forming the main operative mechanism underlying

learning and remembering. The macromolecules,
RNA and proteins are the most likely candidates
to mediate storage of information in the brain.
They have recognition sites. The neurons are
rich in RNA and proteins. No other somatic cell
can compete with the neuron as a RNA producer (7).

At this point, I would like to stress that
no data support the view that brain cells contain mechanistically taping "memory molecules"
that store information. This is biological nonsense. In current literature, such views are
seen, but they do little more than add to the
anatomy of confusion. As will be seen, RNA and
proteins in brain cells do respond at the establishment of new behavior, but the mechanism and
its regulation remain for the future to be elucidated.

Proteins, synthesized in neurons and glia at
learning, are probable candidates as executive
molecules in a mechanism for acquisition and retrieval of memory. They have a high specificity
and could respond rapidly in millions of brain
cells on a trigger mechanism.

In this discussion, I would like to give
some results of biochemical analysis of brain
cells at learning and some data which deal with
orderliness and age in brain cells. This is in-
timately linked with learning and memory.

2. Macromolecular synthesis in brain cells
at the establishment of a new behavior:

A main question has been: do macromolecular
changes occur in brain cells which characterize
the learning process but which do not occur at
only increased neural activity? Another line of
research has been to interfere with synthesis of
macromolecules in the brain and to observe the
effect on behavior.

When a new behavior is established, brain
cells respond with a synthesis of RNA of a highly

specific base compostion and a synthesis of several types of acidic proteins (8-12). When rats are induced to switch over from the left to the right paw in reaching down in a narrow glass tube to retrieve food pills, very small amounts of new RNA are formed in the nerve cell nuclei in a part of the motor-sensory cortex, which serves as a control center of transfer of handedness. The base composition of this RNA is characterized by high adenine and uracil values, that is, a DNA-like RNA.

A long series of studies has revealed that increase in sensory and motor stimulation (13) and certain drugs (14, 15) easily give rise to an increased synthesis of RNA in engaged nerve cells. But there occurred no change of the proportions of the adenine and uracil components of the newly synthesized RNA as it did at the establishment of new behavior. Neither did such changes occur in brain cells during stress experiments. also the nerve cells in the hippocampus respond. This is an area of the old part of the brain necessary for formation of long-term memory.

In the closely arranged nerve cells of the hippocampus, the synthesis of two fractions of acidic proteins occurs at a 100% increased rate at learning.

These proteins are interesting because some of them are produced only in the brain and in no other organ (16, 17). They have a molecular weight of 30,000 and a composition characterized by a high percentage of glutamic acid and also aspartic acid, but are almost devoid of tryptophane. Proteins of such a small size could be assumed to consitute an electrogenic protein (18) which could respond on electrical fields in 10^4 sec and undergo conformational changes, activate transmittors and be incorporated in membranes in a more stable configuration.

A question for the future is whether such acidic proteins are the specific executive molecules which operate at the biochemical differentiation of the brain and are also used in mechanisms for storage and retrieval of information.

Another type of learning experiment which involves motor and sensory functions has been performed on goldfish by Shashoua (19). A piece of plastic foam fastened below the jaw turns the fish around and lifts the head out of the water. In a while, the fish has learned to swim correctly. During this learning, new RNA was synthesized in the brain of the same type as we had found during learning in rats, and with similar ratios of the uracil/cytosine values. No such RNA was produced when the fish performed the same amount of swimming but did not learn, and, most important, stress did not produce these RNA changes. If protein synthesis in the goldfish brain was inhibited by puromycin, no learning occurred.

Another way to elucidate the mechanism would be to catch the primary RNA species by so-called hybridization with DNA, and to determine the activity of the enzyme copying the DNA sections, the RNA-polymerase. See the article by Pelc (34). Such analyses are now going on in several laboratories. On the other hand, to show simply increased RNA synthesis in behavioural experiments without specification of RNA is without value for assessing possible mechanisms specific for learning.

At this point, I would like to return to the question of the relationship between macromolecules and establishment of behaviour and to conclude: at learning, small amounts of nuclear RNA with a highly specific base composition are synthesized in brain cells. Similar RNA responses have not been found up to now at other types of physiological stimulation in brain cells in mammals. The synthesis of certain acidic proteins at learning may also be a specific response.

The RNA response has been interpreted as reflecting an activation of hitherto silent gene areas
in brain cells when the animals were faced with
a situation they had not encountered before, and
which required learning (8, 20). The task in
both of the quoted examples was obviously within
the capacity of the species. As learning involving problem solving, the tasks were not difficult to judge by performance and time course.

3. Interference with macromolecular synthesis in brain cells at the establishment
   of new behaviour:

In another line of research, the effect of
certain antibiotics has been studied on behaviour.
They have all aimed at blocking the production of
protein or nucleic acids of brain in a selective
way (21, 22).

Dr. Agranoff and Dr. Barondes will discuss
this field. The conclusion of these experiments
has been that brain protein synthesis during
training is necessary for the formation of long-
term memory, as is intact nuclear DNA and RNA-
mediated protein synthesis in the brain cell
bodies. Short-term memory, on the other hand,
is not dependent on protein synthesis. It can
persist concomitantly with the long-term memory
and for as long as six hours. You can also read
more about the formation of protein in brain at
learning in the papers by Dr. Bogoch and Dr. Ungar in this volume.

4. The gap between electrical and biochemical phenomena in brain cells:

How can the gap in knowledge between the
electrophysiological and biochemical data on
brain cells be bridged? This is a question of
dundamental importance. There is no straight
way to be seen, but we move in the best circles!

Some recent data seem to open up a new lead.
Adey and his colleagues have studied intra- and

extracellular electrical phenomena as well as the
EEG during learning experiments (23,24). They
have advanced the view that there exist current
pathways outside the neurons in the extracellular
spaces which may modulate the neurons. They have
suggested that information can be processed in
a parallel fashion by a primary system of waves
generated in the neurons (50 to 100 µV). These
waves are supposed not to be dependent on synaptic connections. The EEG reflects partly such
non-linearly, graded waves. The pulse-coded
nerve impulses represent for Adey a secondary
system for information processing. It is interesting that clear changes of the EEG pattern
have been found to accompany the establishment of
a new behaviour. There was a decreasing scatter
of phase relations in the wave trains of the
hippocampus at higher level of performance. The
20% extracellular space in the brain is probably
filled with mucopolysaccharides and mucoproteins.
Such molecules can bind calcium ions which would
then modulate the flux of sodium and potassium
according to Adey's views. However this may be,
it is no doubt that experiments have to encompass
and to integrate data from the three compartments
in specific sites of the brain: the neuron, its
glia, and the activity in the extracellular space.

5.   The neuron and its cellular environment
     in learning:

Can the glia take part jointly with the neuron in the learning process? In commenting recently on brain correlates of learning, Galambos
(25) observed that some investigators give something less than enthusiastic support to this
idea. True, some investigators cannot accept the
idea that even secondarily, learning is tied to
macromolecular changes in brain cells.

For ten years we have analysed the neuron
and its glia in mammals and have come to the
conclusion that they together constitute a metabolic and functional unit. When the demand on
energy is increased in the neuron, the capacity

of the electron transferring system in its respiratory chain increases (26).   In the glia surrounding the neuron, there occurs a switch-over
to the less-efficient anaerobic glycolysis to
cover the glial energy-demand (27).   In a pathological process there occurs first an alteration
of the polynucleotide pattern of the glia (28).
Later on, a similar change can be seen in the
neurons.   Only then do the nervous symptoms become overt.   Furthermore, acidic proteins are
localized in nuclei of neurons and in cell bodies
of glia (29).   There is what we consider good
evidence that governing molecules are transferred
between the glia and its neuron.   RNA seems to
flow between these cells (20).   During learning,
the glia responded in a similar way as did the
neurons with respect to synthesis of specific
nuclear RNA (9).   Therefore, the glia may be as
potent in their function as are the synapses.

A view I have tried to substantiate by data,
is that neuron and its glia represent a two-cell
metabolic and functional collaboration, a stable
system from a cybernetic point of view.   The
glia may stabilize, influence and program synthesis in the neurons by transfer of molecules, and
may modulate electrical properties of the neuron.

6.   Some aspects of future brain research:

The macromolecular response in brain cells
when a new behaviour is started and consolidated
was interpreted as reflecting gene activation of
brain cells.   Motivation and the set of environmental factors had served as triggers and the
intraneuronal genetic mechanisms responded according to their set of rules.

One may then ask what possible consequences,
if any, in the future such neurobiological data
may have if we turn to the problems in development and early experiences of children.   More
specifically, one could ask whether this type of
research could have implications for the way in
which the realization of an individual's poten-

tialities should be planned.

The concept of intelligence, the expression of brain capacity, has changed greatly since the early 1930's. At that time, the capacity to succeed in the environment was largely looked upon as a fixed individual characteristic. It was believed to show little fluctuation from early childhood to maturity, since it was related to parental genetic traits and was relatively uninfluenced by the impact of environment. This view is no longer upheld due to the results of longitudinal studies of children. I would like to quote from the results of most important studies by Skeels (30) which covered a period of 30 years. Skeels has convincingly shown how children around one year old and classified by conventional test methods as clearly mentally retarded, were transformed to mentally normal children. This was achieved by increasing the amount of developmental stimulation and the intensity of relationships between the children and mother-surrogates. During a two year period, these children made a gain of around 30% IQ points. They were then placed in foster homes, grew up and showed emotional, educational and occupational achievements during 21 years time which compared favorably with the 1960 U.S. Census figures for the United States in general. A contrast group of small children, initially higher in intelligence than the experimental group, were transformed to mentally retarded individuals by being exposed to a relatively non-stimulating environment over a prolonged period of time. The 12 children of the contrast group, all except one, underwent a tragic fate over the years, institutionalized or in mental hospitals. It can no doubt be postulated that if the children of the contrast group had received the same stimulation during the critical early period of their lives, they would have achieved within the normal range of development, as did the experimental children.

By inference, I would like to advance the view that a critical stimulation and interaction

with the environment during the right period of
early life produces a gene activation of brain
cells as to give an additional activation to
those gene areas which are already active at the
time of birth. This leads to a biochemical,
structural and functional differentiation of the
brain.

From research on other species, it is also
known how animals living in a restricted and
poor milieu will develop a brain cortex which
is thinner and biochemically less developed than
the cortex of their litter mates living in a
stimulating environment (31). If young animals
are reared in the dark, the retina and its ganglion cells will not differentiate functionally,
or, biochemically (32,33). For further details,
see the article of Utina in this volume.

For the future, there could be foreseen definite consequences of research in molecular
neurobiology for early development of man, even
for teaching and planning of education. We
ought to be able, eventually to answer primary
types of questions,such as: which is the critical age for training for definite achievements,
abstractions and problem solving and complicated
sensory-motor tasks? Which is the time-inte-
grated pattern and type of stimulation to realize
maximally the potentialities of an individual?

7. Neurons, orderliness and age:

The last comment may deal with orderliness
and age of brain cells. The nerve cells share a
rare characteristic with a few other cells of the
organism. They do not divide. Over a life-
cycle, therefore, no order is added to the cells
by means of cell division. There will be an increasing danger of errors at synthesis in the
cells which means errors in function. Evidence
has been presented that neurons may renew their
DNA without preparing for division (34). Future
studies have to elucidate what happens in aging
neurons, since hybridization analyses have sug-

gested that DNA from an old animal may not be
identical with DNA from a young animal.(35).

Old animals do not learn and consolidate the
the memory with the same alacrity and efficiency
as do young animals (36).

At old age, some organs or systems may be
in good shape, including circulation, although
the brain function reflected as memory and higher
intellectual function may falter. In such cases,
can we increase orderliness in the brain cells in
order to harmonize the functional output of the
various organs? If learning and the expression
of memory ultimately reflect potentialities of
the genome, then, addition of extraneous gene
material with high intrinsic orderliness to such
aging brain cells would be a logical attempt. We
have, therefore, as a first step, prepared brain
DNA of varying degrees of purity. This DNA was
injected in the ventricles of the brain in other
animals. In one hour, protein synthesis was
found to increase by more than 100% (37). Biochemical analyses showed that the DNA had been
incorporated in the recipient's brain cells in a
polymerized state. It is not surprising per se
that brain cells can take up macromolecules like
DNA, since it is by now a well-known fact that
many other somatic cells have this capacity (38).
It was made highly probable by a series of control tests that the incorporation of the DNA had
caused the stimulation of protein synthesis in
the brain cells, and RNA seemed to mediate the
effect. This does not mean that an elixir of
life has been found. More hard facts are needed.
What is the nature of the protein formed? Is it
functionally valuable or nonsense protein? How
long does the effect last, for hours, days, or
years?

How could the observation be utilized in
future brain research? I would like to suggest
that a way may eventually be found to add orderliness to brain cells of one individual by incor-

poration of gene material from another. The
question is how this would be accomplished.
There are several possibilities. The most direct
would be to infect the brain with genes attached
to a harmless virus entering the brain, for example, from the mucosa of the nose. Viruses
have the capacity of penetrating into host cells,
and should thus act as transporting agents, as
Dr. Kornberg has suggested. A successful counter-
acting of entropy increase in brain cells could
change the whole structure of our society.

REFERENCES

1. Dahlström, A., Fuxe, K., Acta Physiol.
    Scand. Suppl., 232, 1964.
2. Hydén, H., in Macromolecular Specificity and
    Biological Memory (F. O. Schmitt, Ed.),
    Cambridge, Mass., 55, 1962.
3. Egyhazi, E., Hydén, H., Life Sci., 5, 1215,
    1966.
4. Harreveld, A. van, Crowell, J., Malhotra,
    S. K., J. Cell Biol., 25, 117, 1965.
5. Plato, Meno (R. S. Buck, Ed.), Cambridge,
    1961.
6. Locke, J., Essay Concerning the Understand-
    ing, Knowledge, Opinion and Assent (B. Rand,
    Ed.), Oxford, 1931.
7. Hydén, H., in The Cell, Vol. IV (J. Brachet
    and A. Mirsky, Eds.), New York, 215, 1960.
8. Hydén, H., Egyhazi, E., Proc. Nat. Acad.
    Sci., 48, 1366, 1962.
9. Hyden, H., Egyhazi, E., Proc. Nat. Acad.
    Sci., 49, 618, 1963.
10. Hydén, H., Egyhazi, E., Proc. Nat. Acad.
    Sci., 52, 1030, 1964.
11. Hydén, H., Lange, P., Proc. Nat. Acad. Sci.,
    53, 946, 1965.
12. Hydén, H., Lange, P., Science, 159, 1370,
    1968.
13. Hydén, H., in The Neurosciences (G. C. Quar-
    ton, T. Melnechuk and F. O. Schmitt, eds.),
    New York, 248, 1967.

14. Hydén, H., in Recent Advances in Biological Psychiatry (J. Wortis, d.), New York, 31, 1964.
15. Hydén, H., Egyhazi, E., Neurology. In Press, 1968.
16. Moore, B. W., McGregor, D. J., J. Biol. Chem., 240, 1647, 1965.
17. Moore, B. W., Biochem. Biophys. Res. Comm. 19, 739, 1965.
18. Schmitt, F. O., Davison, P. F., Neurosci. Res. Progr. Bull., 3, 55, 1966.
19. Shashoua, V. E., Nature, 217, 138, 1968.
20. Hydén, H., Lange, P. W., Naturwiss., 3, 64, 1966.
21. Flexner, L. B., Proc. Amer. Phil. Soc., 111, 343, 1967.
22. Agranoff, B. W., in The Neurosciences (G. C. Quarton, T. Melnechuk and F. O. Schmitt, eds.), New York, 756, 1967.
23. Adey, W. R., in The Neurosciences (G. C. Quarton, T. Melnechuk and F. O. Schmitt, eds.), New York, 637, 1967.
24. Elul, R., Adey, W. R., Nature, 212, 1424, 1966.
25. Galambos, R., in The Neurosciences (G. C. Quarton, T. Melnechuk and F. O. Schmitt, eds.), New York, 637, 1967.
26. Hydén, H., Lange, P. W., J. Cell Biol., 13, 233, 1962.
27. Hamberger, A., Hydén, H., J. Cell Biol., 16, 521, 1963.
28. Gomirato, G., Hydén, H., Brain, 86, 773, 1963.
29. McEwen, B. S., Hydén, H., J. Neurochem., 13, 823, 1966.
30. Skeels, H., Monographs of Soc. for Res. in Child Development, Vol. XXXI, No. 3., 1966.
31. Diamond, M. C., Law, F., Rhodes, H., Lindner, B., Rosenzweig, M. R., Krech, D., Bennett, E. L., J. Comp. Neurol., 128, 117, 1966.
32. Riesen, A. H., Science, 106, 107, 1947.
33. Brattgård, S.-O., Acta radiol. Suppl., 96, 1952.
34. Pelc, S. R., J. Cell Bio., 22, 21, 1964.
35. Hahn, H. P. v., Gerontologia, 12, 18, 1966.

36. Oliverio, A., Bovet, D., Life Sci., 5, 1317, 1966.
37. Egyhazi, E., Hydén, H., Proc. Symp. on Nucleic Acids and Proteins in the Neuron, Prague, May, 1967.
38. Robins, A. B., Taylor, D. M., Nature, 217, 1228, 1968.

SOME CONSIDERATIONS OF THE EFFECTS OF SHORT TERM
LEARNING ON THE INCORPORATION OF URIDINE INTO RNA
AND POLYSOMES OF MOUSE BRAIN.

DR. GLASSMAN: There is an extensive literature that indicates that behavioral experiences
can have marked effects on the macromolecules of
the nervous system.   Unfortunately, a number of
difficulties mar unequivocal interpretation.

First, all too often the behavioral task
lasts for too long a time or is too complicated
to allow clearcut conclusions concerning the
critical behavioral components responsible for
the chemical response. Second, all too often there
is no comparable examination of chemical changes in
tissues other than the nervous system to be sure
the response is specific to it. Third, all too
often either the entire brain is homogenized or
only a very small part is examined, and an extensive inventory of all the major portions of the
brain and even of the cells involved in the response
is not available. A lack of consideration of these
problems has led many investigators to conclude
that their findings are relevant to memory research,
when in fact, there is often a lack of critical data
to establish such relevance. These considerations
have made it difficult, if not impossible, to draw
any conclusions concerning the possible cause and
effect relationships between the experience or the
behavior and the chemical change. Nor are there
any data to indicate the role that the chemicals
are playing in the responding cells.

In an attempt to approach this problem, an interdisciplinary research program was initiated in
collaboration with Dr. Kurt Schlesinger of the Psychology Department to develop a behavioral paradigm
in which the animal could be approached by biochemical methods. We chose the mouse because it has been
used extensively in behavioral, biochemical and
genetical studies. To minimize genetic variation,
we use only 6 to 8 week old males of strain C57Bl/
6J supplied by the Jackson Laboratories. Many of
the methods and the data have been published elsewhere and we shall confine ourselves here mainly to
an evaluation of these results.

The training apparatus:

Dr. Schlesinger researched the "jump box" and we have been using it ever since (1). It consists of a box divided into two sections with a common electric grid floor. One mouse is placed in each section. A light and a buzzer are attached to the outside of the box so that each mouse receives equal stimulation. The sections are identical except that one side has a shelf onto which the mouse in that section can jump. The light and buzzer are presented for three seconds after which the electric shock is applied. Initially, both mice jump in response to the shock and the animal that has the shelf uses it as a haven. The shock is terminated as soon as he does so. The mouse is then removed from the shelf, placed on the grid floor and another trial commences. The training lasts for 15 minutes and between 30 and 35 trials are carried out in this interval. The mouse that has the shelf will usually start to avoid the shock in response to light and buzzer by the fifth trial and is performing to a criterion of 9 out of 10 by 5 minutes. It should be noted that when the trained mouse avoids the shock, the animal on the other side also does not receive one. The untrained mouse also receives equivalent handling at random during the training. Thus with respect to lights, buzzers, shocks, handling, and injections, the untrained mouse is yoked to the trained mouse.

Biochemical methodology:

We originally planned to carry out extensive experiments using radioactive precursors of RNA, protein, lipids, etc., to ascertain whether there is altered incorporation into these macromolecules during the training experience, but our results involving RNA have been so interesting that we have not yet worked with the other compounds. We are planning to do so in the near future, however.

During the course of this part of the work we were privileged to be associated with two excellent graduate students, Mr. John Zemp, who did the biochemical research on RNA, and Mr. William Boggan, who did the training. Their zeal and talent con-

tributed much to the research. The experiments which
were finally worked out after more than a year of
trial and error research consisted of an extension
of the double isotope labeling method, then used
extensively to study phage and bacteria. One mouse
of a pair was injected intracranially with ^{14}C-uri-
dine, the other with ^{3}H-uridine. Thirty minutes
later, one of the mice was trained for 15 minutes
in the jump box while the other served as the yoke
animal. After the training period, the brains of
both mice were homogenized simultaneously in the
same homogenizer, after which the homogenate was
fractionated into nuclei, ribosomes and a supernatant fraction (2). UMP was isolated from the
supernatant fraction while RNA was extracted from
each of the subcellar components and the amount of
^{14}C and ^{3}H in each was determined (2).

The purpose of using two labels in this way
is to avoid the problem of differential losses of
RNA that occur during the complicated manipulations
we had to employ to isolate the RNA. The amount of
^{14}C or ^{3}H in the RNA indicates the extent of incorporation of uridine into RNA of the mouse that was
injected with ^{14}C or ^{3}H, respectively. The ratio
of ^{3}H to ^{14}C in the UMP is a useful indication of
the relative efficiency of the injection and of the
relative amount of uridine that entered brain cells,
and we have used it to correct the observed ratio
of ^{3}H to ^{14}C in the RNA. This double isotope labeling method was used in all of the experiments described in this paper to measure the incorporation
of radioactive uridine into polysomes or into RNA
extracted from nuclei or ribosomes.

The incorporation into RNA:

Preliminary experiments indicated that more
radioactive uridine was incorporated into the RNA
extracted from brain nuclei or from brain ribosomes
of the trained mouse, and coded blind experiments
were carried out thereafter. Only after the biochemical analysis was completed and all the data
were computed was it revealed which isotope (^{3}H or
^{14}C) the trained mouse received. In 25 of such
blind double labeling experiments, all trained mice
incorporated more radioactivity into RNA than the

untrained mice (2). The average increase in RNA
isolated from nuclei was 38% and the range was 6.5
to 119%. The average increase in RNA isolated from
the ribosomal pellet was 64% and the range was 7 to
180%. It is of interest that if the mice are killed
30 minutes after training instead of immediately
after, then the differences are no longer observed.
This, however, is probably due to the fact that the
total radioactivity in the RNA increases greatly
during this time, and the small differences due to
the experience are obscured.

This difference between the incorporation into
brain RNA of the trained and untrained mice cannot
be unequivocally ascribed to increased synthesis
of RNA in the trained animal. It could, for example, be due to decreased destruction of RNA, to
increases in permeability of the cells or their
nuclei to uridine, to a decrease in the synthesis
of endogenous uridine, or to any one of a number
of other alternatives. At present, we have not
pursued this problem directly because of technical
difficulties. Thus we refer only to the change in
incorporation and do not specify a mechanism. It
is of interest that a similar uncertainty exists
for most experiments involving the incorporation
of a radioactive precursor into any molecule, whether it is in brain or in any other tissue. Usually, however, this problem is not taken into account.

To test whether there is an increased incorporation into brain RNA in the trained mouse or a decreased incorporation in the yoke animal, we compared the incorporation into brain RNA in the yoke
animal with the incorporation in a quiet mouse;
that is, a mouse that had been returned to its home
cage after the injection of radioactive uridine.
We felt that if the yoke animal was undergoing a
decrease in incorporation due to its peculiar experience, we should observe the same decrease when it
was compared with the quiet animal. On the other
hand, if the trained animal did have an increased
incorporation into RNA, then we might expect that
the yoke animal, because of its greater activity,
could show slightly more incorporation than the
quiet one. The data indicated that the yoke and
the quiet animal had similar amounts of incorpora-

tion of radioactive uridine into brain RNA (2). Two
conclusions can be drawn. First, it seems clear
that the increased amount of radioactivity in the
RNA from the brain of the trained mouse when compared to the yoke mouse is due in fact to an increased incorporation of uridine in the trained
mouse, and not due to a decreased incorporation in
the yoke mouse. Second, since the yoke and quiet
mice give similar chemical responses, it demonstrates
that the lights, buzzers, shocks, and handling that
the yoke mouse receives are not sufficient to have
any effect by themselves. To test this second idea
further we compared the incorporation of radioactive
uridine into RNA taken from the brains of quiet
mice with the incorporation in mice that were subjected to 30 electric shocks given at random over
the 15 minute period. The data clearly show that
the radioactivity in the RNA was similar in both
mice giving further credence to the idea that mere
stimuli and activity were not responsible for the
effect we observed. It is possible, of course,
that such stimuli can cause such effects, but that
the stress of the injection of solutions into the
brain that all the mice undergo is large enough to
mask the effects of other input stimuli.

The response of other tissues:

The question arose as to whether this difference in incorporation into RNA was specific to the
brain or whether it occurred generally throughout
the animal. By giving intraperitoneal injections
of radioactive uridine as well as the intracranial
ones, we were able to show that neither liver nor
kidney showed any differences between the trained
and untrained animals in the incorporation of radioactive uridine into RNA or polysomes, while these
very same animals showed pronounced differences in
the brain (2, 3). We conclude, therefore, that
while tissues other than the liver and kidney might
be responding, it seems likely that the effect may
be specific to the brain.

Localization in the brain:

Not only does the effect seem to be specific
to the brain, it is specific to limited areas. The

localization within the brain was investigated by
two methods: first, by applying the double labeling method to grossly dissected brain parts (4),
and second, by autoradiography following the injection of ^{3}H-uridine (5).   The results of the gross
dissection revealed that the increased incorporation
of uridine into RNA took place entirely in the di-
encephalon and associated structures (4).   These
studies also showed a small but significant decrease
in the cortex.   The results of an autoradiographic
study carried out by Dr. Martin Krigman and Mr. Barry Kahan showed that this was correct.   There are
regions in the outer 2 to 3 layers of the neocortex
of the untrained mice where the cell bodies are
covered with the silver grains produced by the radioactivity incorporated into RNA; the corresponding
regions in the trained mouse are practically free
of grains.   On the other hand, there are many regions throughout the limbic system, especially in
the amygdaloid area that contain cells and cell
clusters that are covered with grains, whereas the
corresponding regions in the yoke animal have few,
if any, grains.   The intensity of labeling varies
from a noticeable increase in the number of grains
to a very large increase.   Most areas are, of course,
equally labeled in the trained and untrained mice.
The details of this work will be published elsewhere (5), but they indicate how important it is to
know the full extent of the involvement of the
brain in chemical responses to behavioral stimuli
in order to gain proper perspective on the phenomenon being observed, and to enable the production of
realistic interpretations.

The function of the RNA:

     The only known function of RNA is that it par-
ticipates in protein synthesis.   Several workers,
however, have suggested that RNA might have a unique function in brain related to memory storage.
To examine this possibility, we sedimented most of
the ^{14}C and ^{3}H labeled RNA mixtures from yoke and
trained animals in a sucrose gradient to see if the
increased radioactivity was located in a single
species of RNA that might have a unique function (2).
In all cases, the increased radioactivity associated
with the RNA of the trained mouse was located

throughout the gradient and was quite heterogeneous with respect to sedimentation rate. The patterns of radioactivity were indistinguishable from the patterns for such RNA from the livers or kidneys of the same animals, and were of the same general shape for both trained and untrained mice. Thus the increases resembled those found after RNA synthesis has been stimulated in liver by hydrocortisone or in uterus by estrogen.

These findings led us to the conclusion that the increased radioactivity was not confined to a single species of RNA, but was distributed among many species, and that we were detecting a general increase in the synthesis of rapidly-labeled RNA due to the stimulation of the cells involved. To further substantiate this idea we found that there also was increased incorporation into polysomes of the brain during the training experience but not during the other behaviors (3). This work was carried out by another talented graduate student, Mrs. Linda Adair. These results suggest that the increased radioactivity is either in messenger RNA or in preribosomal RNA. Thus we find no evidence for an RNA with a function that does not involve protein synthesis; indeed the RNA we do find is similar to RNA extracted from other tissues. Whether this RNA is involved in the syntheses of new proteins or in the replenishment or increase in the amount of proteins already present needs further work. It is of interest that the base ratios of the rapidly labeled RNA of the trained and yoke animals is the same as determined by ^{32}P incorporated into the nucleotides (Wu and Wilson, unpublished). We are now attempting DNA-RNA hybridization to see if more subtle differences can be detected between the RNA synthesized in the trained and untrained animals.

Behavioral components:

Another question concerns which part of the training experience, or which part of the response of the trained mouse, triggers the increased incorporation of radioactive uridine into RNA of those brain cells which participate in this phenomenon. Clearly there are many differences between the trained mouse and the yoke mouse in addition to

learning. For example, the trained mouse has a
change in cue and his attention is now directed at
a stimulus that the untrained mouse probably views
as benign, Also, the stresses on the trained mouse
are different, as are his responses to them. Finally, the trained mouse jumps more often than the
yoke mouse, which does not get a shock if the
trained mouse avoids. We know that the increased
number of jumps is not responsible for the chemical
change since a mouse given 30 shocks in 15 minutes
does not show any difference in radioactivity in
RNA or polysomes when compared with the quiet or
yoke mouse (2). There is, however, a difference
in the quality of the jump, in that the trained
mouse quickly learns to organize his locomotion to
reach the shelf, while the untrained mouse jumps
with random purpose. This could be an important
factor.

To examine some of these alternatives we used
a mouse that had been previously trained for two
successive days with 15 minute sessions in the jump
box. On the third day, this mouse was injected with
radioactive uridine using the double labeling method
for polysomes and was then required to perform in
the jump box for 15 minutes. This mouse did not
show any differences in the incorporation of radioactive uridine into brain polysomes when compared
with a yoked mouse (6). Thus we conclude that the
organized locomotion, the changed cue and attention,
and those stress factors which operate after the
animal has learned and is performing the task are
not responsible for the chemical changes we observe.
It is, of course, possible that the prior trained
mouse had habituated to the training apparatus and
the handling, and thus experienced reduced stress.
This cannot be ruled out completely, but the animals
do show fear reactions when placed into the training
apparatus. Furthermore, mice which had been previously exposed to the training situation as yoke
mice for 15 minute sessions on two successive days,
learned the task perfectly well on the third day
and showed the increased incorporation into brain
polysomes. Thus habituation is probably not playing
a role here. Finally, the effects of stress as related to adrenal activity can be eliminated, since
adrenalectomized mice learned to jump to the shelf

and showed the same increased incorporation into brain polysomes as did the normal mice (6).

Other behaviors related to the jump box were also tested and the results are clear; when the animal learned to jump to the shelf in response to a conditioned stimulus, the increased incorporation of radioactive uridine into brain polysomes was observed. If the behavioral experience did not involve avoidance learning, then the increased incorporation into polysomes did not take place. For example, when the mouse was conditioned to avoid the floor in a 15 minute training session, an increased incorporation was observed in the brain polysomes. When a mouse that had previously been through this training procedure was made to _perform_ this task, no such increase was noted. Thus, the process of changing cue (i.e. learning) seems to be relevant to the chemical change we observe (6).

It is still not possible to reach an unequivocal conclusion concerning cause and effect relationships or the biological significance of these results. We do not have enough information to do this On the one hand, we have a behavioral experience that can cause both a change in behavior (learning) and a change in chemicals in the brain. On the other hand, we have no data that indicate whether either of these changes is the cause of the other, or that they are even induced by the same component of the experience. Indeed, they might be completely unrelated responses to two different input stimuli. This dilemma plagues all research on the nervous system where an experience exerts one or more behavioral, biological, or chemical responses, and it does not help to clarify the situation by taking a monolithic approach to interpreting the phenomenon Only by keeping an open mind can we hope to generate useful hypotheses that can lead to experiments that answer non-trivial questions.

Our own approach has been mostly descriptive thus far in trying to elucidate the parameters of the chemical response that we must know and deal with in order to understand this aspect of nervous system function. Whether it has anything to do with

the learning process per se, or with a response in-
cidental to this process, does not seem to be a relevant question at this time. What does seem important is that an experiential input stimulus has e-
voked a consistent chemical response. It is up to
us now to analyze the details of the response and
to explain its role and importance in the brain.

Recently we have started using rats in the expectation that we can insert cannulas into the corresponding areas in the brain in the hope that we
can monitor the electrical activity in trained and
untrained animals to see if significant differences
exist. Of equal importance will be experiments to
see if we can interfere with the function of these
areas through minute lesions, with chemicals, and by
other means, and to see whether such treatments will
affect the learning of the animal and the chemical
response in the other parts of the brain. For example, one can produce a small lesion or add actinomycin to the cells in one region that incorporates
the radioactive uridine during the training experience. How will such treatment affect the behavior
and the ability to learn? If the animal still
learns, does a normal pattern of incorporation still
exist in other parts of the brain? Experiments such
as these may shed a great deal of light on the significance of the processes we are studying.

We believe that one hope for unequivocal assess-
ment of work such as this will come from the analysis
of even simpler paradigms in simpler organisms where
one can control the input stimulus better, and where
one can monitor or regulate the response with greater
precision. To this end we have been carrying out
similar single and double isotope labeling experiments using the headless cockroach preparation originated by Horridge (7). Preliminary data show that
significant differences in the incorporation of radioactive precursors into RNA and protein can be observed in the thoracic ganglion that innervates the
trained and untrained legs. These experiments in
the cockroach may eventually help us to understand
the process in the mouse brain.

REFERENCES

1. Schlesinger, K. and Wimer, R., Genotype and
    conditioned avoidance learning in the mouse,
    Jour. Comp. Physiol. Psychol., 63, 139-141,
    1967.
2. Zemp, J.W., Wilson, J.E., Schlesinger, K.,
    Boggan, W.O. and Glassman, E., Brain Function
    and Macromolecules. I. Incorporation of uridin
    into RNA of mouse brain during short-term trai
    ing experience, Proceedings of the National
    Academy of Sciences, Washington, 55, 1423-1431
3. Adair, Linda B., Wilson, J.E., Zemp, J.W. and
    Glassman, E., Brain Function and Macromole-
    cules. III. Uridine incorporation in polysomes
    of mouse brain during short-term avoidance con
    ditioning, 1968. In preparation.
4. Zemp, J.W., Wilson, J.E.,and Glassman, E.,
    Brain Function and Macromolecules. II. Site
    of increased labeling of RNA in brains of mice
    during a short-term training experience, Pro-
    ceedings of the National Academy of Sciences,
    Washington, 58, 1120-1125, 1967.
5. Krigman, M.R., Kahan, B., Wilson, J.E. and
    Glassman, E., Brain Function and Macromole-
    cules. V. Autoradiographic studies on the ef-
    fect of a short-term training experience on
    the incorporation of radioactive uridine into
    brain RNA, 1968. In preparation.
6. Adair, Linda B., Wilson, J.E. and Glassman,
    E., Brain Function and Macromolecules. IV. Uri
    dine incorporation into polysomes of mouse
    brain during different behavior experiences,
    1968. In preparation.
7. Horridge, G. A., Learning of leg position by
    the ventral nerve cord in headless insects.
    Proc. Roy. Soc., London, ser. B, 157, 33-52,
    1962.

HYBRIDIZATION OF NUCLEIC ACIDS

DR. SINEX: I am not sure there is a true difference in DNA-DNA hybridization between DNA from young and old animals. This is a complicated question because of difficulties in preparing DNA completely free of protein, the great complexity of mammalian DNA compared to bacterial DNA, and the greater probability for non specific inter an intra molecular interaction. Attempts at hybridization between mammalian as opposed to bacterial DNA result in much more intramolecular reannealing in mammalian systems rather than intermolecular interaction. The relative amounts of protein present and the degree of interaction with protein with DNA varies between young and old a animals. Some protein comes off more readily, some less readily in older animals.

DNA-RNA hybridization experiments are more hopeful. Even here there is a technical problem. The hybridization of true messenger is a very slow process compared to the hybridization of other RNA's of less theoretical interest. Richard Cutler at Brookhaven National Laboratory probably has the best data on the effect of age on DNA-RNA hybridizability. In theory, hybridization should give some hint as to how many types of messengers are being produced in what concentrations during learning and whether the messenger being produced in the neurone is identical to the messenger being produced in the glia. Only a portion of the messenger-like RNA found in the nucleus ever gets to the cytoplasm as true messenger.

Indeed, true messenger is the most difficult of all to hybridize, and is difficult to isolate from the nuclei. Polysomes are a better source of messenger.

In theory you could block DNA sites with unlabeled RNA from control animals in a hybridization study, make labeled RNA in a learning situation and look for evidence of new species of RNA with which the control RNA did not compete.

Hybridization studies might indeed be helpful in learning whether the RNA which is being stimulated in learning is a unique species.

This information is now in the range of modern biochemical technology.

One of the problems technically is to move from the very elegant microtechniques of the Hyden laboratory where great regard has been paid to the uniqueness and complexity of the brain as an organ with great emphasis on individual cell groups to hybridization studies which for the moment require much larger amounts of tissue.

EFFECT OF AGE ON BRAIN DNA AND PROTEIN

DR. SINEX: I think the question that was asked this morning of where does protein go, is a sensible one.

A small portion appears to go into neuronal age pigment. The old brain appears to accumulate a lipoprotein garbage. The ability of the brain to dispose of old protein has apparently some limitation and there is a suspicion that such deposits become harmful.

Other proteins look like they are going down the axone. Where do they go after they get to the tip?

About half-lives. Implicit in the idea of protein recognition is the thought that if memory has a life and that if recognition sites are formed from protein, the half-life of those proteins should approximate the half-life of memory.

If you believe that there are allosteric modifications of proteins at sites of ion transport or electrical activity following the synthesis of a particular species of protein, some might be catabolized and some deposited in a more permanent form which would have a longer life.

In aging, the amount of histones associated with brain protein does not change much during the total course of development. What does change is the amount of non-histone protein. This non-histone protein is high in the newborn animal, low in the mature animal and at least in the mouse, appears elevated in the senescent animal.

The temperature of melting of the DNA double helix of brain chromatin is some six or eight degrees higher in the newborn mouse than it is in the mature adult.

These chromatin associated non-histone proteins are not histones in the sense that they are not acid soluble. The maximum amount of non-histone protein is found when the chromatin is rich in RNA.

DR. HYDEN: There are only suggestions of a difference between DNA of young and old animals, and I mentioned studies by von Hahn (Hahn, H. P. von, Gerontologia, 12, 18, 1966).

DR. SINEX: Dr. Wang has shown there is lipid in residual protein fractions from the nucleus. The problem becomes one of how to define and prepare chromatin. In modern papers on chromatin, few refer to the presence of lipid. This is not to say that it is not worth checking every so often.

To discuss the total structure of the chromatin would take longer than we have. In addition to DNA, chromatin contains about 15 subspecies of associated RNA. The amount of RNA and protein in brain varies during development and aging is influenced by age. The variation in learning has never been studied to my knowledge.

Bacterial repressor from Gilbert and Ptashne's studies appears to be an ordinary protein.

In bacteria the role of these histones are taken by polyamines. The protein content of bacterial chromatin compared to mammals is low -- it is not clear that the Jacob theory literally applies

as well to mammals. The whole grand edifice of
repressor is built on the Pasteur Institute hypothesis about bacteria. It is a bacterial theory.
We assume that mammalian genetics operate the same
way and we assume that the effect of protein on
the mammalian chromosomes is to turn it off. This
is not necessarily a proven fact.

SYNTHESIS OF MACROMOLECULES

DR. SINEX: The concentration of precursor
may affect both species activity and absolute rate
of synthesis. Investigators who do experiments
with labeled precursor may sometimes make errors
on interpretation concerning total synthesis when
what they are actually measuring is the specific
activity of the precursor pool at the time they do
the experiment. It is hard in studying learning
in old and young to make sure that you have controlled the absolute specific activity of the nucleosides when you wish to study polynucleotide
synthesis.

At the same time the absolute concentration
of precursor may affect the rate of polynucleotide
or protein synthesis. There are these fascinating
experiments in which large amounts of polynucleo-
tides are injected and observers report that patients benefit or animals improve their performance.
There are several interpretations of such experiments so that to be able to say that the injection
of a polynucleotide can produce an effect as an intact polynucleotide in the brain requires some
pretty careful controls to be sure there isn't some
sort of a nutritional or pharmacological intervention at another level in the basic control mechanism of the brain.

I would like to comment on this problem of the
genetic versus quantum models. Two things to be
noted are time and permanence. To synthesize a
large molecule takes time, more than milliseconds.

Concepts of time and permanence can be extended clear up through the evolutionary memory of the
species. It is not just the brain which must evolve

and be made to work. The brain helps define not only the individual but also the species.

You have to have a system which produces an organism, undergoes evolution and works in milliseconds. These are distinctions you have to make to fit the fast quantum models and the longer synthetic models at the nuclear level.

REDUNDANCY OF DNA NUCLEOTIDE SEQUENCES

DR. HYDEN: Lately most interesting findings have been reported of a redundancy of DNA nucleotides. In mouse DNA, for example, there is a great redundancy of DNA nucleotide sequences which are similar enough to react the same in hybridization experiments (Britten, R. J. and Kohne, D. E., Science, 161, 529, 1968; Walker, P. M., Biochem.J. 108, 293, 1968). As these authors point out, a sudden occurrence of such a redundancy of a certain frequency may have had a decisive effect for a species during evolution. The protein pattern resulting from such a DNA-nucleotide redundancy could be of importance for storage of information in the nervous system.

EFFECTS OF STRESS

DR. HYDEN: I am glad Dr. Mandel took up the question of stress. Handling of laboratory animals easily leads to stress, and it is difficult to avoid a stress factor in any learning experiment. Therefore, we asked ourselves whether stress can give rise to RNA changes in nerve cells similar to those which we have observed in learning experiments. The answer is no. Some years ago, we found that a typical stress experiment could result in an increased RNA content of nerve cells, but the nuclear RNA synthesized had the same base composition as that existing in the nuclei before the experiment (Hyden, H. and Egyhazi, E., Proc. nat. Acad. Sci., 52, 1030, 1964).

FATE OF SYNTHESIZED PROTEIN

DR. HYDEN: Proteins are constantly being synthesized in nerve cell perikarya and transported periferad with velocities differing from 2 mm to 200 mm per day. The break-down of neuronal protein is presumably considerable, and I assume Dr. Lajtha will later comment on this. However, specific protein capable of undergoing conformational changes and incorporated in, for example, membranes in the perikarya and the synapses and with a slow turnover, may be assumed to serve the long-term memory function.

SPECIFICITY OF NUCLEIC ACID CHANGES OBSERVED

DR. AUSTIN: I would like to ask Professor Hyden if there are any changes in the acidic proteins in the hippocampus, secondly if you control for the effect of stimuli alone, and third whether these changes are limited to the sensory cortex in the paw area, to the motor area or more generalized?

DR. HYDEN: In response to Dr. Austin's questions, first, we studied the synthesis of acidic proteins in the pyramidal nerve cells of the CA3 hippocampal area, but we have not studied whether any qualitative changes occurred in these protein fractions.

Secondly, we controlled for unspecific stimulus effect by allowing rats to perform the same number of reaches with the preferred paw as did the experimental rats with the non-preferred paw, and did not observe any change in the specific activites as we did in the case of rats training with the non-preferred paw. We analysed the hippocampal CA3 area of both hemispheres.

Thirdly, there was a trend to localization of the greatest changes in the specific activities of the protein fractions 4 and 5 to the hippocampus contralateral to the training paw. It was only a trend and we did not pursue this line in the investigation. The main result of the study was that both left and right side of the hippocampus showed higher protein synthesis than did the corresponding areas in the controls.

Our RNA study was performed on nerve cells from a 1mm^3 area in the sensory motor rat cortex, 1.5 mm cranially and 2.7 mm laterally from the bregma. If this area is destroyed, the capacity of handedness reversal is also destroyed (Peterson, G. M. and Devine, J. V., J.comp. physiol. Psychol., 56, 752, 1963). Since our analyses take considerable time, we kept to this area. It is interesting, however, that Booth has recently reported that he found an increased grain density over the handedness control area we studied after training of the rats (Booth, D. A., Psychol. Bull., 68, 149, 1968). He found the response extending also cranially from this area and, interestingly enough, he also found a trend of localization to the area contralateral to the training paw.

In the paper published in Science two months ago, we showed that there occurs an increased synthesis of two protein fractions in the pyramidal nerve cells of the hippocampal CA3 area when rats are trained in handedness reversal.

This raised the question whether this protein response was specific for the learning process or not. Therefore, we have studied the incorporation of ^{3}H-labeled leucine in the 4 and 5 protein fractions of the hippocampal CA3 region during one month of intermittent training in handedness reversal. The rats were trained for 5 days with two 25 min. periods per day, trained again briefly after a fortnight, and, finally, trained for 3 days (two 25 min./day) one month after the initial training. This indicates that the protein response, the increased synthesis of the fractions 4 and 5 over that of controls, is specific for the learning process. In this study we also got evidence that the acidic, brain specific protein S 100 became synthesized in increased amounts during training.

DR. BOGOCH: I want to draw attention to what I think is a very important point that was brought up by Dr. Mandel when he showed that changes occur in some cell layers, but not in others, in nucleic acid base ratios.

Utina and Byzov (Biophysica, Moscow, 10, 1965) showed some years ago by rather crude methods that there was a marked increase in RNA and DNA in ganglion cells in the retina during flickering light stimulation.

This basic fact, I think, is of crucial importance because at the very first entry, at the place where transduction occurs, at the site of the first coding, there are changes one can compare with reported changes in the deepest nuclei in the brain.

Does this mean that the coding which occurs is finalized at the stage of the transduction at the sensory organ, and changes no further? How should one interpret the changes in deep nuclear structures in the light of the changes one sees at the periphery?

In other words, wherever you look at nerve cells along the path from sensory receptor to the deepest cells there are changes.

Are these changes due just to general cellular activity or is it possible that one can specify the nature of the relationship of the changes to what we are interested in in terms of memory?

DR. MANDEL: The answer is that if you have cell stimulation, you induce a higher amount of some messengers to produce more proteins and several enzymes which are involved in the activity of the cell. When you stimulate that cell you have a whole stimulation but you have also specific stimulation.

Now the problem is how to separate this fraction of specific stimulation from the overall stimulation. This is up till now a technical problem for RNA.

DR. HYDEN: Over the years, we have performed around 15 experiments using various types of stimulation and studied amount and base ratios of nuclear and cytoplasmic RNA of mammalian nerve cells engaged during stimulation. As examples, I would like to mention intermittent rotatory stimulation and vestibular nerve cells, caloric stimulation and Purkinje nerve cells, thirst and NaCl and hypothalamic nerve cells. In all experiments we could show increase in the RNA content per cell following stimulation. But in no case we found nuclear RNA changes after stimulation of a type observed in the training experiments.

This DNA-like RNA may be a messenger type of RNA. It is interesting that if the template RNA for the brain specific acidic protein is calculated, it turns out to have base ratios similar to those we found in the nuclear RNA emerging during training in the cortical nerve cells.

Since the nuclear RNA characteristics could not be found in control experiments, and since the recent analysis of protein response in the hippocampus indicated specificity for the response for the learning process, I would like to suggest that during learning there occurs a synthesis of protein in nerve cells which is the consequence of a gene activation and which protein may be needed for the formation of a long-term memory.

DR. AUSTIN: Might I ask how it was that learning is known to occur in the transfer of handedness experiments?

DR. HYDEN: The transfer of handedness experiment in rats is well known in the literature. It was first described by Peterson in 1934, (Peterson, G.M., Comp. Psychol. Monogr., 9, 67, 1934) and has been used in extensive studies. We have sharpened the experiment by letting the a animals reach down in a narrow glass tube to retrieve food pills, one by one. The performance curve (number of reaches per 25 min. training period) is linear up to the 6th-8th day. The

advantage with the experiment is that no surgical
or mechanical handling is needed and the animals
are not forced, neither is an electric shock given.
One control is, of course, the case where animals
are allowed to use the preferred paw for the same
number of reaches as do the experimental rats with
the non-preferred paw.

DR. MANDEL: I should like to make an additional short remark and I think Dr. Hydén will agree
with this.

We should not have the simple scheme, one messenger, one engram, since if we extrapolate this, tha
will lead us to absurdity. Every messenger RNA has
to be transcribed from DNA. Thus, the hypothesis,
one engram, one messenger, means that everything we
will learn in our life already exists in our DNA.
That is why I think we have to look for another
mechanism.

It is reasonable to assume that we are in
agreement with basic molecular biology if we say
that macromolecules such as RNA or proteins may be
involved in behaviour or memory phenomena. But we
cannot ignore the fact that we can produce only
that kind of RNA for which a complementary strand
exists in DNA. The hypothesis that proteins are involved in memory storage processes is in agreement
with what was said by Dr. Von Foerster as well as
by Dr. Eccles. If transmitters participate in this
phenomenon, the immediate event will be interaction
between a protein and a transmitter.

C. NEUROPSYCHOLOGICAL STUDIES

PERSPECTIVES OF THE RESEARCH ON THE
INTEGRATIVE ACTIVITY OF THE BRAIN

Jerzy Konorski

It should be realized that the contemporary research of the brain functions has so far proceeded along two separate lines, each of them following a different aim and using different experimental methods. The object of the first line of research, which may be called <u>analytical</u>, is to elucidate the fundamental properties of the brain on the basis of its electrophysiological and histological examination. Owing to the increasingly exact electrophysiological methods, such as recording the activity of single neurons, and neuro-anatomical methods, such as electron microscopy, the advancement of analytical studies is both rapid and promising. The main disadvantage of these studies is, however, that they are carried out on anaesthetized or curarized animals, that is in a condition in which a normal activity of the brain is excluded.

The other approach, which may be called <u>synthetic</u>, consists in examining correlations between the manifold stimuli impinging upon receptive surfaces and their effects produced in the <u>waking animal</u> with the aim of understanding on the basis of these correlations the cerebral mechanisms controlling animal behavior.

However important is the close cooperation between these two lines of investigation, they developed in the past almost independently from each other. The synthetic approach to the study of brain functions was founded at the turn of this century by Pavlov who called it physiology of higher nervous activity. Then it became obsolete for many years because at that time the general neurophysiology could not supply an adequate basis for its development. The Pavlovian heritage was, however, overtaken by behavioristic psychology which utilized only the empirical achievements of Pavlov and his coworkers, completely ignoring his physiological theory, which for him was the gist of the whole matter. As the result a somewhat anomalous situation arose in which the "true" neurophysiologists were not concerned with the problems of the central mechanisms of animal behavior, considering them too complex to be dealt with by precise electrophysiological methods. On the other hand, the study of animal behavior was exclusively in the hands of psychologists who were not concerned with physiological interpretation of its mechanisms, regarding this interpretation as an unjustified speculation.

Only in the last decades this rather abnormal split of two kindred scientific disciplines began to be overcome. Many a neurophysiologist began to feel that a pure study of intercentral connections in the brain without any reference to their functional role is somewhat sterile. On the other hand, the gradual penetration of neurophysiological methods into behavioristics, in particular, removal of various parts of the brain and implanting electrodes into the brain both for stimulation and recording purposes, made a purely behavioristic attitude untenable. You cannot observe the effects of ablation or stimulation of various cerebral structures upon behavioral responses and at the same time claim that the brain is a black box inaccessible to any scientific analysis.

Therefore, what I think is that we are witnessing now an extremely important period in the development of brain research in which these two lines of investigation, the analytical and the synthetic one, begin to merge. In parallel with this process we observe a real explosion of new research work in which most refined neurophysiological methods are utilized in experimentation on waking animals. The best sign of this explosion is the amount of new scientific journals which have emerged in last years in this field. To conform to the old tradition this branch of brain research keeps the root "psycho"--psychophysiology, neuropsychology, physiological psychology, etc.--but to my mind it is nothing else but simply the <u>physiology of integrative activity of the brain</u>.

Entering into our discussion of further perspectives of this discipline we should first emphasize an important aspect of integrative activity of the brain. This is that the brain of the waking animal, in contradistinction to that of the anaesthetized preparation, is endowed with <u>plasticity</u>, that is with the ability to change its reactivity by previous stimulations. In other words, a quantum of the input impinging upon the brain not only elicits an appropriate response, but also changes the structure involved in such a way that the next identical input produces a different response. In this way such phenomena as learning and memory, which cannot be detected in the anaesthetized brain, are the most essential attributes of the function of the waking brain.

This being so, the first problem to be discussed is that of the physiological mechanisms of memory in general. It is generally accepted that there are two different types of memory traces, namely the long-term memory and the short-term memory. The aim of today's discussion will be to characterize briefly these two types and their mutual relations. Such a discussion seems to me important because in order to promote further our research in this field, it is indispensable to have a clear idea about the problems we have to solve. Indeed, I believe that in order to obtain meaningful answers one must ask correct questions.

The long-term memory is the basis of learning

in the broad sense of the word. Although its intimate mechanism is still unknown, it certainly consists in some microstructural and/or biochemical
changes which occur in the neurons involved, probably in the synaptic contacts. Since there are excellent specialists on this subject in our conference,
I shall not deal with it now, but instead I would
like to turn to the behavioral level of the problem.

It is important to realize that at this level
we should discern two different types of learning,
which, although possessing probably the same intimate mechanism, serve different purposes. These
are <u>perceptual learning</u> and <u>associative learning</u>.

Perceptual learning consists in the integration of particular stimulus-patterns presented to
the animal and their memorization, i.e. the ability of their recognition when they are presented
again. It seems to me that recent studies of Hubel and Wiesel (1965) showing that particular
relatively simple, visual patterns are represented
by particular units in the non-striate visual cortex may be the point of departure for a new physiological theory of perceptions, which I have developed in my recent publication (Konorski, 1967). According to this theory all known stimulus-patterns
of various modalities, perceived by a single act of
attention, are represented by <u>particular units</u> in
the gnostic fields of various analysers. This sort
of representation is accomplished owing to the principle of convergence of messages arriving at the
units of a given gnostic field from the lower levels of the given afferent system, plus the principle
of lateral inhibition, providing the selectivity
of input to thses units. The intimate nature of
perceptual learning is assumed to consist in the
formation of functional synaptic contacts between
the terminals of axons reaching the given gnostic
field from the lower part of the afferent system,
and the units of that field.

Whether this theory is right or wrong it

opens up a tremendous field of future investigations according to the following schedule. The animals are kept from birth in a surrounding very meagre in respect to a given modality of stimuli. For instance, the rats are exposed to only very few visual stimulus-patterns (e.g. white oval objects with horizontal long axis) or auditory stimulus-patterns (e.g. only some sounds imitating those emitted by other rats). It is predicted by the theory that in such animals the majority of units in the corresponding gnostic fields will selectively react to those patterns. If this appears to be true it will provide a new step for our understanding of perceptual processes. I should add in parenthesis that to my mind the main weakness of the contemporary studies of these processes is that instead of using stimuli actually encountered in normal life, they make profit of stimuli which are artificial and biologically quite unusual.

Turning now to associative learning, it consists in establishing functional synaptic contacts between the units of different gnostic fields representing particular stimuli, if these stimuli coincide in time. Of course these connections are unidirectional leading from the transmitting unit to the recepient unit. In many cases, however, associations between stimuli are bidirectional, but then they are based on different connections each leading in an opposite direction.

In humans, associative learning is best manifested by introspection when hearing a known person speaking we visualize his appearance, or seeing a given dish we imagine its taste. In experimentation on animals such an approach is of course impossible and therefore we must recur to some less direct methods. The most popular one is that of classical conditioning. Briefly it consists in the recipient unit being <u>labelled</u> by some overt response. For instance, taste of food in the mouth produces salivation, and electric shock to the paw produces flexion of the

limb. Accordingly, if some external stimulus is
paired with taste of food or noxious stimulus to
the paw, and connections between the corresponding
units are formed they are manifested by the external stimulus eliciting salivation or leg flexion
respectively.

Therefore, we may consider a conditioned reflex as an experimentally privileged case of association. Unfortunately, it led to a supersti-
tion held by many experimentalists that this is
almost the only form of associative learning.
This is certainly not true. We have plenty of
evidence to show that not only in humans, but in
animals, too, intercentral connections are formed
between the so-called neutral stimuli. The only
condition of the formation of such connections is
that the corresponding afferent systems be in a
state of arousal produced by particular drives
(e.g. curiosity), i.e. that the subject pays attention to the presented stimuli. Now, when we
have at our disposal a new method of labelling
the states of central excitations, much more universal and precise than the previous one--the
method of recording action potentials from the
brain--the realization of this fact is most important. There is no doubt that in not a very remote future the associative function of the brain,
and in particular of the cerebral cortex, will be
much better understood than it is now on the basis of the conditioning experiments.

To end our considerations on associative
learning it should be mentioned that this phenomenon may be explained by two alternative hypotheses. On the one hand we may assume that pairing
of two stimuli simply leads to the formation of
functional synaptic contacts between the appropriate units. On the other hand it may be conceived
that there exist plurimodal "associative" units
representing each of the paired stimuli (cf. Fessard and Gastaut, 1958). It seems very important
to elucidate which of these two hypotheses is correct.

We shall now turn to the other phenomenon under discussion, that of short-term memory. Most authors assume that it is based on throwing into action reverberating circuits of neurons which may be either silenced gradually within a lapse of time or inhibited by antagonistic influences. The functional role of the short-term memory is to preserve temporarily a given information for its utilization in a not very remote future.

A typical experimental model of short-term memory is provided by the delayed response test. According to our experimental set-up a dog being on a leash at the starting platform hears a buzzer sounding from one of the three feeders situated in the three corners of the compartment. After one or a few minutes he is released, and if he runs to the feeder from which the buzzer sounded, he receives food there. Accordingly, in order to solve the problem the dog must remember the direction of the signal during the delay period.

The delayed response technique is obviously a test for short-term memory of spatial kinesthesis. A test for short-term memory of exteroceptive stimuli may be devised in the following way (Konorski, 1959). We present a pair of successive stimuli within a certain modality separated by several seconds. The stimuli change from trial to trial but in a single trial they are either different or the same, the former pair being reinforced by food and the latter not. As a result, in order to solve this problem the animal must remember the first stimulus of the pair when the second one is presented.

As seen from these tests, which imitate many natural situations in human and animal life, the short-term memory has two properties quite distinct from long-term memory. One is that the cues the subject has to remember are not new cues (as is the case with respect to the material to be memorized by long-term memory), but they are

already well known to the animal. The second
property is that the short-term memory serves only for tasks requiring temporary memorizing,
otherwise the memory of a subject would be over-
burdened with completely superfluous information.

Here I approach a very important and controversial point which urgently requires clarification because it seems to prevent the proper devising of experimental work in this domain. After the pioneering experiments of Duncan (1948),
Gerard (1953), Thomson (1957) and others on the
obliteration of memory traces by ECS or anoxia,
it was thought that short-term memory serves for
consolidation of memory traces, i.e. for their
transformation into long-term memory traces. I
fully agree with Dr. Agranoff who has shown experimentally on goldfish that this view is incorrect (Agranoff et al, 1965, Davis and Agranoff, 1966). No less convincing are data obtained
by Kleinsmith and Kaplan (1963, 1964) on man.
Taking into consideration this criticism of the
previous view (to which, by the way, I adhered
for a long time), most studies concerning the
preventing of consolidation of memory traces produced by ECS, anoxia and other agents should be
revised in order to clarify whether these agents
affect short-term memory, or long-term memory, or
both.

Of course, it was far beyond my possibility
to depict the future of the physiology of integrative activity of the brain in a broader aspect. Therefore, I have concentrated only upon
some selected problems, simply because they are
familiar to me and because I see clearly what
should be done in the near future to contribute
to their solution. Looking into the future is
always an intriguing affair because the accuracy
or fallacy of our predictions reflect the accuracy or fallacy of our present concepts. But in
this context it is perhaps useful to quote a wise
saying which I heard not once from Professor
Ralph Gerard: "Young man, beware lest you find
what you are looking for".

REFERENCES

1. Hubel, D. and Wiesel, T., J. Neurophysiol.
    28, 229-289, 1965.
2. Konorski, J., Integrative Activity of the
    Brain, Chicago, University of Chicago Press,
    1967.
3. Konorski, J., Bull. Acad. Pol. Sci., Ser.
    Sci. Biol., 7, 115-117, 1959.
4. Fessard, A. and Gastaut, H., Correlations
    neurophysiologiques de la formation des re-
    flexes conditionnels. In: M. A. Fessard, H.
    Gastaut, A. N. Leontiev, G. de Montpellier,
    and H. Pieron, (Eds.), Le conditionnement et
    l'apprentissage, Paris, Presses Universi-
    taires, 15-90, 1958.
5. Duncan, C. J. Comp. Physiol. Psychol., 42,
    32-44, 1948.
6. Gerard, R., Sci. Amer., 189, 118-125, 1953.
7. Thomson, R., J. Comp. Physiol. Psychol., 50,
    397-400, 1957.
8. Agranoff, B., Davis, R. and Brink, J.,
    Proc. Nat. Acad. Sci., 54a, 788-793, 1965.
9. Davis, R. and Agranoff, B., Proc. Nat.
    Acad. Sci., 55, 555-559, 1966.
10. Kleinsmith, L. and Kaplan, S., J. Exp.
    Psychol., 65, 190-193, 1963.
11. Kleinsmith, and Kaplan, S., J. Exp. Psy-
    chol., 67, 124-126, 1964.

WHAT IS AN ADEQUATE STIMULUS IN A PERCEPTUAL LEARNING SITUATION?

DR. GALAMBOS: To open the discussion of the papers just presented I would like to comment on one of Professor Konorski's points--the question of what constitutes an adequate stimulus in a perceptual learning situation. In auditory studies the use of "simple" stimuli--pure tones, noise and clicks--has been justified for a long time with three good arguments. First, they are easy to generate and control. Second, their physical properties can be more accurately specified than those of "natural" stimuli. Third, all natural stimuli, however complex, can at least in theory be generated by mixing one or more of the simple ingredients in proper proportions.

As long as fifteen years ago, however, the idiosyncratic behaviors of certain medial geniculate [1] and auditory cortical [2,3] units toward the "simple" stimuli raised some questions about their appropriateness. Some units do indeed respond like well-behaved physical devices to noise, tones and clicks, but others in the same animal respond only to two or even one of these. Still other cells turn out to be utterly indifferent to the "simple" stimuli but nevertheless react to a miscellany of sounds or noises that defy systematic classification.

One such sound is frequency modulation (FM) of a tonal signal, such as is produced in the laboratory by turning an oscillator dial rapidly in one direction or the other, and in nature by certain bats as they echo-locate obstacles or the insects they seize in flight. Using FM signals resembling natural bat cries, Nobuo Suga has recently shown that certain single cells in the bat auditory brain respond better to FM signals than to simple ones, while others in fact actually respond only to FM sounds [4,5,6,7,8]. The direction of frequency modulation, its speed, and the number of octaves swept are all important variables, and in several instances the optimal

stimuli for driving the brain cells were those
most like the natural cry these bats emit.

Suga's auditory cells are the counterpart in
the auditory sphere of the visual cells that respond better to complex features of visual stimuli (e.g., form, direction and speed of movement)
than to the "simple" hue and brightness variables
so widely used in the past. Both the auditory
and the visual systems seem from such studies to
have the equivalent of filters built into them--
filters that pass and process signals we infer to
have real or potential significance for the animal,
and that de-emphasize or even exclude alterations
in stimulus dimensions not used by the animal in
making his discriminations. Students of animal
behavior (psychologists, ethologists) describe
these filters when they show behavior to be aroused
by limited and specific features of complex stimuli,
and neurophysiologists are just beginning to investigate and appreciate some of the ways a brain
is organized so that it can exhibit these filter
properties. Stimuli "adequate" for perceptual
learning must, if these investigators are correct,
vary widely from species to species in directions
corresponding to the selective analyses the animal
performs on the stimuli that bombard it.

REFERENCES

1.   Galambos, R., Rose, J.E., Bromiley, R.B. and
     Hughes, J.R., Microelectrode studies on medial
     geniculate body of cat, III, Response to pure
     tones, J. Neurophysiol., 15, 381-400, 1952.
2.   Galambos, R., Studies of the auditory system
     with implanted electrodes, Neural Mechanisms
     of the Auditory and Vestibular Systems, Grant
     L. Rasmussen and W. Windle (eds.), Charles C.
     Thomas, Springfield, Ill., 10, 137-151, 1960.
3.   Hind, J.E., Unit activity in the auditory sys-
     tex, Neural Mechanisms of the Auditory and
     Vestibular Systems, Grant L. Rasmussen and
     W. Windle (eds.), Charles C. Thomas, Spring-
     field, Ill., 14, 201-210, 1960.

4.   N. Suga, Recovery cycles and responses to fre-
     quency-modulated tone pulses in auditory neu-
     rones of echo-locating bats, J. Physiol., 175,
     50-80, 1964.
5.   N. Suga, Analysis of frequency-modulated sounds
     by auditory neurones of echo-locating bats,
     J. Physiol., 179, 26-55, 1965.
6.   N. Suga, Functional properties of auditory neu-
     rones in the cortex of echo-locating bats, J.
     Physiol., 181, 671-700, 1965.
7.   N. Suga, Neural responses in the inferior col-
     liculus of echo-locating bats to artificial
     orientation sounds and echoes, J. of Cellular
     Physiol., 67, 319-332, 1966.
8.   N. Suga, Analysis of frequency-modulated and
     complex sounds by single auditory neurones
     of bats, J. Physiol., 198, 51-80, 1968.

DR. CHOROVER: I would like to respond with a strong affirmation of two points made by Professor Konorski in what I thought was an exceptionally interesting paper. His presentation contained much that is of crucial importance to people interested in working on memory. The first point is one to which Professor Galambos just referred: the impor- tance of using stimuli of a sort that animals have evolved to cope with. Two obvious examples of this point are well-known to all of us. Namely, the elegant experiments of Hubel and Wiesel and of Lettvin and Maturana and their colleagues.

Another example is the one that Professor Galambos just gave, In studying the auditory sys- tem of the bat, it is necessary to use stimuli that bats are interested in. Let me add one additional example from my own laboratory. For quite a long time students of brain and behavior have recognized the difficulty in specifying the nature of stimuli that are most relevant for various animals. Be- cause I too am interested in using echo and logic- ally significant forms of stimulation in the study of learning and memory, I was intrigued by a phenom- enon in mouse behavior called the Bruce-Parkes ef-

fect. This is the pregnancy-blocking effect obtained by exposing a recently impregnated female mouse
to a male mouse she has never met before. If this
is done within the first several days after insem-
ination, the fertilized ova fail to become implanted in the uterus. Pregnancy is thus terminated, and
within a few days she becomes sexually receptive to
the "strange" male. This effect can be reproduced
on the basis of olfactory cues alone by introducing
the odor of urin from a strange male into the female's cage. Introducing the urine of the male
with whom she has mated has no effect upon the
progress of the pregnancy.

In some initial experiments on unit activity
in the mouse olfactory bulb, Foteos Macrides and I
have been using stimuli which are very different
from the sort that most people who work in olfactory
stimulation have used in the past. Rather than Re-
agent-grade, pure chemicals, we have simply been
using mice themselves or urine from different mice
as olfactory stimuli.

We are beginning to find that units in the olfactory system of the mouse respond very nicely and
very selectively to such stimuli. I believe there
is a general lesson to be learned from the use of
stimuli of this type.

RELATIONS BETWEEN BEHAVIORAL AND OTHER BIOLOGICAL DISCIPLINES IN INVESTIGATION OF BRAIN PROCESSES IN LEARNING AND MEMORY*

Mark R. Rosenzweig

In attempting to forecast the progress of the brain sciences, let me start with a general proposition which I will then apply to the study of the brain mechanisms of learning and memory. The proposition is this: <u>The brain exists only for purposes of behavior.</u>† Perhaps this assertion will

*For helpful comments and suggestions on the original draft of this paper, I wish to thank Drs. Edward L. Bennett, James A. Dyal, Leo J. Postman and Walter H. Riege.
†At an interdisciplinary conference, it is necessary to define the term "behavior". This was made clear to me by comments of two physiologists who supposed that I had meant to contrast overt muscular behavior with such activities as perceiving and thinking. On the contrary, I was employing "behavior" to include both overt and covert activities, in conformity with current psychological usage. Thus, Harriman in his 1947 New Dictionary of Psychology gives this as part of his definition of behavior: "human behavior includes mental activities, consciousness, muscular functions, and the like" (p. 48). What I had meant to do was to distinguish between instrumental part-function of the brain (such as synthesis of biochemical compounds, conduction of nerve impulses, and the like) and their end product--behavior.

not arouse much debate, but perhaps I can provoke
some by drawing from it the following conclusion:
<u>Study of the brain as an organ can progress in the
long run only in collaboration with the study of
behavior</u>. In the short run, we may be able to describe a chemical compound that is specific to the
brain or a specialized anatomical structure without
yet knowing "what is it there for?" Or we may be
able to describe a property or capacity of the be-
having organism without yet know in neural terms
"what makes it work?" Thus at any time biological
findings may outstrip behavioral work in a particular area, or behavioral study may advance even where
biological mechanisms have not yet been revealed.
But our understanding of the brain takes a major
step forward only when we are able to link structure
and function closely together. A clear consequence
of this is that other biological scientists cannot
afford to be unaware of behavioral findings nor to
understimate the complexities of behavioral research.

        The request made of me was to survey research
on learning and memory within the context of this
conference--and all in 20 minutes. Since it is obviously impossible to squeeze a meaningful survey
of such a broad field into so limited a time, I
will attempt to fulfill only a small part of the
assignment. In doing this, I will comment briefly
on the current status of certain subfields and note
unevenness of development among them. A few references will be given for each area; these may help
the interested reader to start his own survey. Five
subfields will be touched upon: (a) human verbal
learning, (b) animal learning, (c) brain mechanisms
of memory storage, (d) brain changes induced by experience, and (e) chemical transfer of memory. Then
an example of the complexities of research involving
memory will be developed in some detail. The comments on the subfields will be brief and the evalu-
ations individual, so the discussants may well wish
to supplement or challenge some of them.

        The first field--the study of human verbal
learning-- is being pursued enthusiastically and
productively by large numbers of investigators--

perhaps more than are engaged in all the four other subfields combined. Recent reviews or surveys of this field or parts of it can be found in the following references: Underwood (1966, Chaps. 11-13), Cofer and Musgrave (1963), Keppel (1968), Waugh and Norman (1965), Cooper and Pantle (1967). According to Keppel (p. 171), "Contemporary research in verbal learning has veered away from the gross functionalistic approach which characterized research efforts in this area ten years ago. Instead, experimental analyses of the acquisitive process generally involve a specification and isolation of component processes which are assumed to underlie the functional relationships established with the typical learning tasks. A reflection of this change may be seen in the large number of multi-process theories which have been proposed..." Active areas of research in verbal learning include the following problems: What factors govern transfer--both positive and negative--from one task to another? Are there separate processes and stores for short-term and long-term memories? (William James discussed this question in his PRINCIPLES OF PSYCHOLOGY in 1890, referring to the two stores as primary and secondary memory.) Does short-term memory have a limited storage capacity, such that one can trade off number of presentations and presentation rate of items if total presentation time is held constant? Can forgetting be explained solely by interference processes or solely by autonomous decay, or are both factors required? At a few points later in this paper we will note where concepts and findings from verbal learning have been or can be applied to the study of brain mechanisms of learning.

In contrast to verbal learning, the study of the second subfield--animal learning--seems at present to be less lively. Perhaps one reason is that certain theoretical issues that once seemed important have faded away and have not been replaced by equally compelling questions. Some of the main problems currently being studied in this area include the following: How is the magnitude of reward related to rate and strength of learning? How do partial and varied reinforcement (as opposed to continuous reinforcement) affect resistance to extinction? How does overtraining on one habit affect transfer to a succeeding habit? How do the capacities for learning and problem-solving compare among species, and at what points in the phylogenetic scale do new capacities emerge? Fundamental questions about the orientation of comparative studies of learning and intellectual performance have been raised by Diamond and Hall (1968). They suggest that a fruitful reorientation of effort will be a union of comparative neurology with comparative psychology directed toward discovering the functions of homologous brain structures that expand in a given phyletic line of descent. Other recent reviews or surveys of animal learning will be found in the following references: Kimble (1961), Underwood (1966, Chaps. 9 and 10), Warren (1965).

The third topic--brain mechanisms of memory storage--really includes the final two topics of brain changes induced by experience and chemical transfer of memory. I plan to make a few remarks about the larger topic, then to go on to the two more specialized areas, and afterwards to return to offer a detailed example of the complexities of behavioral research on consolidation of memory.

Research on brain mechanisms of memory storage has drawn an increasing band of investigators, and it represents a promising area of interdisciplinary research. This subject is an obvious focus of attention at the present conference. Some of the main questions being investigated are the following: In what form is memory stored--molecular changes, synaptic changes, or other? Are there separate neural mechanisms for short-term, intermediate-term and long-term storage of memory? More specifically, can protein synthesis be demonstrated to be requisite for long-term storage? In a recent review, Flexner (1967) concluded that puromycin, used to inhibit protein synthesis, blocked the exp ession of memory rather than maintenance of the memory trace. Other reviews and summaries of research in this area will be found in these references: Booth (1967), Bower (1966), Deutsch and Deutsch (1966),

Hydén (1967), Gaito (1966), Rosenzweig and Leiman (1968).

Concerning the fourth subfield, a few laboratories are attempting to find chemical or anatomical changes induced in the brain by training or experience. Several promising leads have been uncovered, and it is clear that the brain is far more plastic than we would have supposed only a few years ago. We have shown at Berkeley that even the adult brain can be altered measurably and significantly by new demands of the environment (Rosenzweig, Bennet and Krech, 1964; Bennet, Rosenzweig and Diamond, 1968). Even a few hours a day of differential experience is sufficient to modify brain chemistry and anatomy (Rosenzweig, Love and Bennett, 1968). It remains to relate any of these changes specifically to memory storage and to rule out completely effects that may be attributed to such tied variables as sensory stimulation, motivational effects, activity, stress, etc. A few general references in this area are the following: Gaito (1966), Hydén (1967), Rosenzweig and Leiman (1968).

The fifth and last subfield to be mentioned is the search for chemical transfer of memory. If specific memories could be transferred in injecting brain extracts from a trained to a naive animal, this would be decisive proof of the chemical nature of at least certain engrams or certain stages in formation of memory storage. For this reason, a number of researchers are devoting themselves to investigating the possibility of chemical transfer. While positive reports at present outnumber negative ones, it is not clear that this is representative of the number of experiments, since many studies have been withheld from publication. No current topic in the brain sciences appears to arouse more differences of opinion. Professor Ernest Chain at the 1968 UNESCO-IBRO meeting vehemently deplored "wasting precious time on such chimerical and ridiculous goals when it should be spent on better founded biochemical investigations." On the other hand, there are some convinced proponents of chemical transfer who are devoting their time and tal-

ents to this research. The rest of us can only
hope that it will soon be possible to tell whether a phenomenon of chemical transfer of memory
does indeed exist. Hopefully, interested investigators will make more attempts to replicate the
experiments that others have claimed to give the
clearest results. Hopefully, the experiments
will be done "blind" to all participants to prevent any biasing of results. And hopefully, as
Dyal and Golub (1967) have urged, "... all experiments, positive and negative, should be reported
in a research area as controversial and important
as the present one" (p. 31). Reviews of research
on this topic have been prepared by McGaugh
(1967), Byrne (1968) and Rosenzweig and Leiman
(1968).

Now let us consider some of the complexities
of research involving behavioral variables. As a
specific example, let us consider the use of electroconvulsive shock to affect the consolidation of
memory storage. In reference to Professor von
Foerster's paper and the discussion this morning,
I should add a note, before proceeding further,
about the term "consolidation of memory storage".
Since this term is commonly used by investigators
in the area that I will be discussing, it will be
convenient for us to employ it too; however, I
believe that it would not make any difference to
the questions to be raised if we were to talk instead about establishing new programs in a brain
computer so that it would be able to produce new
responses.

Consolidation of memory is a concept that
was originated by investigators of verbal learning
(Müller and Pilzecker, 1900) at the start of this
century. It was later taken over by physiological
psychologists (e.g., Hebb, 1949; Duncan, 1949)
and by other biologists (e.g., Gerard, 1953). The
concept is that a permanent form of memory storage develops over a period of time after the
training or experience. Before the permanent
storage is elaborated, the memory is presumably
in a more labile form.

The concept of consolidation could be tested and its temporal course measured if we could interrupt the proposed process at some particular time after the completion of learning trials. There was already clinical evidence that could be interpreted in this way: It was known that a severe blow to the head often led to amnesia for the period just before the accident (Russel & Nathan, 1946). Electroconvulsive shock (ECS) used in therapy had also been found to lead to retrograde amnesia. Various investigators then attempted to obtain retrograde amnesia in animal experiments--by passing an electrical current through the head, or by administering convulsant drugs or anesthetics or antimetabolites. In earlier experiments of this sort, the training consisted of a number of trials given one per day; an electroconvulsive shock was given at a fixed time interval after each daily trial. This procedure led to difficulties of interpretation, since a course of electroshock treatments seemed to lead to qualitatively different effects from those of a single shock. Multiple cerebral shocks became clearly punishing or aversive (Madsen & McGaugh, 1961). Therefore, most current research employs a one-trial learning design. While this simplifies problems of interpretation, it limits the complexity of the learning tasks. Also, as we shall consider later, it ignores the possibility that a second combination of training and ECS might not produce the same amnesic effects as did the first combination.

Two one-trial learning tasks frequently employed with rats and mice are the step-down and the step-through procedures; both involve passive avoidance of punishment. In the step-down procedure, the animal is placed on a small pedestal a few inches above a grid floor. The animal usually steps down from the pedestal within a few seconds, whereupon it can be given a strong foot shock through the grid floor. When tested 24 hours later, such an animal will usually be much slower to descend than an animal that was not punished when it stepped down. In the step-through procedure, the animal is placed on a

small platform outside a hole that leads to a
black box. When the animal promptly steps through
the hole it can be given foot shock just as in the
step-down procedure.

To attempt to interrupt consolidation, ECS
can be given to different groups of subjects at
different time intervals after foot shock. Numerous investigators have reported that the longer
the head shock is delayed, the larger the percentage of animals that will recall the punishment and
that will refuse to leave the safe perch. This
increasing resistance to disruption of memory is
exactly what the consolidation hypothesis predicts, so these findings have been taken as strong
support for the concept of consolidation. Thus
far the story seems straightforward, but this is
just the introduction. Now the plot thickens and
the complications begin to appear.

In the first place, there are rival claim-
ants for the status of the rightful hypothesis.
Various investigators have proposed quite differ-
end hypotheses to account for the same basic observations as does the consolidation hypothesis.
And the quest seems to have become more popular,
since three new contenders, to my knowledge, have
appeared during the first four months of 1968.

The first of the 1968 crop was proposed by
Neilson (1968); he holds that ECS alters retrieval of memory rather than consolidation of memory
Basic to this argument is the observation that a
single ECS can alter both locomotor activity and
brain excitability levels for as long as several
days after the shock. Neilson hypothesizes "that
the neurological aspect of learning may involve
changes in levels of brain excitability as reflected in the thresholds of functional neural
systems, that retention implies a maintenance or
reconstruction of these modifications of brain excitability, and that failure of retention occurs
whenever brain excitability is modified away from
that established by the training procedure" (p. 3).
While the findings offered in support of this new
contender do not appear to be sufficient to estab-

lish its claims, some of them do challenge predictions of the consolidation hypothesis, and the new retrieval hypothesis clearly warrants further investigation.

Secondly, Schneider and Sherman (1968) have proposed an alternative that places stress on the arousal after-effects produced by the foot shock rather than on the learning produced by the foot shock. In their experimental design, two foot shocks were given, one (contingent) immediately upon step-down and the other (non-contingent) at some later time. Amnesia occurred when ECS was administered promptly after the second foot shock, even if this occurred six hours after step-down trial. These investigators conclude, "Although the after-effects notion is in need of further tests, the ease with which it accounts for both the time-dependent and recovery data suggests that it is the most parsimonious explanation available" (p. 221).

A third hypothesis was proposed in April by Kohlenberg and Trabasso (1968). It was based on Estes' (1955) stimulus fluctuation theory and predicted recovery over time of the amnesia produced by a single ECS. Such recovery was observed, adding to a controversy that we will touch upon later concerning the permanence of amnesia induced by ECS. It should be noted that stimulus fluctuation theory was developed by Estes in the early 1950's to give a mathematical formulation of animal learning and conditioning; it was a purely behavioral account with no concern for brain mechanisms of learning.

Even without challenges from alternative explanations, the consolidation hypothesis has been running into internal difficulties. That the experiments designed to test this hypothesis are not simple is shown by the fact that competent investigators--some of them taking part in this conference--have reached quite different estimates of the time course of consolidation. According to some, consolidation has run its course

in 10 seconds (Chorover & Schiller), while other
investigators find that several hours are required for completion of consolidation (Kopp, Bohdanecky & Jarvik, 1966; Alpern & McGaugh, 1968;
McGaugh, 1968). Not only do these discrepant
findings alert us to methodological and perhaps
conceptual difficulties; they also are inadequate to suggest what biochemical processes may
underlie learning and memory storage. That is,
the time range of seconds to hours leaves too
much latitude to offer significant guidance to
the researcher who wonders which biochemical
processes to test in relation to learning and
memory.

An ingenious way to account for the discrepant temporal results has been proposed recently
by Cherkin (1968). He suggests that submaximal
ECS may slow but not completely stop consolidation. It has generally been assumed that retention tested 24 hours after ECS provided a measure
of the consolidation achieved at the time ECS
was administered. If, however, some consolidation has continued to occur between ECS and the
time of testing, then this will lead to an over-
estimation of the amount of consolidation that
had occurred prior to ECS.

It would appear that one could test Cherkin's
suggestion readily by giving behavioral tests at
different numbers of hours after ECS, instead of
waiting for 24 hours to elapse. This sort of
testing had, in fact, been done by Geller (reported by Jarvik, 1968) even before Cherkin had
stated his hypothesis. We have been considering
the possibility that retention might continue
to improve slowly after the administration of ECS.
In Geller's results, on the contrary, retention
was at a high level one hour after ECS, but it
was clearly lower at 2 hours and still further
diminished at 6 hours, practically disappearing
at 24 hours. One might interpret these results
by saying that a short-term memory was established in spite of the administration of ECS, but
that there was no consolidation of long-term mem-

ory, so that apparent retention declined steadily
after training and ECS. These results of Geller
are rather similar to results obtained with chemical
inhibitors of protein synthesis. Thus, Barondes and
Cohen (1966) found evidence that consolidation but
not initial learning was affected when mice were
given intra-cerebral injections of puromycin 5 hours
before training. The animals were able to learn a
Y-maze problem, and 15 minutes after reaching criterion they still showed savings of 80 per cent, but
thereafter successively tested groups showed declin-
ing effects. Davis and Agranoff (1966) found that
when puromycin was given to goldfish immediately after training, savings scores remained at about 100
per cent during the first 6 hours and then declined
to zero in three days. The studies of Geller are
being replicated with a number of controls for post-
convulsive effects, since, as Neilson has pointed
out in his retrieval hypothesis, the behavior of the
animal is modified in a number of ways by ECS. It
is certainly clear that testing Cherkin's hypothesis
is complicated both by the possibility of an inter-
vening short-term memory and by post-ictal phenomena.

The site of administration of ECS poses a
number of problems. Often ECS is administered
to rats or mice through ear clips. This probably
causes peripheral pain which may affect later behavior. Neilson (1968) has shown that simply attaching ear clips to rats decreases their ambula-
tion in an open field and it increases their rate
of defecation, which is often taken as a measure
of emotion. Stimulation through the eyes is another frequent route; while this involves lesser
resistance, it also may produce pain. It is possible that strain differences in thresholds and
responses to ECS may be related to the differences
in the conformation of the head with resultant
differences in distribution of current through the
brain. Kesner and Doty (1968) have found that
direct electrical stimulation of the brain causes
different effects on later memory tests according
to the precise site employed. This technique
of direct stimulation of the brain offers consid-

erable promise, both in eliminating peripheral
pain factors and in allowing investigators to
pinpoint the brain regions important to the behavior in question.

When ECS after training eliminates any sign
of the learned behavior 24 hours later, it is
generally considered that the memory has been
abolished--and permamently. Recently, it has
been suggested that the memory may reappear under
certain circumstances (Zinkin & Miller, 1967),
again indicating that the problem may be one of
retrieval rather than consolidation. This study
was criticized for the use of repeated tests on
the same subjects. Other investigators (Luttges
& McGaugh, 1967; Peeke & Herz, 1967) using separate groups for each time interval have since reported that the retrograde amnesic effects of ECS
do not dissipate but are long lasting. Schneider
and Sherman (1968) have added another factor--
strength of training. They reported that the
amnesia was temporary if conditioning was strong,
but permanent if conditioning was weak. Since,
however, the recovery was found in repeated testing of the same subjects, those who claim permanence of amnesia after ECS will probably insist
on seeing replications done with independent
groups at the various test periods.

Even if the effects of a single administration of ECS do last indefinitely, as shown by
some tests of memory, this does not necessarily
mean that there is no residue in memory of the
initial training. Kesner (1968) has recently
given animals repeated cycles of one-trial training, each trial followed by ECS. The later ad-
ministrations of ECS did not produce as complete
retrograde amnesia as did the first one. This
brings to mind interesting findings by Hebb
(1961), confirmed by Melton (1963), on a verbal
learning task. The subject had to recall a nine-
digit number immediately after presentation, and
there was a long series of such nine-digit numbers. It is usually considered that the short-
term memory for one string of digits is complete-

ly wiped out by the interference of the succeeding numbers. In Hebb's experimental design, however, the same nine-digit number was given as every third item throughout the list. This repeated nine-digit number was recalled better and better as the list went on, although the subjects did not improve in performance for the other numbers. Hebb suggested that his results demonstrated the cumulative automatic effects of repetition even under extremely adverse conditions involving large amounts of retroactive inhibition. Melton found similar results with a more elaborate design in which both frequency of repetition and spacing of numbers were varied. Melton added that the evidence implies a continuity between short-term and long-term memory and does not support the concept of different mechanisms for two sorts of memory. It should be noted, however, that other students of verbal learning have produced evidence that they believe supports a distinction between short-term and long-term memory mechanisms (Broadbent, 1963; Waugh & Norman, 1965).

Limitation of time has allowed me to present only a part of the catalog I had prepared on the complexities of experimentation on memory consolidation using the technique of electroconvulsive shock. Others present at this conference could undoubtedly add to the list that I prepared. A similar catalog could probably be drawn up of the problems involved in employing chemical agents that block protein synthesis in order to study learning and memory storage.

Although I may appear to be approaching the conclusion that experimentation on cerebral mechanisms of behavior is impossibly complex, this is not my intention. Rather, it seems to me that the example we have considered brings out the following points: that alternative interpretations of brain-behavior relations are frequently possible, that testing among them usually requires a battery of approaches, that a manipulation introduced to affect one variable frequently also affects other (and sometimes as yet unrecognized) variables, and that valuable clues may often come from quite different areas of behavioral study. For all of these reasons, experiments on brain and behavior are unlikely to be short and simple. There is little likelihood that those of us in this field will shortly be faced by technological unemployment, as an enthusiastic molecular biologist predicted a decade ago. There is, however, some danger that an impatient or insufficiently informed investigator may fail to carry on the thorough testing of alternative hypotheses that is necessary to establish functional brain-behavioral relations. Such a researcher may nevertheless provide helpful clues to others, or he may drag false indications across the trail. But only an investigator who is prepared to persevere through the complexities of behavioral research is likely to be present at the denouement. I conclude that the future of the brain sciences is inextricably bound to the future of the behavioral sciences.

REFERENCES

1.   Alpern. H. P. and McGaugh, J.L., Retrograde amnesia as a function of duration of electro-shock stimulation. J. comp. physiol. Psychol, 65, 265-269, 1968

2.   Barondes, S. H. and Cohen, H. D., Delayed and sustained effect of acetoxycycloheximide on memory in mice. Proc. Natl. Acad. Sci., 58, 157-164, 1967.

3.   Bennett, E. L., Rosenzweig, M. R. and Diamond, M. C., Time courses of effects of differential experience on brain measures and behavior of rats, in W. L. Byrne (Ed.), Molecular approaches to learning and memory, New York, Academic Press, 1968. In press.

4.   Booth, D. A., Vertebrate brain ribonucleic acids and memory retention, Psychol. Bull., 68, 149-177, 1967.

5.   Bower, G. H., Neurophysiology of learning, in E. R. Hilgard and G. H. Bower (Eds.), Theories of learning, New York, Appleton-Century-Crofts, Chap. 13, 1966.

6.   Broadbent, D. E., Flow of information within the organism, J. verb. Learn. verb. Behav., 2, 34-39, 1963.

7.   Byrne, W. L. (Ed.), Molecular approaches to learning and memory, New York, Academic Press, 1968. In press.

8.   Cherkin, A., Retrograde amnesia effect of flurothyl in chicks, Fed. Proc., 27, 437, 1968.

9.   Chorover, S. L. and Schiller, P. H., Reexamination of prolonged retrograde amnesia in one-trial learning, J. comp. physiol. Psychol., 61, 34-41, 1966.

10. Cofer, C. N. and Musgrave, B. S. (Eds.), Verbal behavior and learning: Problems and processes, New York, McGraw-Hill, 1963.

11. Cooper, E. H. and Pantle, A. J., The total-time hypothesis in verbal learning, Psychol. Bull., 68, 221-234, 1967.

12. Davis, R. E. and Agranoff, B. W., Stages of memory formation in goldfish: evidence for an environmental trigger, Proc. Natl. Acad. Sci., 55, 555-559, 1966.

13. Deutsch, J. A. and Deutsch, D., Physiological psychology, Homewood, Ill., Dorsey Press, 1966.

14. Diamond, I. T. and Hall, W. C., Evolution of neocortex and intelligence. In preparation.

15. Duncan, C. P., The retroactive effect of electroshock on learning, J. comp. physiol. Psychol., 42, 32-44, 1949.

16. Dyal, J. A. and Golub, A. M., An attempt to obtain shifts in brightness preference as a function of injection of brain homogenate, J. biol. Psychol., IX, 29-33, 1967.

17. Estes, W. K., Statistical theory of spontaneous recovery and regression, Psychol. Rev., 62, 145-154, 1955.

18. Flexner, L. B., Dissection of memory in mice with antibiotics, Proc. Amer. Philos. Soc., 111, 343-346, 1967.

19. Gaito, J., Molecular psychobiology: a chemical approach to learning and other behavior, Springfield, Ill., Charles C. Thomas, 1966.

20. Gerard, R. W., What is memory? Sci. American, 189, 3, 118-126, 1953.

21. Hebb, D. O., The organization of behavior,

New York, Wiley, 1949.
22. Hebb, D. O., Distinctive features of learning
    in the higher animal, in Brain mechanisms and
    learning, Springfield, Ill., C. C. Thomas, 37-
    51, 1961.
23. Hydén, H., Behavior, neural function, and
    RNA, Prog. Nucleic Acid Research, 6, 187-218,
    1967.
24. Jarvik, M. E., Associative interference and
    consolidation, Int. Symp. on Recent Advances
    in Learning and Retention, Atti Accad. Naz.
    Lincei, 1968. In press.
25. Keppel, G., Verbal learning and memory, in
    Annual review of Psychology, 19, 169-202,
    1968.
26. Kesner, R. P., Personal communication, 1968.
27. Kesner, R. P. and Doty, R. W., Amnesia produced
    in cats by local seizure activity initiated
    from the amygdala, Exper. Neurol., 21, 1968.
    In press.
28. Kimble, G. A., Hilgard and Marquis' condition-
    ing and learning, Second edition, New York,
    Appleton-Century-Crofts, 1961.
29. Kohlenberg, R. and Trabasso, T., Recovery of
    a conditioned emotional response after one or
    two electroconvulsive shocks, J. comp. physiol.
    Psychol., 65, 270-273, 1968.
30. Kopp, R. Bohdanecky, Z. and Jarvik, M. E.,
    Long temporal gradient of retrograde amnesia
    for well-discriminated stimulus, Science, 153,
    1547-1579, 1966.
31. Luttges, M. W. and McGaugh, J. L., Permanence
    of retrograde amnesia produced by electrocon-
    vulsive shock, Science, 156, 408-410, 1967.
32. Madsen, M. C. and McGaugh, J. L., The effect
    of ECS on one-trial avoidance learning, J.
    comp. physiol. Psychol., 54, 522-523, 1961.
33. McGaugh, J. L., Analysis of memory transfer
    and enhancement, Proc. Amer. Philos. Soc.,
    111, 347-351, 1967.
34. McGaugh, J. L., A multi-trace view of memory
    storage processes, Intl. Symp. on Recent Ad-
    vances in Learning and Retention, Atti Accad.
    Naz. Lincei, 1968. In press.
35. Melton, A. W., Implications of short-term
    memory for a general theory of memory, J. verb.

memory for a general theory of memory, J. verb.
Learn. verb. Behav., 2, 1-21, 1963.

36. Müller, G. E. and Pilzecker, A., Experimen-
telle Beiträge zur Lehre vom Gedächtnis, Z.
Psychol., Ergbd. I, 1900.

37. Neilson, H. C., Evidence that electroconvulsive shock alters memory retrieval rather than
memory consolidation, Exper. Neurol., 20, 3-
20, 1968.

38. Peeke, H. V. S. and Herz, M. J., Permanence
of electroconvulsive shock-produced retrograde amnesia, Proc. 75th Ann. Conv. Psychol.
Assoc., 1967.

39. Rosenzweig, M. R., Bennett, E. L. and Krech,
D., Cerebral effects of environmental complexity and training among adult rats, J. comp.
physiol. Psychol., 57, 438-439, 1964.

40. Rosenzweig, M. R., Love, W. and Bennett, E.
L., Effects of a few hours a day of enriched
experience on brain chemistry and brain weights
Behav. & Physiol., 1968. In press.

41. Rosenzweig, M. R. and Leiman, A. L., Brain
functions, in Annual Review of Psychology,
19, 55-98, 1968.

42. Russell, W. R. and Nathan, P. W., Traumatic
amnesia, Brain, 69, 280-300, 1946.

43. Schneider, A. M. and Sherman, W., Amnesia: a
function of the temporal relation of footshock
to electroconvulsive shock, Science, 159, 219-
221, 1968.

44. Underwood, B. J., Experimental Psychology,
second edition, New York, Appleton-Century-
Crofts, 1966.

45. Warren, J. M., Comparative psychology of
learning, Ann. Rev. Psychol, 16, 95-118, 1965.

46. Waugh, N. C. and Norman, D. A., Primary memory,
Psychol. Rev., 72, 89-104, 1965.

47. Zinkin, S. and Miller A. J., Recovery of memory after amnesia induced by electroconvulsive
shock, Science, 155, 102-104, 1967.

MEMORY CONSOLIDATION AND THE DISTINCTION BETWEEN
SHORT-TERM AND LONG-TERM MEMORY

DR. GERARD: Thank you Dr. Rosenzweig, for
that thoughtful and balanced statement of some very
difficult issues.

You have either disproven my generalization
that no one can take a balanced view of intensely
felt problems, or else these aren't as intensely
felt as sometimes they seem to be in the heat of
an argument.

It is always a good idea for a speaker who
hasn't managed to say all he wants to say to
leave hanging a sentence or two at the end, which
encourages someone to get up and ask the right
questions--as Heinz did this morning. If you
want to quickly say one or two or three things
that you didn't get said so that we can direct
questions to them, I will invite you to do so in
a moment.

I can't resist one additional comment dealing
with the things that I said, and Konorski, too.
The really inexcusable schism between psychology
and neurophysiology is illustrated in the dynamic
structural consolidation mechanism, referred to
by Dr. Rosenzweig as the hypothesis that had been
introduced by Hebb in 1949; psychologists almost
invariably referred to it that way. Since it is
now on the way out, perhaps it won't seem ungra-
cious when I point out that physiologists are
rather likely to refer to it as the Gerard hypothesis because it was published in the same year
by each of us; but neither group apparently reads
the other's literature.

The consolidation time, or as I have committed myself to in my thinking, "fixation time",
certainly can be very different for different
things, and I have no doubt it must be different
from individual to individual. But I do remind
you that it can have an extremely sharp end point

in certain circumstances, and that end point is essentially the same for all members of the species for a particular kind of fixation.

I am referring back to the rat spinal cord experiments in which 45 minutes of time for physiological impulses to come down the cord was sufficient to give an enduring postural change and less than that was not; and this fixation time was amazingly constant (see Chamberlain, T.J., Rothschild, G.H. and Gerard, R.W., Proc. Nat. Acad. Sci., 49, 918, 1963).

DR. ECCLES: I must be a little old-fashioned about this because I still believe in the two categories. There seems to be a fundamental difference between on the one hand the concept of a purely dynamic memory, with its physiological basis of impulses running through complex spatiotemporal pathways that are postulated during perception and short-term memory, and on the other hand long-term memory, with its anatomical and physiological changes not at any particular synapses, but in the vast assemblages of synaptic pathways that form the basis of the specific patterned engrams of Lashley. However, it must be recognized that such changes have not yet been observed by electronmicroscopy. But of course they can be presumed to be displayed indirectly by the specific patterns of neuronal operation that form the neuronal mechanism of memory recall and the retrieval of information.

DR. CHOROVER: A confusion which has become fixed in many minds is that short-term memory is the initial process in long-term memory. Dr. Konorski distinguishes between the short-term memory functions which mediate the demands of life on the part of animals and the initial stages in the processes responsible for long-term memory. This distinction is a very important point. I think that we tend to over-value and hold too tightly and too long to our favorite dichotomies. The distinction between long-term and short-term memory is a case in point. Because of a preoccupation with this distinction, those of us working in the area of memory consolidation have, for too long, been concerned with the temporal characteristics of consolidation. Is "consolidation" finished in a matter of seconds? Does it take a matter of hours?

I think it may be misleading to speak of memory as a process which has a sort of terminal point. The process that is responsible for maintaining memory, which Professor Pribram so beautifully characterizes, is a dynamic one. I doubt whether a "memory trace" ever comes to a point where one could say that it's finished and that it's consolidated. Every event that involves reintegration of previous events obviously changes memory. Every memory trace that is going to be at all useful to the animal must be one which is sufficiently dynamic to be changed as a function of its subsequent usage.

When we do autopsies on animals we never find anything in the brain that has the requisite structural characteristics of a consolidated entity. As Professor Lajtha very neatly pointed out to us, the stuff of which the brain is made is constantly turning over.

I think we should begin more and more to look for dynamic activity patterns which are more likely to represent the information that we and other animals use as memory.

I didn't mean to say that reverbatory activity was necessary or sufficient to maintain long-term information. I did, however, try to suggest that there is an alternative to the view that long-term memory is stored macro-molecularly, or that structural changes in membranes and synapses were the sort of things that were built as the result of initial reverbatory activity. My point is that whatever the mechanism, it is one which must be continuously active and dynamic. It must be, I think, necessarily subject to further alteration.

It's not a question of distinguishing between
a transient and a stable system; it's a system
which is dynamic at all levels, although different
mechanisms may be involved at different stages.

I just want to respond quickly to two points
that Professor Rosenzweig raised: first, to reiter-
ate something that I tried to say this morning, but
I guess I didn't say very clearly then. I want to
repeat it to see if people will find it something
to disagree with; and it's something that Dr. Mc-
Gaugh might at some point in his talk want to speak
to as well, if he is going to be talking to these
issues.

The question relates to whether among all the
experiments concerning the problem of brain function in learning, there are any that tell us anything about the time required for consolidation.

For example, as one of the people involved in
experiments involving short-term effects of ECS,
I never did, nor did I intend to, imply that these
experiments define the time course of consolidation.
I don't believe any such experiments can do that.
All that we do define is the time intervals during
which one or another particular treatment is effective in interfering with retention.

The implication should not be drawn that at
times thereafter memory is no longer susceptible
to interference. On the contrary such interference is clearly shown in other experiments in which
other agents are used under somewhat different
conditions.

I suspect that the process is sufficiently
continuous, if not an indefinite one, so if you had
the right problem at any point you could interfere
with it.

Just one other point.

On the recent experiments of Schneider and
others that you have mentioned, recent results

in our lab suggest that, indeed, something unforeseen to me, at least, is true about ECS experiments. The effects of an electroconvulsive shock upon the brain are very much altered by the immediately previous history of the animals. Animals that have received footshock or some similar treatment immediately before the electroconvulsive shock don't show the same patterns of seizure.

I would just like to add to what Dr. McGaugh has said by reminding people that one of the manifestations of the memory disorder to the investigator, whether he be a clinician or a student of animal behavior, is that it's very difficult to distinguish between memory disorders which are specific and non-specific for time. The latter type often follow a series of ECS treatments. The patient manifests what looks like a retrograde amnesia but the impairment may be very non-specific with respect to time. It is only the former type - the sort that are specific with respect to time - that are relevant to what Dr. McGaugh has described.

I want to point out that one other possible clinical context for the study of retrograde amnesia is in spontaneously occurring seizures. If spontaneously occurring petit mal seizures in patients are monitored by EEG and if these are studied in connection with some serial task, one can often observe memory lapses comparable to those which follow ECS. That is, one can study the patient's ability to recall events immediately preceding the seizure.

DR. ROSENZWEIG: In regard to the point about a continuity in progression of development of memory processes, it is important to note that some of the investigators are attempting to indicate what processes take place at different time intervals. So, for example, in the work of Flexner, you have the statement that at one point electroshock is no longer effective, so that at this point memory could no longer be stored as circulating impulses but must already have entered one form of chemical storage. But he has other evidence for believing that the chemical trace

is not yet consolidated, since it is still susceptible to attack by antibiotic agents. Therefore while one may say that there is a long
continued process of consolidation, it may,
nevertheless, be worthwhile to ask what are the
different stages in the process and how long does
each stage last. Investigators are certainly
working on the problems of differentiating stages
in consolidation.

D. NEUROPHARMACOLOGICAL STUDIES

PROTEIN SYNTHESIS AND MEMORY FORMATION

DR. AGRANOFF: Various antibiotic antimetabolites have been shown to specifically affect one or another growth process such as membrane synthesis without disturbing the other macromolecular processes or energy metabolism. It was of interest to see what effect these agents had on higher brain function. Our laboratory has employed the goldfish, an animal that is suitable for such studies for many reasons. The goldfish has a relatively simple brain, can easily be trained and has good memory of its training weeks after a brief training session. Furthermore the brain resides in a rather large cranial cavity into which drugs can be easily injected into the unanesthetized animal by means of a microsyringe. The brain itself is not touched by the needle, which penetrates the skull in a region just above the tectum. We found that the antibiotic puromycin (170 μg) or acetoxycycloheximide (AXM, 0.2 μg) rapidly produced inhibition of protein synthesis to about 20% of normal with recovery within a day. We found that while these drugs were acting, animals showed no evidence of neurological damage or of gross behavioral disturbance. When injected immediately before or within an hour after training, these drugs produced a loss of the trained response as tested 3 days later. The same amount of drug produced no effect if injected at later times. In order to score an avoidance response, fish must swim over a barrier within 20 seconds after a light goes on and avoid a punishing shock administered through the water.

Our experiments had certain characteristics in common with the "consolidation" phenomenon reported in other animals. There was a time-dependent effect on memory which terminated shortly after the training session. The experiments differed in two ways from the usual consolidation experiment. First, animals did not lose consciousness as a result of the amnestic agent and showed no gross behavioral disturbance, unlike electroconvulsive shock, an agent commonly used in memory

studies. Secondly, we found that the drug could
be injected before the actual training session.
In this instance it had no effect on acquisition
but nevertheless produced a deficit observed on
retraining days later. None of the agents we have
used affect that memory involved in acquisition--
what we term short-term memory. In further studies
we found that the amnesia produced by pre- or
post-trial injection of drugs was not immediately
apparent, but developed over a period of 2-3 days.
This we have referred to as the decay of short-
term memory. Animals may achieve a higher responding level and yet not convert this memory into a
long-term form if one of the inhibitor agents is
acting. Smaller doses of puromycin or AXM injected in the standard 10 µl volume produced no effect.
The drugs and not the injection itself were responsible for the amnesia.

During the course of this work a phenomenon
was observed, which may have implications for consolidation studies in general. Dr. Roger Davis
found that animals will not fix memory if they remain in the training apparatus following training.
Even though puromycin will ordinarily block memory only if injected into animals who have been
returned to their home tanks within an hour following a training session, animals remaining in the
shuttlebox without further training trials remain
susceptible to the drug. It has been possible to
disrupt memory even 3 hours following training when
animals remain in the training apparatus during
this period.

In addition to puromycin and AXM, actinomycin
D also blocks memory formation, implicating RNA as
well as protein synthesis in memory formation.
Arabinosyl cytosine, an inhibitor of DNA synthesis,
does not appear to block learning or memory formation. Our experiments implicate but do not
prove that protein synthesis is required for memory formation. We have looked into the possibility
that these drugs may act in some other way.

It has been suggested that puromycin is a convulsant agent and that this property accounts for

its amnestic properties. We have found that metrazol convulsions are potentiated by intracranial
puromycin injections of the fish as has been observed with intracerebral injections of puromycin
in the mouse. AXM does not potentiate metrazol
convulsions nor does it prevent the puromycin potentiation of metrazol convulsions. We have also
found that puromycin aminonucleoside, a substance
which does not block protein synthesis or memory
in the goldfish, does nevertheless potentiate
metrazol convulsions in the fish. These studies
suggest that the convulsant activity of puromycin
is not responsible for its amnestic effect in the
fish. It further suggests that peptidyl puromycin
is not the convulsant agent, since pretreatment
with AXM did not protect the fish against convulsions. AXM should prevent peptidyl puromycin
formation.

The issue of whether protein synthesis is required for memory formation is still open but in
the fish, at least, agents that block protein synthesis also block memory. While we cannot at
present say much more than this, we are hopeful that
our approach adds support for the conviction that
conventional biological tools are potentially
powerful ones in the study of behavior.

REFERENCES

1. Agranoff, B.W., Davis, R.E. and Brink, J.J.,
   Memory fixation in the goldfish, Proc. Nat.
   Acad. Sci., U.S., 54, 788-793, 1965.
2. Agranoff, B.W., Davis, R.E. and Brink, J.J.,
   Chemical studies on memory fixation in goldfish, Brain Res., 1, 303-309, 1966.
3. Agranoff, B.W., Davis, R.E., Casola, L. and
   Lim, R., Actinomycin D blocks formation of
   memory of shock avoidance in the goldfish,
   Science, 158, 1600-1601, 1967.
4. Bohdanecka, M., Bohdanecky, Z. and Jarvik, M.
   E., Amnesic effects of small bilateral brain
   puncture in the mouse, Science, 157, 334-336,
   1967.
5. Casola, L. and Agranoff, B.W., Studies on
   RNA in goldfish brain. I. Isolation and in

    vivo labeling, Proc. Nat. Acad. Sci., U.S.,
    60, 1389-1395, 1968.
6.   Cohen, H.D. and Barondes, S.H., Puromycin
     effect on memory may be due to occult seizures,
     Science, 157, 333-334, 1967.
7.   Davis, R.E. and Agranoff, B.W., Stages of
     memory formation in goldfish: evidence for an
     environmental trigger, Proc. Nat. Acad. Sci.
     U.S., 55, 555-559, 1966.
8.   Potts, A. and Bitterman, M.E., Puromycin and
     retention in the goldfish, Science, 158, 1954-
     1956, 1967.

IMPLICATIONS OF STUDIES WITH INHIBITORS ON CERE-
BRAL PROTEIN SYNTHESIS FOR "MOLECULAR" AND "MOLAR"
VIEWS OF MEMORY

    DR. BARONDES: Studies of memory storage are
typically oriented by a "molecular" or a "molar"
point of view. In the "molecular" view, memory
represents a change in interneuronal relationships
based on a change in the composition of the brain.
Inquiry is directed to identifications of the substance which is changed or synthesized. Some of
the alternative possible mechanisms for this change
have been discussed previously (Barondes, 1965).
The "molar" view concerns itself with the details
of the changes in the behavior of the organism as
a result of a learning experience. The existence
of phenomena like "consolidation" and "forgetting",
discussed elsewhere in this symposium, indicate
that the behavioral changes are, to some extent,
a function of the time since learning.

    Studies of the effects of inhibitors of cere-
bral protein synthesis on learning and memory have
contributed to our understanding on both the "molecular" and the "molar" levels. The results of experiments done thus far suggest that cerebral protein synthesis is required for the development of
the new intercellular relationships which are believed to represent "long-term" memory. This represents a beginning towards our understanding of
the molecular basis of memory. These studies have
also contributed to our understanding of the "molar"
phenomenology of memory. They support the suggest-

ion from other studies that memory is stored in a different manner at different times after training. They also provide a technique for dissecting the "short-term" process from the "long-term" process. Experiments done in my laboratory with Dr. Harry D. Cohen will be used to illustrate these points.

Intracerebral or subcutaneous injections of acetoxycycloheximide in mice produce profound but reversible inhibition of cerebral protein synthesis.

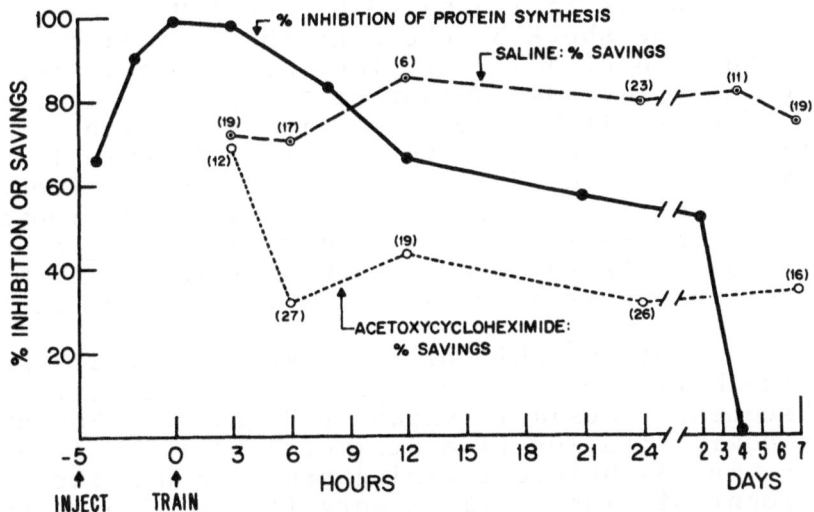

Fig. 1. Effect of intracerebral acetoxycycloheximide on cerebral protein synthesis and memory. Mice were injected in both temporal regions of the brain with a total of 20 microliters of 0.15 M NaCl with or without 20 micrograms of acetoxycycloheximide. Five hours later they were trained to escape shock by choosing the correct limb of a one choice maze to a criterion of 3 out of 4 consecutive correct responses. Retention was determined at the indicated times. Protein synthesis inhibition was estimated in 3-6 mice at the indicated times. There was no significant difference in the savings of the two groups 3 hours after training but thereafter acetoxycycloheximide-injected mice had significantly poorer savings (P<.05, or less) than saline-injected mice, (from Barondes and Cohen, 1967).

After intracerebral injection (Figure 1) there is a gradual onset of inhibition and a rather prolonged duration of inhibition. After subcutaneous injection of larger doses, 95% inhibition of cerebral protein synthesis is achieved within 10 or 15 minutes of injection of the drug and its duration of action is substantially briefer (Barondes and Cohen, 1968). Even briefer inhibition can be achieved by subcutaneous administration of cycloheximide, a less potent derivative of acetoxycycloheximide (Cohen and Barondes, 1968 b). When mice, whose cerebral protein synthesis is inhibited by about 95%, are trained to escape shock by choosing the left limb or the lighted limb of a T-maze, they acquire this response in exactly the same number of trials as saline injected controls. When tested for retention three hours later, their memory is normal (Figure 1). However, when tested for retention six hours after training or thereafter, they have markedly impaired memory (Figure 1). Tests conducted as long as six weeks after training also show that the impairment of memory persists for at least this period of time.

The relationship of the time of establishment of inhibition of cerebral protein synthesis to the subsequent amnesia is shown in Figure 2. Subcutaneous administration of acetoxycycloheximide five or more minutes before training produced a marked impairment of "long-term" memory (tested 7 days after training in this experiment). If the drug was given immediately after training, there was a significant but less marked impairment of memory 7 days after training. Administration of the drug 30 minutes after training had no effect on memory.

These and related experiments suggest the following conclusions: 1) There is no evidence that protein synthesis is required for either learning or for "short-term" memory (for at least three hours after training in the situations studied). Presumably, some other mechanism is responsible for memory storage for this period. 2) Acetoxycycloheximide, a potent inhibitor of cerebral protein synthesis, impairs "long-term" memory. This suggests that protein synthesis is required for "long-term" memory storage. However, it remains possible that

some other action of acetoxycycloheximide is responsible for its amnesic effect. 3) The protein
synthesis apparently required for "long-term" memory storage occurs during learning and/or within

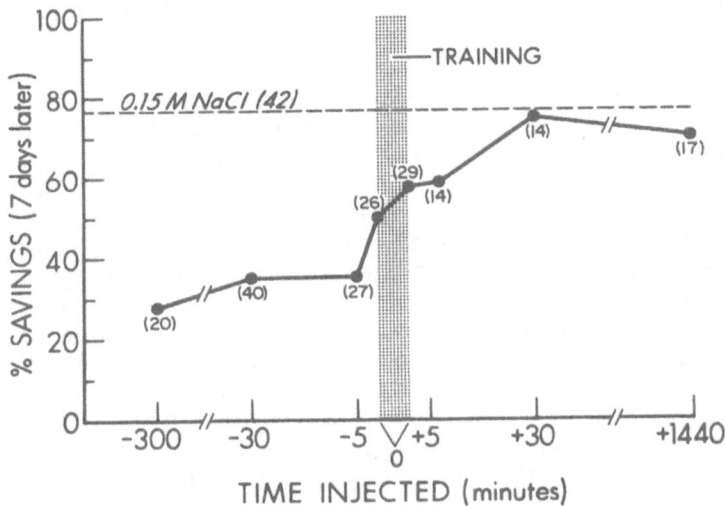

Fig. 2. Effect of subcutaneous administration of
acetoxycycloheximide at various times before or after training on memory. Mice were injected subcutaneously with 240 micrograms of acetoxycycloheximide at the indicated time relative to training.
They were trained to escape shock by choosing the
lighted limb of a T-maze to a criterion of 5 out of
6 correct responses (Cohen and Barondes, 1968a).
Training took an average of 8 minutes. Approximately 90% of cerebral protein synthesis was inhibited
within 10 - 15 minutes of subcutaneous injection of
acetoxycycloheximide. All mice were tested for retention 7 days after training, long after they had
recovered from the drug. The mice injected before
or within 5 minutes after training all had significantly less savings (P<.05 or less, Mann-Whitney U
test) than saline controls. The mice injected 5 or
more minutes before training had significantly less
savings than those injected immediately after training. Injections 30 minutes or more after training
had no effect on memory (from Barondes and Cohen,
1968).

minutes thereafter. 4) Since the amnesic effect
of acetoxycycloheximide is not observed until three
to six hours after training, the "short-term" mechanism, which is apparently independent of protein
synthesis, may survive until this time and temporarily obscure the effect of the inhibitor on the
"long-term" process.

Inhibitors of protein synthesis have, therefore, been useful in memory research in two regards.
From the "molecular" point of view, they suggest
that the development of new functional interneuronal
relationships may be established initially without
the mediation of cerebral protein synthesis but
ultimately requires that cerebral protein synthesis
occur. From the "molar" point of view, they have
added to our understanding of difference between
"short-term" and "long-term" memory, have indicated
the duration of "short-term" memory in the absence
of "long-term" memory, and have made possible studies (in progress) on the effect of manipulations
of the "short-term" memory process on "long-term"
memory storage.

REFERENCES

1.   Barondes, S. H., The Relationship of Biologi-
      cal Regulatory Mechanisms to Learning and
      Memory, Nature, 205, 18, 1965.
2.   Barondes, S. H. and Cohen, H. D., Delayed and
      Sustained Effect of Acetoxycycloheximide on
      Memory in Mice, Proc. Nat. Acad. Sci., U.S.,
      58, 157-164, 1967.
3.   Barondes, S. H. and Cohen, H. D., Memory Im-
      pairment after Subcutaneous Injection of Ace-
      toxycycloheximide, Science, 160, 556, 1968.
4.   Cohen, H. D. and Barondes, S. H., Acetoxycyclo-
      heximide Effect on Learning and Memory of a
      Light-Dark Discrimination, Nature, 218, 271,
      1968 a.
5.   Cohen, H. D. and Barondes, S. H., Cycloheximide
      Impairs Memory of an Appetitive Task, Comm.
      Behav. Biol, 1968 b. In press

TRAINING, NEURONAL CELL PROCESS GROWTH, AND PROTEIN SYNTHESIS

DR. ROSENZWEIG: Dr. Barondes has said part of what I wanted to say. I would like to add that this is one of the reasons why it seems important to pin down the temporal aspects of different phases of consolidation. Even granting, as Dr. Chorover mentioned this morning, that consolidation may be a long continuing process, we want to know what's going on in the different parts of that process. And if protein synthesis comes in, where does it?

Along that line, I would like to mention that some evidence is beginning to accumulate which suggests that stimulation or experience may cause changes in synaptic number. For example, Cragg recently published in Nature (215, 251, 1967) some indications that after visual stimulation of only an hour or so of animals that had been raised totally in the dark, there was an increase both in synaptic number and in the size of certain synapses. I would also like to note preliminary evidence of synaptic changes following one month in our enriched or impoverished environments. Dr. Albert Globus has found evidence, in two experiments with us, that enriched-experience rats have more dendritic spines than their impoverished littermates, and their difference appears to be restricted to one part of the dendritic tree of the cells studied. Replication and extension of such experiments on synaptic number may furnish valuable evidence of changes in the brain caused by training. It appears that we cannot rule out the possibility that there is protein synthesis which is related to outgrowth in fairly distant parts of the neuron.

INTERPRETATION OF PUROMYCIN INHIBITION EFFECT

DR. GLASSMAN: One of the difficulties in interpreting this type of work is that one does

not know whether the effects are on specific proteins involved in the process of consolidation, or
whether the memory deficit is caused by the inhibition of processes that are generally required
for the health of the cell. The function of the
neurons is not so impaired as to prevent learning,
but they are sick enough so that consolidation does
not take place.

Would you please discuss any data that would
indicate that one or the other was more likely?

DR. AGRANOFF: The animals are not sick at
the time of re-testing.

DR. GLASSMAN: I did not mean sick animal.
I meant sick neurons, i.e., neurons whose functions
are impaired.

DR. AGRANOFF: If I can rephrase your question,
is it possible that a cell that's "half sick" in
some way can learn perfectly but cannot remember?

DR. GLASSMAN: I would like to continue the
discussion of the interpretation of the inhibitor
experiments by bringing in the work of Dr. Flexner
and his associates. Mice treated with puromycin
one day after training will show a large memory
deficit when tested later. If the mouse is injected with a small amount of saline up to 30 days
later, it will perform as though it always knew the
task. Flexner interprets these results to mean
that consolidation went on normally in these mice,
but puromycin interfered with retrieval by inducing
the formation of abnormal puromycin peptides.
These interfere with the action of normal peptides
that are presumably responsible for the maintenance
and retrieval of the conditioned avoidance. The
saline supposedly removes the abnormal puromycin
peptides. Thus, it is clear that under these
conditions, puromycin does not interfere with
consolidation. This raises a number of questions
concerning the mode of action of other inhibitors.
Are they acting specifically on consolidation, or
on the other processes?

I should like to ask Dr. Agranoff if he has ever tried saline or other agents to bring back the memory which he no longer sees after the administration of certain antibiotics to goldfish.

DR. CHOROVER: I would like to add to the previous remarks an observation made in my laboratory by Foteos Macrides that puromycin injections in the sites that Flexner and others have used provoke seizures.

Now, in our laboratories the seizures have been seen to last for periods that long outlast the period during which protein synthesis is inhibited by puromycin, in one experiment, as long as three weeks after the injection.

In quite different experiments we have seen in our laboratory that seizure activity in the brain in cats can be arrested and occasionally reversed by changing the balance in the brain by saline injections, and the suggestion I would like to make--and hear Dr. Agranoff's comment on it-- is the possibility that the so-called washout effect might be due to a blocking of abnormal activity, induced initially by puromycin injection, now reversed by the saline injection, permitting a return to more normal patterns of behavior and brain activity. This would suggest, it's true, the possibility that puromycin effect is due less to interference with consolidation that to a long-term proactive interference with performance.

DR. AGRANOFF: We have not done the washout type of experiment, and I don't think we really can because in our experiments we do not make a hole in the brain, which itself can produce a convulsive effect.

I think that in terms of retrieval that one should also mention the experiments of Deutsch with inhibitors, where apparently the effect is completely on some sort of retrieval mechanism as far as we can tell. I am attracted to Chorover's interpretation of the washout effect.

I am somewhat dumbfounded that there is some sort of clot in the brain that you could wash out with a slug of saline. I have heard proposed an explanation based upon state dependent learning. It wouldn't be a typical type of state dependent learning but the injection could serve as some sort of reminder.

ON THE NEED TO BETTER UNDERSTAND PROTEIN SYNTHESIS IN BRAIN

DR. DAVISON: I would like to direct my remarks toward protein synthesis in the nervous system; the subject has been touched several times in this conference but without discussion of the classes of cells involved. Some of this interest in protein synthesis in brain tissue has been aroused by the experiments of Flexner, of Agranoff and of Barondes and their associates; their experiments have demonstrated that systematically or locally applied drugs, among other effects, largely block protein synthesis, interfere with the consolidation of memory.

Hydén and Egyhazi (Proc. Nat. Acad. Sci., 52, 1030, 1964) and Shashoua (Nature, 217, 238, 1968) have shown that in the brains of animals learning new skills significant RNA changes occur. These findings make it plausible that with the acquisition of learning, new proteins are built into the brain. An increase in the brain size of animals held in enriched as compared to a "deprived", dark monotonous environment has in fact been observed.

On closer examination, however, the direct relationship between learning and protein synthesis appears less convincing. The brain as a whole is the site of rapid protein synthesis and catabolism. Over 70% of the proteins in the brain of a mouse are turned over in 30 days (Lajtha and Marks, Protides of the Biological Fluids, 13, 1965); only a very small fraction of the brain proteins have long-term stability and among these must be many such as histones that probably play no direct role in nervous function. What is the nature of this large synthetic response that training stimulates and how are the changes wrought in the brain maintained when most of the proteins have been catabolized?

It is questions such as these that pinpoint the need for deeper investigation into protein synthesis in the brain. Ultrastructural studies suggest that both neurons and glia have well developed endoplasmic reticula that coud subserve rapid protein synthesis, but how many species of proteins are turned out by either class of cell and what is their function? The glial cells of the brain divide infrequently, the neurons probably not at all; nor is there a secreted product--other than neural transmitter--from either species of cell (if we except the relatively small population of neurosecretory cells). Where is it utilized, the protein that is turned over so rapidly and consistently, whether the brain is stimulated or unstimulated? Much of the output of neuronal protein may be consigned to the cellulofugal axoplasmic transport (Weiss, 1967) but the nature and function of these proteins too are unknown. Hopefully a better understanding of protein synthesis in the brain may help us to unravel the complex functions of the brain.

FACILITATION OF MEMORY STORAGE PROCESSES

James L. McGaugh

If we are to achieve an understanding of the neurobiological bases of memory, several conditions must be satisfied. First, as Rosenzweig has pointed out at this conference, we must have a clear under-standing of the processes of memory at a behavioral level. Second, we must understand the mechanisms of neural plasticity. And most importantly, we must be able to show in detail how the findings of neural plasticity can explain in precise detail the phenomena of learning and memory observed at a behavioral level. Although considerable theoretical speculation concerning neural bases of learning and memory has been proposed in recent years, pre-cise empirical support for the favored theories is lacking.

One of the major problems facing theorists in this area is that we do not yet know how many systems are involved in producing the behavioral phenomena of memory. It would seem unlikely that the mechanisms of memory storage can be understood unless we are first able to specify the number of processes and the nature of their inter-relation-ships.

For example, earlier in this conference Bogoch suggested that in certain instances some individuals may have a very rapid consolidation as when the "secret agent" needs only to read the "serial number" in order to store it. It is not entirely clear that such ability is based upon the kind of change generally referred to when we think of long-term memory. Perhaps there is more than one memory system. It may be that the evolution of memory involved the selection of a system of memories; one system allowing us to retrieve and use information recently experienced so that we can behave adaptively in the present environment. Another system might serve to accumulate experiences.

The distinction between short-term memory and long-term memory is, of course, not a new one. It is only in recent years, however, that the distinction has seriously influenced research strategy. Figure 1 outlines several simplified representations of the way in which the memory for recent events might relate to long-term memory and long-term memory are but different ends of a continuum based on the same process or mechanism. Or, as others have suggested (e.g. 12, 11) the processes underlying recent memory might be different from, but essential for the formation of long-term memory. And as I have suggested elsewhere (19) it might be that there are several independent processes. If so, they might operate in parallel or in a serial fashion; that is, each process might be essential for producing the subsequent one. The point is that we do not know at the present time which conception most adequately fits the facts of learning and memory. And it is essential that we know if understanding of the bases of memory storage is to be achieved.

The view that there is more than one stage of memory storage is suggested by some common experiences. For example most of us have virtually perfect memory for very recent experiences. We can repeat sentences just heard and complete sentences just planned. We can do this accurately and rapidly within a period that would seem to be too brief for molecular-synthesis to be significantly involved.

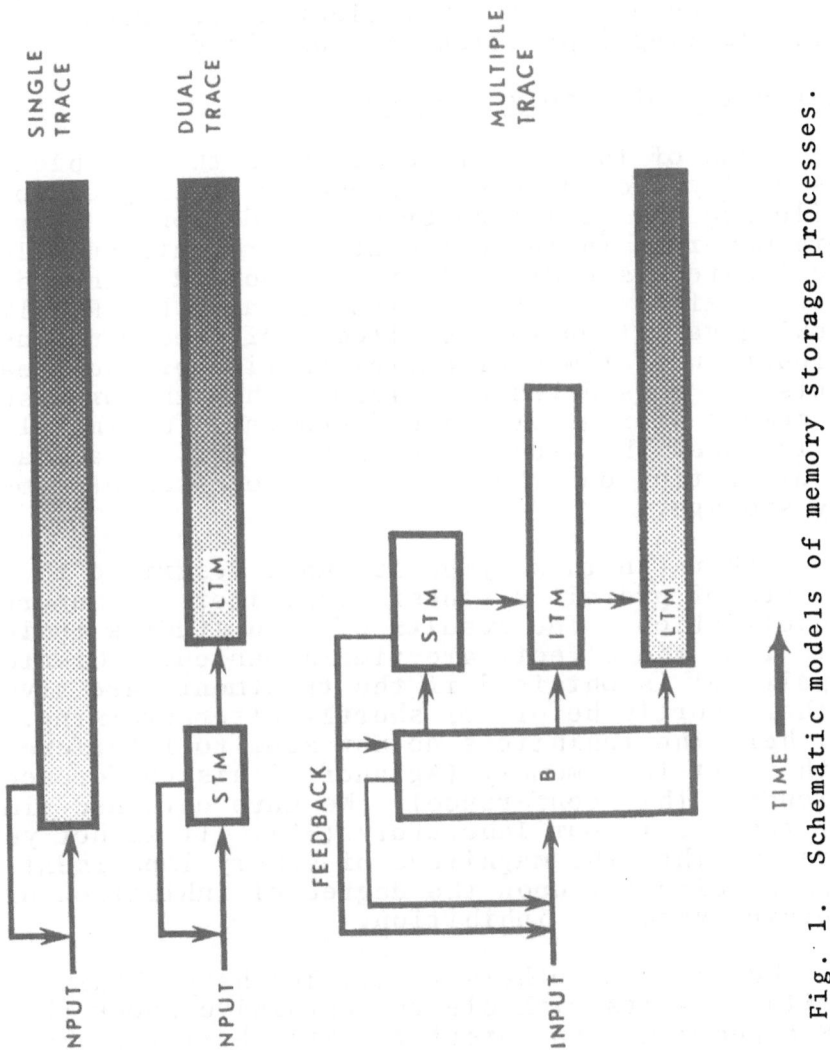

Fig. 1. Schematic models of memory storage processes.

However, such memories are fleeting. Durable
memories do not appear to be formed within seconds.
The problem of memory storage is largely that of
discovering how it is that fleeting experiences
leave lasting impressions of some kind.

Disruption of Memory Storage

Much of the recent research on this problem
has employed experimental procedures designed to
influence the formation or consolidation of long-
term memory. In these recent experiments animals
are treated (see Fig. 2) at some period either before or after they are trained on a task. Results
showing variation in the effects of treatments as
a function of times of administration of the treatments are considered as evidence that the processes
of memory storage are time dependent. Ultimately
it is hoped that such studies will provide a means
of dissecting out the stages or processes of memory storage.

Others in this symposium have described the
effects of protein synthesis inhibitors on memory
consolidation. The results of such studies indicate that the effects are time-dependent. Greatest
impairment is obtained if the treatments are given
either shortly before or shortly after training.
Further, the inhibitors do not seem to interfere
with short-term memory (Agranoff, this conference;
Barondes, this conference). We have obtained similar results in our laboratory (32). It is not yet
clear whether the magnitude of memory impairment
depends directly upon the degree of inhibition of
protein synthesis inhibition.

We (24) and others (e.g., 10) have obtained
similar results with electroconvulsive shock (ECS).
ECS appears not to interfere with short-term memory,
but is quite effective in impairing the consolidation of long-term memory (18). With ECS as well
as with protein synthesis inhibitors amnesia does
not occur immediately after the training and treatment but, rather, develops gradually over a period
of hours. These findings strongly suggest that
the processes of short-term memory are different

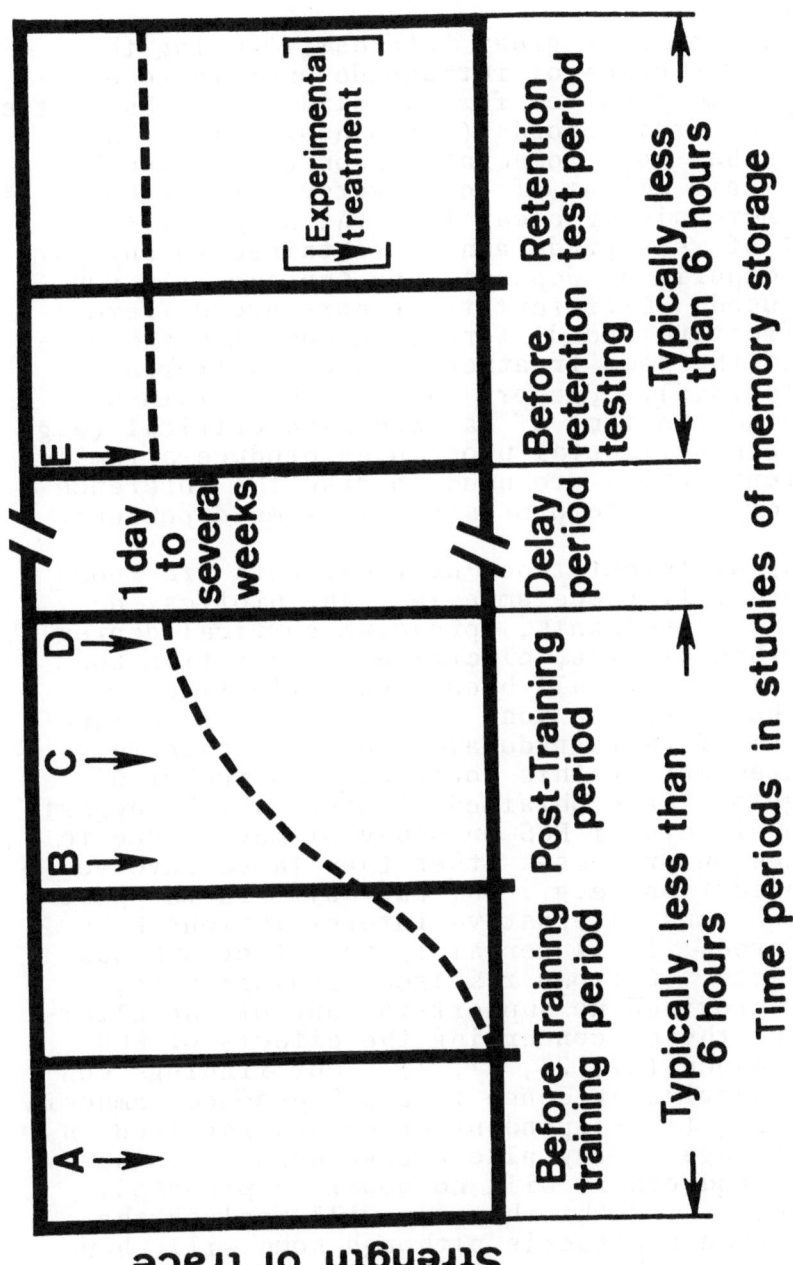

Fig. 2. Times of experimental treatment in experimental studies of memory storage.

from those underlying durable memory.

There has been great interest in using the findings of studies of retrograde amnesia to estimate the time required for consolidation. Estimates have ranged from seconds (6) to hours (14). Such variation has been somewhat disconcerting and has led to a fair amount of controversy. It is becoming increasingly clear that the slope of the gradient of retrograde amnesia obtained in any particular experiment depends upon the specific conditions used. Different treatments are differentially effective and different intensities and durations of the same treatment produce different gradients (3,5). Factors such as the strain of animals tnd the time of day are also critical (e.g 31). The experimental procedures produce a family of gradients which are used to draw the inference that memory storage processes are time dependent.

Many different types of treatments are known to produce retrograde amnesia. Antibiotics, depressants, convulsants, spreading cortical depression, carbon dioxide, electrical stimulation and brain lesions have all been found effective (22). ECS is, however, the most commonly used technique in studies of retrograde amnesia. As Rosenzweig has pointed out at this conference, a number of investigators have obtained findings which suggest that the effects of ECS on behavior may be due to influences on processes other than those involved in consolidation (e.g., 30, 29, 33). In the past decade numerous alternative interpretations have been offered. It is certainly true that ECS has many effects. Our own research findings have, however, provided no support for any of the alternative hypotheses concerning the effects of ECS on performance (27, 21, 17, 7). Our findings continue to provide evidence that ECS produces amnesia by affecting time-dependent processes involved in memory storage. Many alternative hypotheses have been made and others will no doubt be proposed. At this point I doubt that any will replace the consolidation hypothesis although some will, hopefully, sharpen the hypothesis by showing its limitations. Such is the history of science. It should

be remembered, however, that current thinking about
memory consolidation processes does not rest solely
upon findings from studies using ECS.

Facilitation of Memory Storage

In recent years an increasing amount of evidence has indicated that it is also possible to
facilitate memory storage by treating animals with
drugs or electrical stimulation during the early
stage of consolidation. Much of the research on
this problem has employed central nervous system
stimulants such as strychnine, picrotoxin, and pentylenetetrazol (26, 20). The assumption guiding this
research is that if there is more than one stage of
memory storage and if these processes are labile
then they might be influenced by drugs known to
stimulate the central nervous system. In our early
studies (25) we injected animals with drugs prior
to training sessions and studied the drugs' effects
on acquisition. Our findings supported earlier suggestions (16) that analeptics can enhance rate of
acquisition. It is difficult, however, to interpret the findings of these studies. Animals dosed
with analeptics might perform better for a variety
of reasons. They may be more alert, they may be
more highly motivated, they may attend better, or
they may be more efficient in storing information.

In order to avoid these difficulties we adopted the procedure of administering the drugs after
rather than before the training sessions. With
this procedure the animals are neither trained nor
tested while drugged. Influences on retention
tested subsequently are assumed to be due to the
drugs' effects on storage processes.

Figures 3, 4, 5 and 6 show results that are
typically obtained from such studies. In these experiments mice were trained (3 trials each day) on a
visual discrimination task. In some experiments the
mice were injected with graded doses of an analeptic
immediately after each day's training. Training was
continued until all animals learned the task (i.e.,
until they chose correctly on 9 out of 10 trials).
The results of a dose response study using pentyl-

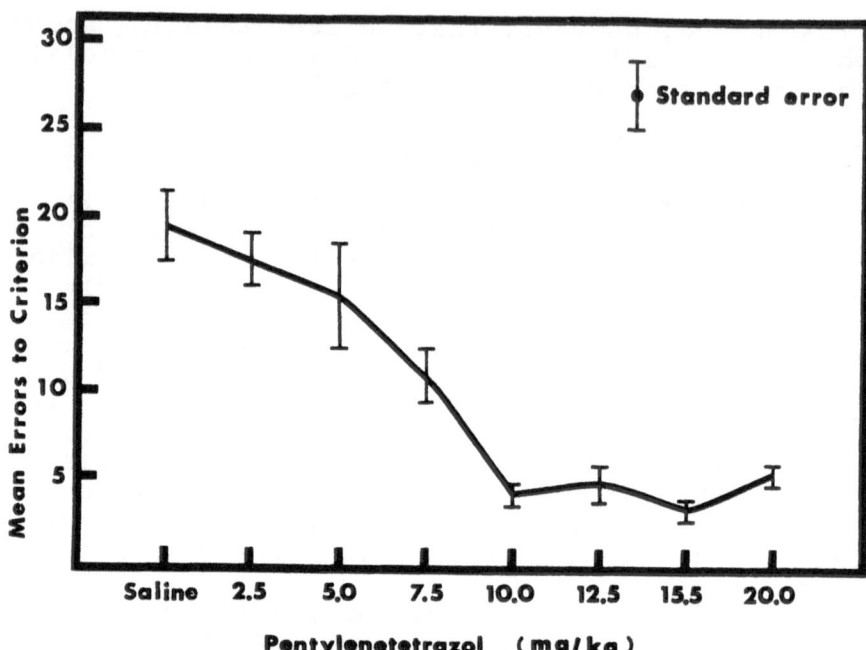

Fig. 3. Effects of post-training administration of pentylenetetrazol on visual learning in mice. Each mean is based on 12 animals (6 males, 6 females). From Krivanek and McGaugh, 1968.

enetetrazol are shown in Fig. 3. As can be seen errors decreased directly with increasing doses up to 15 mg/kg. In other experiments we have varied the time of drug administration. If the drugs act by influencing time-dependent memory storage processes then their effectiveness should vary with time of administration. The results shown in Figure 4 show that this is indeed the case. Facilitation is obtained if the drug is administered either shortly before or shortly after training (i.e., within 15 minutes).

We have obtained similar results with other drugs, including strychnine. Fig. 5 shows effects of varying doses of strychnine given after each

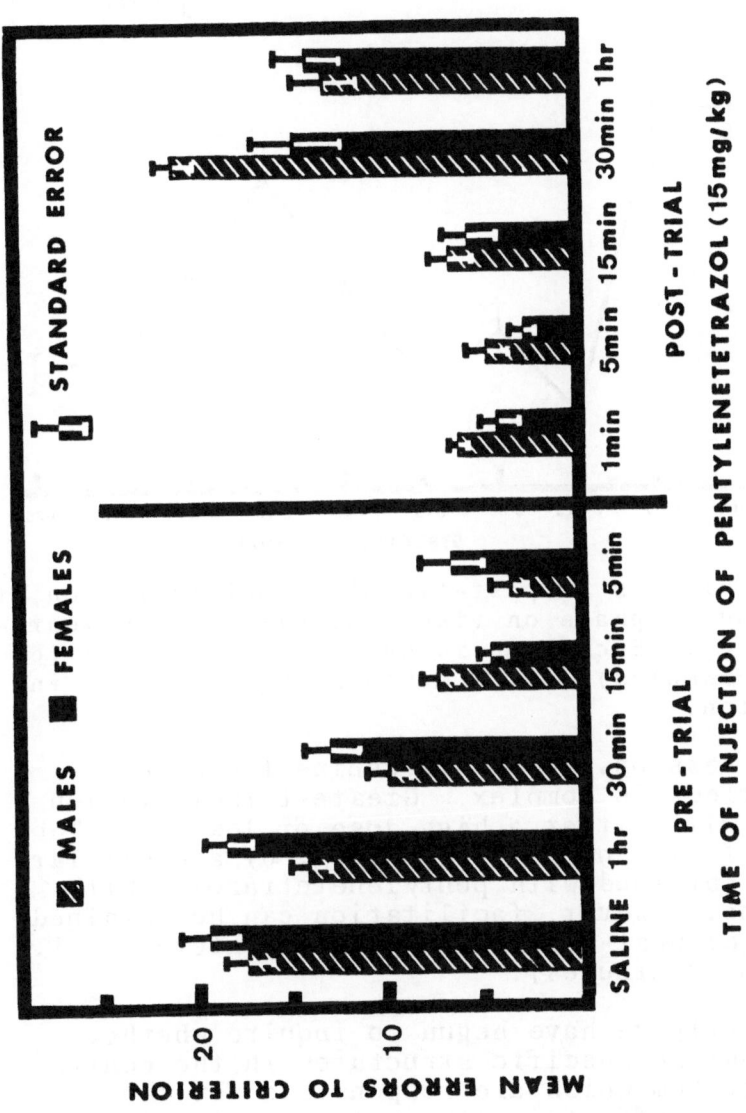

Fig. 4. Effects of time of administration of pentylenetetrazol on visual discrimination learning in mice. Each mean based on 6 animals. From Krivanek and McGaugh, 1968.

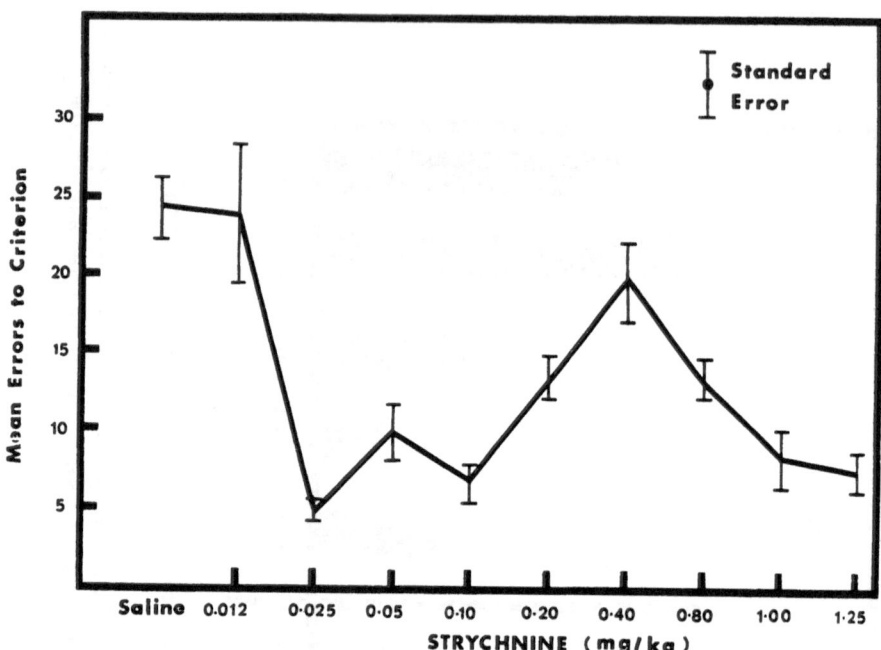

Fig. 5. Effects of post-training administration of strychnine sulphate on visual discrimination learning in mice. Each mean is based on 12 animals (6 males, 6 females). From McGaugh and Krivanek, in preparation.

training session. With strychnine the dose response effect is complex. Greatest facilitation is found with either a high dose or low dose. The effects of time of injection (Fig. 6) are similar to those obtained with pentylenetetrazol. With strychnine, however, facilitation can be obtained with longer post-training injection intervals (i. e., up to 30 minutes).

Recently we have begun to inquire whether there might be specific structures in the central nervous system which are responsible for the facilitating effects. Alpern (2) has found that in rats, discrimination learning can be enhanced by implanting strychnine crystals bilaterally in the mesencephalic reticular formation. A comparable dose implanted in the caudate nucleus or

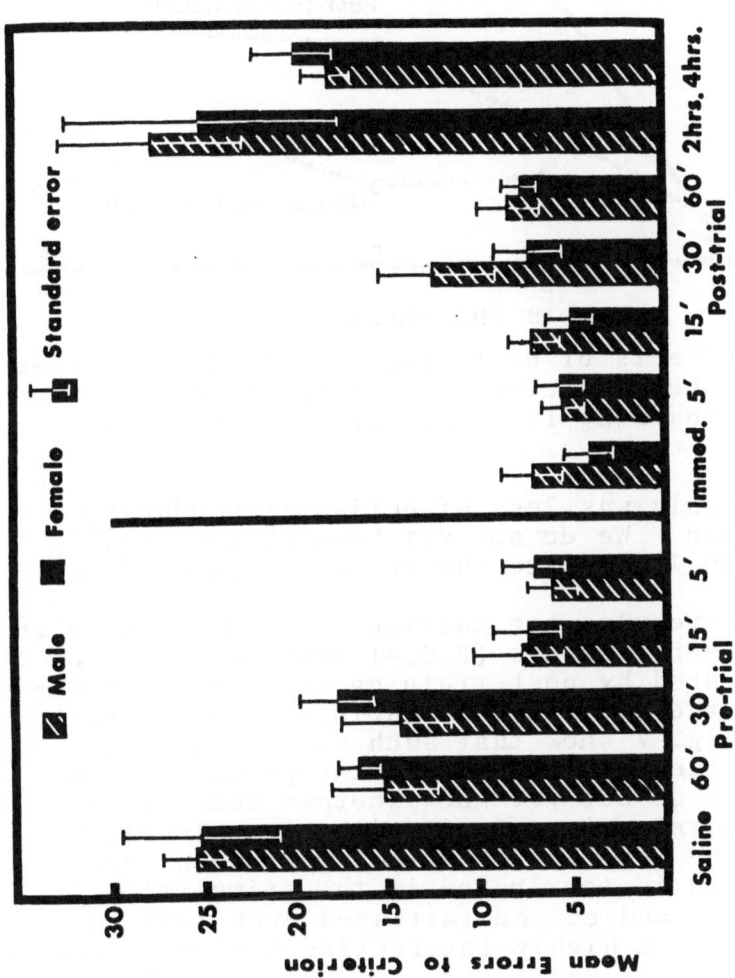

Fig. 6. Effects of time of administration of strychnine sulphate on visual discrimination learning in mice. Each mean based on 6 animals. From McGaugh and Krivanek, in preparation.

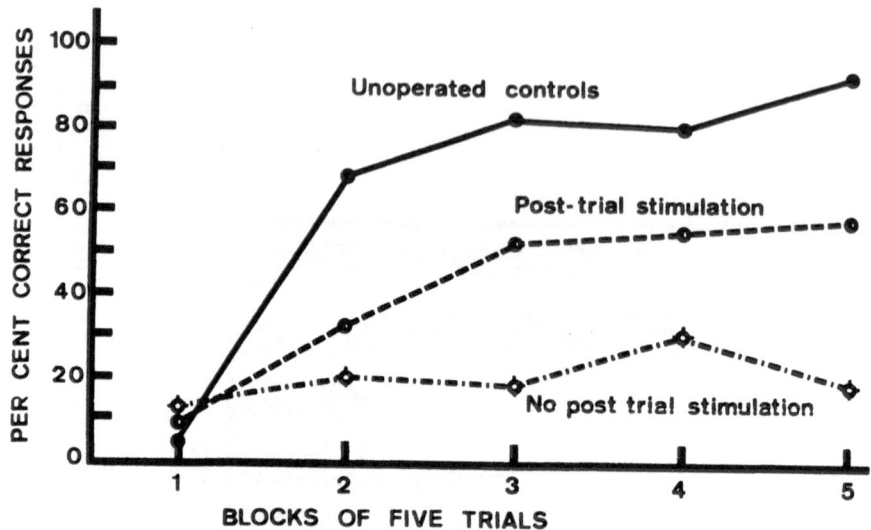

Fig. 7. Effects of post-training electrical stimulation of the mesencephalic reticular formation on avoidance learning in rats. From Denti, et al.

subcutaneously was less effective in producing facilitation. We do not yet know whether such effects are limited to the reticular formation.

These results are particularly interesting in view of other findings (8,9,4) that learning can be facilitated by post-training electrical stimulation of the reticular formation. The results shown in Fig. 7 show that such stimulation can alleviate the deficit in learning produced by a lesion. We do not yet know whether such effects enhance learning in animals without an impairing lesion. The results shown in Fig. 8 show that facilitation is terminated if the stimulation is discontinued and can be initiated even late in training. In a highly interesting series of experiments Albert (1) has found that memory storage processes can be influenced by polarizing currents applied to the cerebral cortices of rats. Impairment is found with cathodal stimulation and facilitation is found with anodal stimulation. It is too early to tell whether these effects are

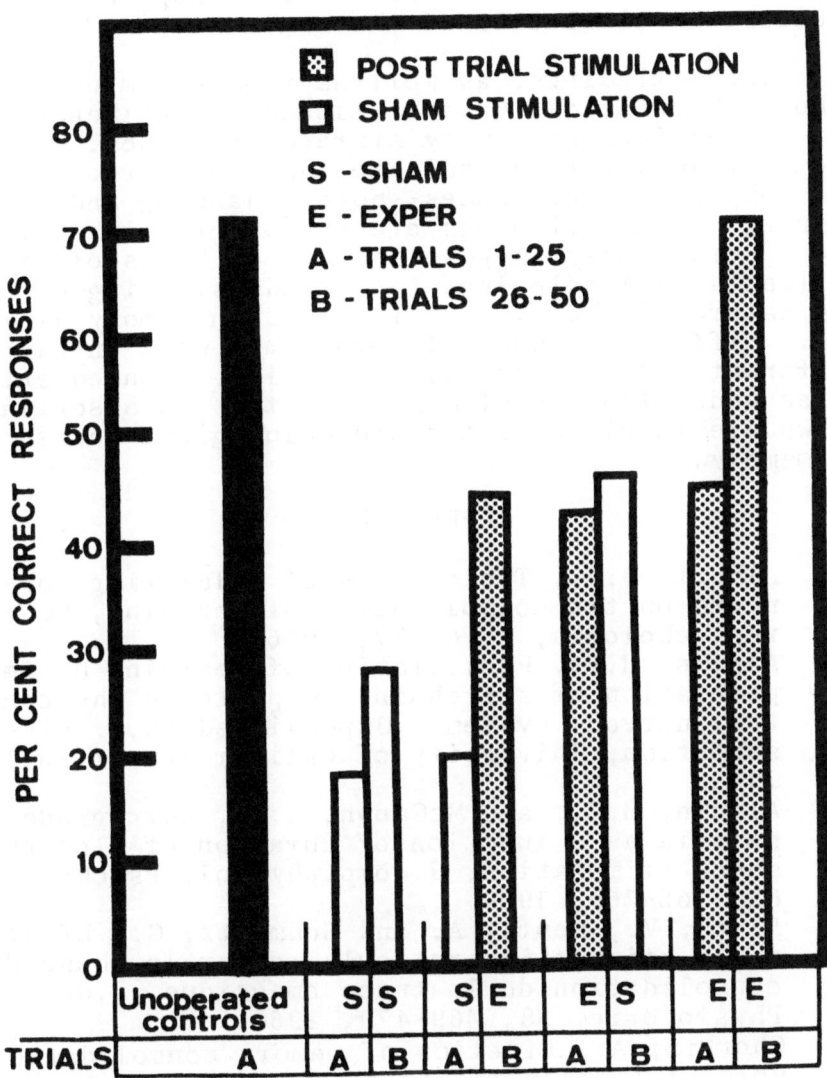

Fig. 8. Effects of initiation and termination of post-training stimulation of the mesencephalic reticular formation on avoidance learning in rats. From Denti, et al.

related to those we have found with drugs and
electrical stimulation of the brain.

These findings, as well as those from numerous
other laboratories, have continued to influence
our conceptions of memory storage processes. At
the very least, they indicate that such processes
are subject to influences--both enhancing and impairing. These findings are encouraging in that
they increasingly suggest that these kinds of experiments will help lead to an understanding of
the nature of the systems involved in memory storage. Different stages of memory may well involve
different kinds of neural plasticity. Consequently an understanding of memory systems is essential
if we are to discover the neurobiological bases
of memory.

REFERENCES

1.　　Albert, D.J., The effects of polarizing currents on the consolidation of learning, Neu-
ropsychologia, 4, 65-77, 1966.
2.　　Alpern, H.P., Facilitation of learning by implantation of strychnine sulphate in the central nervous system. Unpublished Ph.D. dissertation, University of California, Irvine,
1968.
3.　　Alpern, H.P., and McGaugh, J.L., Retrograde
amnesia as a function of duration of electroshock stimulation, J.comp.physiol. Psychol.,
65, 265-269, 1968.
4.　　Bloch, V., Denti, A. and Schmaltz, G., Effets
de la stimulation reticulaire sur la phase de
consolidation de la trace amnesique, J.de
Physiologie, 58, 469-470, 1966.
5.　　Cherkin, A., Kinetics of memory consolidation:
Role of amnesic treatment parameters. In
preparation.
6.　　Chorover, S.L. and Schiller, P.H., Reexamination of prolonged retrograde amnesia for a
well-discriminated stimulus. J.comp.physiol.
Psychol., 61, 34-41, 1966.

7. Dawson, G. and McGaugh, J.L., A further examination of the effects of electroconvulsive shock on a reactivated memory trace. In preparation.

8. Denti, A., Facilitation of conditioning by reticular stimulation in the fixation phase of memory. Unpublished dissertation, University of Paris, France, 1965.

9. Denti, A., McGaugh, J.L., Landfield, P. and Shinkman, P., Facilitation of learning with post-trial stimulation of the reticular formation. In preparation.

10. Geller, Anne and Jarvik, N.E., The time relations of ECS-induced amnesia, Psychon.Sci., 12, 169-70.

11. Gerard, R.W., Biological roots of psychiatry, Science, 122, 225-230, 1955.

12. Hebb, D.O., The organization of behavior, New York: Wiley & Co., 1949.

13. Hudspeth, W.J., McGaugh, J.L. and Thomson, C. W., Aversive and amnesic effects of electroconvulsive shock, J.Comp.Physiol.Psychol., 57, 61-64, 1964.

14. Kopp, R., Bohdanecky, Z. and Jarvik, M.E., Long temporal gradient of retrograde amnesia for a well-discriminated stimulus, Science, 153, 1547-1549, 1966.

15. Krivanek, J. and McGaugh, J.L., Effects of pentylenetetrazol on memory storage in mice, Psychopharmacologia, 12, 303-321, 1968.

16. Lashley, K.S., The effect of strychnine and caffeine upon rate of learning, Psychobiol., 1, 141-170, 1917.

17. Luttges, M.W. and McGaugh, J.L., Permanence of retrograde amnesia produced by electroconvulsive shock, Science, 156, 408-419, 1967.

18. McGaugh, J.L., Time-dependent processes in memory storage, Science, 153, 1351-1358, 1966.

19. McGaugh, J.L., A multi-trace view of memory storage processes, Accademia Nazionale Dei Lincei, 365, 13-24, 1968.

20. McGaugh, J.L., Drug facilitation of learning and memory. In press.

21. McGaugh, J.L. and Alpern, H.P., Effects of electroshock on memory: Amnesia with convulsions, Science, 152, 665-666, 1966.
22. McGaugh, J.L. and Herz, M.J., Controversial issues in consolidation theory. In press.
23. McGaugh, J.L. and Krivanek, J., Strychnine effects on discrimination learning in mice: Effects of dose and time of administration. In preparation.
24. McGaugh, J.L. and Landfield, P., Evidence of a temporary short-term memory in mice following ECS. In preparation.
25. McGaugh, J.L. and Petrinovich, L.F., The effect of strychnine sulphate on maze learning, Am.J.Psychol., 72, 99-102, 1959.
26. McGaugh, J.L. and Petrinovich, L.F., Effects of drugs on learning and memory, Intern.Rev. of Neurobiology, 8, 139-196, 1965.
27. McGaugh, J.L. and Petrinovich,L.F., Neural consolidation and ECS reexamined, Psychol.Rev., 73, 382-387, 1966.
28. Melton, A.W., Implications of short-term memory for a general theory of memory, J.Verb. Learn.Verb.Behav., 2, 1-21, 1963.
29. Misanin,J.R., Miller, R.R., and Lewis, D.J., Retrograde amnesia produced by electroconvulsive shock after reactivation of a consolidated memory trace, Science, 160, 554-555, 1968.
30. Nielson, H.C., Evidence that electroconvulsive shock alters memory retrieval rather than memory consolidation, Exp. Neurol., 20, 3-20, 1968.
31. Stephens, Gwen, McGaugh, J.L. and Alpern, H.P., Periodicity and memory in mice, Psychon. Sci., 8, 201-202, 1967.
32. Swanson R. and McGaugh, J.L., Effect of acetoxycycloheximide on retention of an inhibitory avoidance response. In preparation.
33. Zinkin, S. and Miller, A.J., Recovery of memory after amnesia induced by electroconvulsive shock, Science, 155, 102-103, 1967.
34. Zornetzer, S., Dawson, G. and McGaugh, J.L., Amnesic effects of electroconvulsive shock: Magnitude and stability as a function of prior footshock and ECS stimulation. In preparation.

DISTINCTION BETWEEN DRUG INFLUENCE ON PERFORMANCE
AND ON LEARNING

DR. MCGAUGH: It is important to note the
specific experimental procedures that we have used.
Animals are first trained and then injected with
one of the analeptic drugs. The animals are then
set aside and retested at some subsequent period,
usually 24 hours or longer, when it can be safely
assumed that the drug has been metabolized. With
these procedures the animals have been neither
trained nor tested while drugged. The drugs are
only effective in facilitating learning when they
are administered within a short time before or
after the critical training period. With these
procedures it is unlikely that these drugs influence
performance by directly influencing behavior at the
time of recall. With the implant studies of Alpern
(1968) it is of course not yet possible to conclude that the drugs facilitate performance by
facilitating storage processes simply because the
drugs are present during all phases of the procedure: training, consolidation and retesting.

In the electrical stimulation studies we have
deliberately kept the amount of current lower than
that which produced observable behavioral arousal.
In order to do this we first obtain the arousal
threshold for each animal by observing subtle behavioral effects of direct stimulation. We then
choose stimulation parameters which are lower than
those necessary for producing the behavioral effects
In the drug studies we are careful to use doses
which are below those which produce any clean behavioral effects. That is, we see no effects on
activity or general reactiveness. It may well be
that the basis of learning facilitation is increased arousal immediately following a training session.
This is an interesting hypothesis and certainly at
the present time cannot be rejected by our data.

EFFECT OF STRYCHNINE ON POST SHOCK AMNESIA

DR. MCGAUGH: Our findings indicate that if a
low dose of strychnine is administered to animals

ten minutes prior to training then electroconvulsive shock administered one minute after training
is not effective in producing retrograde amnesia.
It appears as though retrograde amnesia can be pre-
vented by pretreating the animals with analeptics.
It is also possible, however, to attenuate the
amnesic effects of electroconvulsive shock by
treating the animals with strychnine immediately
after electroconvulsive shock.   Such drug treatments
are effective when administered within three hours
following the electroconvulsive shock treatment.
Over this period the effect is graded (McGaugh, 1968).

Several years ago we discovered (Hudspeth, Mc-
Gaugh and Thomson, 1964) that there had been a confusion in the experimental literature with regard to
the effects of electroconvulsive shock.   The effect
of a single electroconvulsive shock is quite different from those produced by a series of convulsive
shocks given over several days.   Prior to approximately 1960 most conclusions concerned with electroconvulsive shock were made on the basis of using
multiple electroconvulsive treatments.   Since that
time most of the conclusions, including our own,
have been based on experiments in which but a single
training session had been followed by a single
treatment.   We do not yet know, of course, whether
the effects of electroconvulsive shock obtained
with animals are analogous at all to those obtained when electroconvulsive shock is used as a treatment for human patients.

NUMBER OF CELLS INVOLVED IN STRYCHNINE EFFECT

DR. MCGAUGH: We have asked a question of ourselves which may be related to Dr. John's question
about number of cells involved in various aspects
of learning.   We have, as I indicated earlier, in-
quired about the structure that might be involved
in producing the learning facilitation that we see
with analeptic drugs.   There are two extreme possibilities.   One possibility is that during the
course of learning connections are formed between
cells as a consequence of experience (synaptic
changes are the kind that we have alluded to), and

that the effect of the analeptic drug is to potentiate the development of changes at each and every
site. Another possibility is that there are mod-
ulating structures in the nervous system whose
activity can alter the functioning of aggregates
elsewhere. That is, without taking a connection-
istic position, one might imagine that there are
various structures, perhaps the caudate nucleus or
perhaps reticular formation, whose function it is to
modulate the information being processed in the cell
(Alpern, 1968). Alpern's work indicates that facilitation can be administered by strychnine directly to
the reticular formation. We cannot yet conclude
that this is the only structure in the brain where
such effects can be obtained. It may well turn out
to be that there is a finite number of structures
in the brain which when so stimulated by analeptic
compounds will serve to modulate the rest of the
nervous system. It is too early to decide, but I
think that the approach that we are taking will
help and directly shed some light on the questions
raised by Dr. John.

CHEMICAL TRANSFER OF LEARNING

DR. UNGAR: Dr. Rosenzweig gave a fair assess-
ment of the situation on chemical transfer of learning. I should like to add a few words just to explain why anyone would want to do such experiments.

This morning Dr. Pribram said that looking for
changes produced in the brain by learning is like
studying the bumps on the grooves of a gramophone
record. I think it is an excellent comparison, not
because I believe for a second that the brain records experience on the same principle as the disc
records sound but because it implies that the best
way of finding out what is on a record is to play
it back on a gramophone.

This is exactly what the chemical transfer experiments are trying to accomplish. Of course, I
immediately anticipate the jocular interpretation
that we are trying to extract the disc and pour the
extract over the record player, as has already been

proposed, in a facetious letter to "Science", to grind up oscilloscopes. It is important, therefore, to state that the transfer approach is valid only if one accepts the possibility that information can be coded in the brain in terms of molecular structure, that is, on a principle completely different from that of the gramophone record or the oscilloscope tracing.

If this is understood, it is evident that the chemical transfer of learning is just as legitimate a bioassay method as those by which Otto Loewi, Dale and others demonstrated the chemical mediation of synaptic transmission. It is, of course, considerably more complicated and some of the failures are probably due to the many unknown variables some of which may be critical. This situation is not unusual at the early stages of a problem.

Successful experiments have now been published from at least fifteen laboratories and it is not the reality of the phenomenon but only its interpretation that remains in doubt. Are the results due to a general stimulation of learning, to transfer of sensitization, stress or other nonspecific factors or to genuine transfer of information?

It may take some time before these questions can be answered to everybody's satisfaction. Since this conference is devoted to the future of brain research, I should like to say that this approach which may sound heretical today may play an important role tomorrow. As Thomas Huxley said and as shown by many examples in the recent history of science--such as, among others, the chemical transmission of nerve impulses--today's heresy becomes tomorrow's dogma.

TERMINOLOGY

(To be read in a relaxed frame of mind ....
H. Klüver)*

Across the moorlands of the Not
　　We chase the gruesome When;
And hunt the Itness of the What
　　Through forests of the Then.
Into the Inner Consciousness
　　We track the crafty Where;
We spear the Ego tough, and beard
　　The Selfhood in his lair.

With lassos of the brain we catch
　　The Isness of the Was;
And in the copses of the Whence
　　We hear the think bees buzz.
We climb the slippery Whichbark tree
　　To watch the Thusness roll;
And pause betimes in gnostic rimes
　　To woo the Over Soul.

*From "A Nonsense Anthology", collected by
Carolyn Wells, Dover Publications, Inc., 1958,
p. 36, author "Unknown". Reproduced with permission of the publisher.

On the Basis of Disorder
in the Nervous System

A. INBORN ERRORS OF NERVOUS SYSTEM METABOLISM

THE ROLE OF PHENYLALANINE IN MENTAL DEVELOPMENT

David Yi-Yung Hsia

The purpose of the present paper is to introduce the concept that with hyperphenylalanemia the retardation of mental development is an expression of the excessive levels of phenylalanine or its related metabolites rather than a manifestation of the specific disease entity which we call phenylketonuria. We shall present first three pieces of clinical evidence which support this concept and then report on a survey carried out to show the distribution of increased phenylalanine levels when ascertained by newborn screening.

During the past two years, a number of reports have appeared emphasizing the importance of maternal phenylketonuria as a cause for mental deficiency in their non-phenylketonuric offspring (1). We have recently surveyed the entire literature on maternal phenylketonuria and found that these women fall into two distinct groups as shown in Table I. Among those who have plasma phenylalanine levels of 20 mg% or higher and who may be viewed as having "classical" phenylketonuria (2), the majority of their unaffected offspring are either mentally retarded or are of borderline intelligence. On the other hand, among those who have plasma phenylalanine levels of 20 mg% or lower and who may be viewed as having "hyperphenylalanemia" without phenyl-

379

I.Q.	Maternal plasma phenylalanine		
	>20 mg%	<20 mg%	All
70	7	1	8
70 - 79	4	0	4
80 - 89	4	0	4
90	0	8	8
Total	15	9	24

$$X = 33.5; \quad 3 \ d.f.; \quad p < .001$$

Table I. Intelligence quotients of offspring of
maternal phenylketonuria based on plasma phenylalanine levels in mother.

ketonuria (3), all but one of their unaffected offspring have normal intelligence.

Next, we turn to the relation between plasma
phenylalanine and intelligence among patients with
phenylketonuria and hyperphenylalanemia. There is
general agreement that there is no correlation between these two factors upon untreated mentally
retarded phenylketonurics. For instance, in one
study, we did a correlation coefficient between
plasma phenylalanine and intelligence quotient and
found the value to be insignificant ($r = +0.060$;
$SE = 0.013$; $p > 0.10$) (4). However, if one takes
all of the atypical phenylketonurics, that is those
with intelligence quotients of 70 or higher, there
is a significant negative correlation ($r = 0.430$;
$SE = 0.027$; $p < .001$) (5). Furthermore, if one
made a list of all of the patients with an intelli-

1. DIET STARTED < 30 DAYS OF AGE

Age at test (mos.)	I.Q. (NO. PATIENTS)						Total	Mean I.Q.
	60-69	70-79	80-89	90-99	100-109	100-119		
12	-	-	2	-	5	-	7	99
12-23	-	4	1	-	2	2	9	92
24-35	3	-	-	1	3	1	7	87
Total	3	4	3	1	10	2	23	92

II. DIET STARTED > 30 DAYS OF AGE

Age at test (mos.)	40-49	50-59	60-69	70-79	80-89	Total	Mean I.Q.
12-23	1	3	-	2	1	7	64
24-35	-	-	1	1	1	3	75
36+	-	-	1	-	-	1	65
Total	1	3	2	3	2	11	68

Table II. Latest intelligence quotients of typical phenylketonuric infants treated with low-phenylalanine diet.

gence quotient of 100 or better, 7 are found to
have plasma phenylalanine levels of 20 mg%, while
only 2 are found to have plasma phenylalanine levels of 20 mg% or more.

Finally, we can measure the effect of reduction of phenylalanine levels by means of treatment
with low-phenylalanine diet. We have recently had
the opportunity of testing the intelligence quotients of cases of typical phenylketonuria with
plasma phenylalanine levels of 20 mg% or higher
born in 1964, 1965, and 1966 which are being followed by the clinics participating in the collabor-
ative study on phenylketonuria (6). In this survey, the mean I.Q. of untreated patients was 61
for 8 index cases and 49 for 18 affected siblings.
In contrast, the mean I.Q. of treated patients
(regardless of the age of start of treatment) was
82 for 34 index cases and 79 for 3 affected siblings. If the index cases treated with low-phenylalanine diet are further subdivided into those
treated prior to 30 days of age and those started
on that diet after that age up to 5 years, the
effect is even more striking as shown in Table II.
It can be seen that the mean I.Q. of the former
group of 23 children is 92, while that of the
latter 11 children is 68.

The widespread use of screening procedures
among newborn infants has led to the recognition
that not all instances of high blood phenylalanine
are caused by phenylketonuria. Because of this,
we have undertaken a survey of all cases with blood
phenylalanine levels of 6 mg% or higher with normal or decreased tyrosine levels born during 1964,
1965, 1966, and in some instances during the first
half of 1967. In Figure 1, we have tabulated the
maximum phenylalanine levels reached without treatment in each of the index cases. It can be seen
that the sample follows a continuous distribution
without segregating into two or more groups suggesting a single homogeneous population. If we
arbitrarily divides the cases into those with a
maximum phenylalanine value of 20 mg% or less and
call them "hyperphenylalanemia" and the remainder

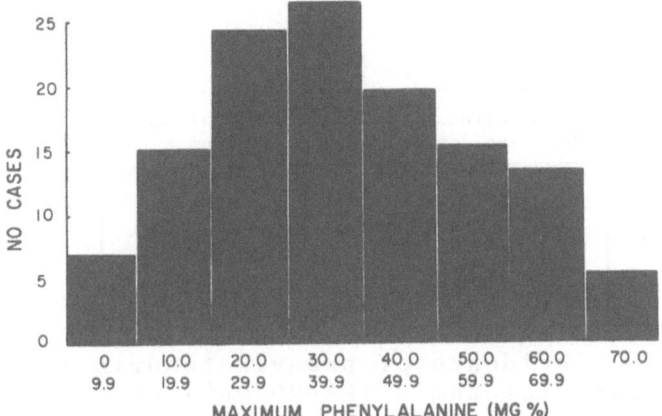

Fig. 1.   Maximum phenylalanine levels of index
cases in survey of all patients with phenylalanine
levels of 6 mg% or higher with normal or decreased
tyrosine levels in six states during 1964-1966.

with a maximum phenylalanine value of 20 mg% or
more and call them "phenylketonuria", we find that
approximately 20 per cent of the patients fall into
the former category. In Table III, we have listed
the number of patients with either phenylketonuria
or hyperphenylalanemia according to years and estimated the incidence of each on the basis of total
births recorded. From these data, we can estimate
the incidence of phenylketonuria to be about 1 in
20,000 and that of hyperphenylalanemia to be 1 in
100,000.

     In Table IV, we have tabulated the distribution
by clinical type and by family size, number of affected, and number of affected expected. Genetic,
or segregation analysis, has been applied to this
data. Only sibships of size two or more can be
groups and the values of expected number of affected sibs are then calculated from the truncated relationship from each sibship size. If one assumes
an autosomal recessive mode of inheritance with a
segregation ratio of 0.25, there is good agreement
between the observed value of 57 and the expected
value of 57.30 in phenylketonuria and between the
observed value of 22 and the expected value of

Year	Total Births	Phenylketonuria No.	Incidence	Hyperphenylalanemia No.	Incidence	All No.	PA Incidence
1964	985,695	35	1:28,161	0	-----	35	1:28,161
1965	925,547	45	1:20,568	8	1:115,093	53	1:17,463
1966	901,936	53	1:17,018	10	1:90,194	63	1:14,316

Table III.   Incidence of phenylketonuria (PA>20 mg%) and hyperphenylalanemia (PA<20 mg%) ascertained in 1964-1966.

Sibship Size	Phenylketonuria No. Fam.	No. Obs.	No. Exp.	Hyperphenylalanemia No. Fam.	No. Obs.	No. Exp.*
1						
2				7	8	8.00
3	7	14	14.70	5	8	6.49
4	9	19	19.88	1	2	11.46
5	2	4	4.65	1	3	1.64
6	2	6	4.92			
7	3	9	7.65			
8	2	5	5.50	1	1	2.23
Total	25	57	57.30	15	22	19.92

Table IV.   Distribution by family size of number of affected children observed and expected for phenylketonuria and hyperphenylalanemia.

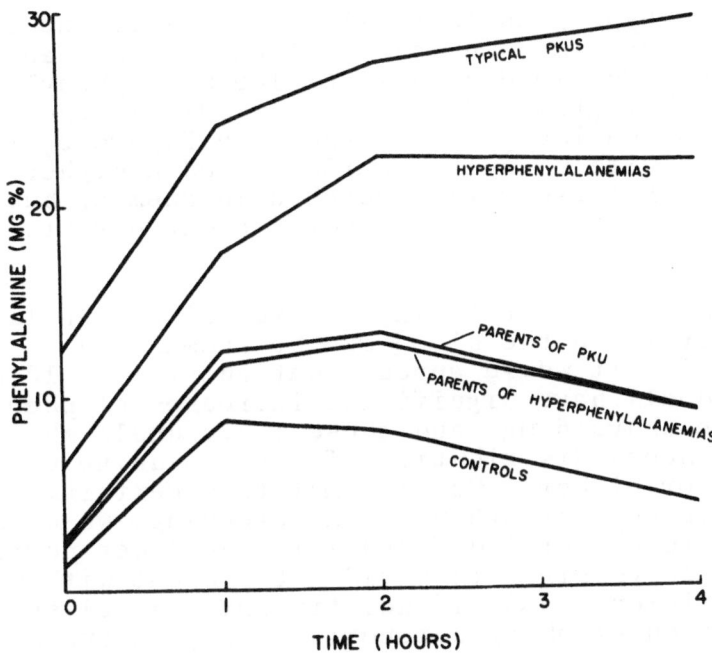

Fig. 2. Plasma phenylalanine levels following 0.1 Gm/kg body weight of L-phenylalanine in phenylketonuria, hyperphenylalanemia, their parents, and controls.

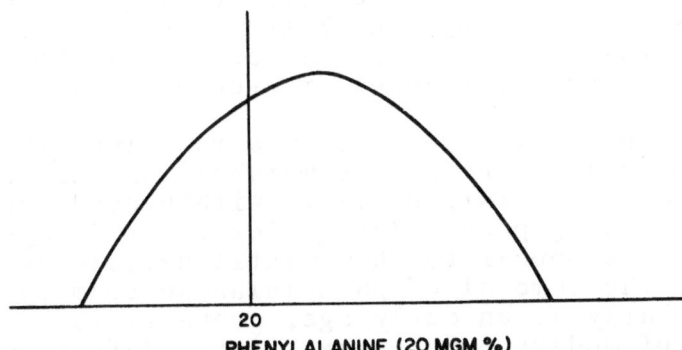

Fig. 3. Diagram illustrating current concept of elevated phenylalanine levels based on complete ascertainment. It is assumed that those below 20 mg% would have been missed prior to newborn screening programs.

19.82 in hyperphenylalanemia. Phenylalanine loading was performed on index patients, sibs, and
parents. The loading dose of 100 mg L-phenylalanine per kilogram body weight was administered in
a polysaccharide gel. As shown in Figure 2, the
patients with phenylketonuria showed a higher
curve than those with hyperphenylalanemia, but
no difference could be detected among parents of
those two groups.

The current thinking on excessive phenylalanine levels in the blood may be summarized in
Figure 3. It would appear that about 1 in 10,000
individuals have significant increases of phenylalanine above 6 mg% and these individuals show
a continuous distribution of phenylalanine up to
80 mg% or higher. Until complete ascertainment
was obtained through newborn screening, we were
unable to detect individuals in the lower portion
of the curve since they did not present with mental deficiency and did not fit into the classical
definition of phenylketonuria. Now, it appears
that the various hyperphenylalanemia syndromes
with phenylalanine levels of 20 mg% or less
represent the lower part of a continuous curve
for phenylalanine. At the present time, there
is no clear evidence that individuals in this
group differ genetically from those with phenylketonuria. However, more detailed segregation
studies with siblings and better biochemical
studies with purified enzyme may reveal the influence of modifier genes or even alleles.

The present studies show a relationship between phenylalanine or its metabolites and mental
development. First, patients with hyperphenylalanemia showing phenylalanine levels of 20 mg% or
less do not appear to show mental deficiency.
Second, the removal of phenylalanine from the diet,
particularly at an early age, prevents the development of moderate to severe mental deficiency in
typical phenylketonuria. Finally, mothers with
phenylalanine levels of 20 mg% or less seldom
show mental deficiency in their unaffected offspring, while those with phenylalanine levels of

20 mg% or more almost invariably do.

In conclusion, we have shown that elevated phenylalanine detected by newborn screening is not synonymous with mental retardation. Rather, there appears to be a quantitative relationship between the level of phenylalanine and mental development. At present, there appears to be no specific indication for treating patients with low phenylalanine levels, but it is not possible to predetermine which infants have typical phenylketonuria and require treatment.

REFERENCES

1. Mabry, C.C., Denniston, J.C. and Coldwell, J.G., Mental retardation in children of phenylketonuric mothers. New Eng. J. Med., 275, 1331, 1966.
2. Jervis, G.A., Phenylpyruvic oligophrenia (phenylketonuria). Res. Publ. Ass. nerv. ment. Dis., 33, 259, 1953.
3. O'Flynn, M.E., Tillman, P. and Hsia, D.Y.Y., Hyperphenylalanemia without phenylketonuria, Am. J. Dis. Child., 113, 22, 1967.
4. Hsia, D.Y.Y., Knox, W.E., Quinn, K.V. and Paine, R.S., A one year controlled study of the effect of low-phenylalanine diet on phenylketonuria. Pediatrics, 21, 178, 1958.
5. Hsia, D.Y.Y., O'Flynn, M.E. and Berman, J.L., The atypical phenylketonuric with borderline and normal intelligence. Am. J. Dis. Child. In press.
6. Berman, J.L., Cunningham, G.C., Day, R.W. and Hsia, D.Y.Y., Causes for high phenylalanine levels in newborn, Abstracts Am. Ped. Soc., Atlantic City, N.J., May, 1968.

SHOULD LOW PHENYLALANINE DIET BE GIVEN DURING
PREGNANCY?   EFFICACY OF THE DIET.

DR. AUSTIN: In regard to Dr. Hsia's interesting paper, he has shown us that mothers who
have plasma phenylalanine levels above 20 milli-
grams per cent have children who are more markedly
retarded.

In contrast, he has shown us that mothers who
have plasma phenylalanine levels under 20 milli-
grams per cent have children who are minimally retarded.

My first question is, were these phenylalanine
levels drawn on these mothers before, during, or
after their pregnancy? Secondly, at what age in
life were the children from these pregnancies tested for their IQ's?

The reason for these questions is that there
is a therapeutic implication in his findings which
leads me to the third question:

Is there any possibility that mothers who had
elevated phenylalanine levels discovered during
pregnancy might be treated with low phenylalanine
diet while pregnant - thus helping to avoid some
degree of mental retardation in their children as
yet unborn?

I realize this may be reaching a bit beyond
the present data, but the basic question is were
these phenylalanine levels determined during the
pregnancy of the same children who had the IQ
tests?

DR. HSIA: I am very grateful to Dr. Austin
for having brought up this area in which I certainly feel we do not know all the answers. We do know
this much. If you take a heterozygous woman and
she carries a phenylketonuric child, the serum
phenylalanine level during that pregnancy does not
become elevated. We have never actually encountered

any instances of a phenylketonuric mother going
through a pregnancy and we do not know what happens
to the blood levels during such a time.

The question may be raised as to whether we
should treat a phenylketonuric mother during pregnancy with a diet low in phenylalanine content.
There is some question in the minds of people as to
whether deficiency of phenylalanine during the pregnancy may result in abortion of the fetus. This is
something which nobody really knows, but something
which we certainly feel appears to justify a clinical trial. I am certainly intrigued by the fact
that the high blood phenylalanine level during pregnancy and during term may have some influence upon
the development of the brain of the fetus in utero
and certainly there is a very clear indication that
the degree of mental deficiency in the offspring of
these women who have maternal phenylketonuria is
more severe than the newborn infant who develops
phenylketonuria after birth. We also have very
good data to show that if you discontinue the diet
even as early as one year of age, it probably does
not have very much effect on the subsequent mental
development of the fetus, so that it is really all
in this period immediately before birth and immediately after birth that I think we are most concerned about.

DR. GAULL: Dr. Hsia gave a very good summary
of the history of phenylketonuria, and I think that
his data were very interesting. I have drawn different conclusions from his data than he has, and I
would like to discuss them and get the reaction of
both Dr. Hsia and the audience.

As all of you know, there is a controversy
raging now as to whether the treatment of phenylketonuria with a low phenylalanine diet for the
prevention of permanent brain damage is efficacious.

When one looks at the data in the literature,
one finds few series that have been reported with
sufficient precision to be able to draw any conclusions whatsoever from them.

Dr. Hsia's series differs from these in that it is reported with a great deal of precision and some conclusions can be drawn. Only I would differ in the conclusions drawn.

For instance, on the slide concerned with maternal phenylketonuria, there were 24 patients culled from the literature. Twelve of these patients had IQ's between 80 and 100 and, of these twelve, eight of them had IQ's between 90 and 100.

If the increased concentration of phenylalanine in the blood, and this level alone, were the cause of the dementia of this disease, it is a little difficult to understand how 12 of 24 mothers with phenylketonuria managed to have normal IQ's in the face of this elevated concentration of phenylalanine.

When you speak of IQ, you also should consider what IQ means. This quotient was first used by Binet as a way of evaluating Paris children for school placement. It was not a measure of intellectual potential but a measure of actual attainment.

I want to make this clear, because I think it enters into the interpretation of Dr. Hsia's data.

If you review the series that have been presented, there are a number of logical fallacies that are easily discerned.

One of the main fallacies in a number of the series is that the cases are ascertained in different ways.

Some of them are ascertained by mental retardation, therefore, they are older when they are ascertained. Some of them are ascertained by screening, therefore, they are started on treatment earlier. The fact, therefore, that those started on treatment earlier do better, may be an artifact of the method of selection.

This series doen't suffer from that fallacy because all of them were ascertained by screening. However, there is one fallacy that seems to me to be inherent in the data that has been presented, and that is the fallacy that the comparison of the untreated siblings to the treated siblings is a valid comparison. I want to tell the biochemists here a little about the treatment of phenylketonuria in this country. It is more than just feeding these children a diet containing hydrolyzed protein from which phenylalanine has been removed.

These children, ascertained by screening, are sent to an "approved center". At the "approved center" they are treated by, not only an astute clinician and biochemist like Dr. Hsia, but a so-called "multi-disciplinary team" which includes nutritionists, visiting nurses, social workers, psychologists. They are tested and retested. And the mother is assured from the very beginning that she doesn't have to worry about this child, like she has had to worry about the previous idiot sibling.

This child is going to have all of the benefits of this diet. She has even heard it on television!

The mother then puts in an enormous amount of work into parceling out the amount of phenylalanine this child has and stimulates this child in a way she never stimulated the poor idiot she had two years ago. Therefore, this child grows up in a completely different environment. A different environment in which one of the factors is the use of this diet, low in phenylalanine.

I am not prepared to say that treatment with low phenylalanine diet is not efficacious. I am only saying that the evidence that treatment with low-phenylalanine diet is salutary does not measure up to the usual criteria for credibility that we apply in science.

I would also like to point out that 20 per cent of this series by Dr. Hsia had hyperphenyl-

alaninemia have been treated but not properly diagnosed. The most valuable thing that has come out
of Dr. Hsia's series is that it doen't matter
whether you treat these patients or not. They go
on to do very well. If you are not as astute as
Dr. Hsia is, you would consider this 20% a great·
therapeutic triumph. Certainly it has biased the
results of series in which they were not carefully
distinguished.

Also, if you just look at the data of the
hyperphenylalaninemics, I don't see how one can
draw the conclusion that having a high concentration of phenylalanine in the blood, per se, has
any necessary causal connection with permanent
brain damage.

I would also like to point out the relevance
to this problem of the beautiful work done by Dr.
Lajtha in which the difference between the concentration and the flux of metabolites was so elo-
quently pointed out. In clinical medicine, we
measure only concentration but maybe what we ought
to be looking at is the flux of metabolites, and
we would draw different conclusions from our data.

DR. HSIA: Dr. Bessman has proposed that the
dietary treatment of phenylketonuria has no effect
on the mental development of the patients with
this disease because: (1) There is a difference
between the patients ascertained by newborn screening based on serum phenylalanine levels and those
ascertained by I.Q. among the mentally deficient,
and (2) The improvement of I.Q.'s observed reflect
the inclusion of a significant population of patients who would have had normal development without dietary treatment anyhow.

The studies which I have presented confirmed
the first of Dr. Bessman's proposals. In our study
approximately one-third of all patients with elevated phenylalanine levels showed serum phenyal-
anine levels less than 20 mg% and fell into the
syndrome of hyperphenylalanemia. As far as can be
ascertained, these patients show mental development

with or without treatment with the low phenylalanine diet. Furthermore, among the remaining two-thirds of the patients with classical phenylketonuria, approximately 20 per cent will show mental development even without treatment with diet. Thus, approximately 43 per cent of all of the patients with elevated phenylalanine levels picked up by newborn screening programs will show normal mental development without treatment.

On the other hand, we have also found that among the patients with classical phenylketonuria there is a clear difference in the I.Q.'s among those treated and not treated by the low phenylalanine diet. Among the untreated individuals, the average I.Q. was 67 while among those treated the average I.Q. was 97. This difference in I.Q. remained regardless of the age when the child was tested.

Thus, it is our current feeling that patients with classical phenylketonuria should continue to be treated with the low phenylalanine diet starting at an early age because of the obvious benefits derived in such patients as compared with their untreated siblings.

The question as to how are phenylketonuric women found and what are their I.Q.'s. Most of the phenylketonuric mothers are married and lead fairly normal lives. They have relatively normal I.Q.'s and can reproduce in a normal way. This certainly makes the sampling biased since the more retarded phenylketonuric women are obviously not going to have children.

Perhaps one way to approach the problem is to take all cases of phenylketonuria with an I.Q. of over 70. In this group approximately 50 per cent of them will have phenylalanine levels below 20 mg%. If you then take the same population and look for women who are pregnant and have children without phenylketonuria, you will find that the offspring of women with low serum phenylalanine levels tend to have normal children and those with high serum phenylalanine levels tend to have retarded children.

DR. SEIFTER: In the formulation of concepts in Dr. Hsia's paper about phenylketonuria and in the preparation of ameliorative diets I wonder whether enough emphasis has been placed on the observed decreased tyrosine levels and on the balance between tyrosine and phenylalanine in the body fluids and in the provided diet.   A decrease in tyrosine levels could result in diminished protein biosynthesis.   Indeed Dr. Hsia did state that tyrosine levels in patients with phenylketonuria were decreased.   I wonder whether we have yet determined the proper balance of phenylalanine and tyrosine in relation to the effects of this clinical condition, or indeed in regard to the formulation of diets for treatment.

DR. STABENAU: I just wanted to say something regarding the aspect of stimulation.

It has already been mentioned in the animal studies how care must be taken to evaluate handling in trying to estimate the variables influencing behavior and learning.   In the phenylketonuric children relative degrees of stimulation by the parents may be of importance.   The only body of data that I know of that begins to consider this point, not regarding retardates but normal children, is a report by Dr. Robert Rosenthal of Harvard. His group tested children in a San Francisco school, grades 1 through 6.   He randomly selected 20% and told their teachers that these individuals had unusual potential for intellectual gain.   When tested 8 months later these children had significant IQ changes in the range of 10 to 30 IQ points over their controls so expectation and stimulation does appear to have its effect.

RECENT STUDIES IN

TWO INBORN ERRORS OF GLYCOLIPID METABOLISM

James H. Austin

INTRODUCTION

Our topic for this conference is inborn errors
of glycolipid metabolism. We are specifically invited at this conference to peer into the future
developments in this field. We can make no attempt
to cover the field as a whole. To recite the chemical names alone of all of the various glycolipids
would consume more time than is at our disposal.
Therefore, the focus of this report will be on sulfated lipids. Several recent reviews of the subject
are now available (1-5).

Sulfated lipids (sulfatides) are distinctive
glycolipids. Like other molecules they are first
synthesized and later broken down. We will present evidence suggesting that important abnormalities involving sulfatides occur in two diseases.
One disease appears to be associated with a disorder in the construction of sulfatides; the other
is associated with an abnormal breakdown of sulfatides. We may expect that certain basic principles
exemplified by these two diseases will be applicable
to other diseases involving other glycolipid molecules.

SULFATED GLYCOLIPIDS (SULFATIDES)

In simple terms, a glycolipid is a lipid which contains a sugar moiety. A sulfatide is a sulfated glycolipid. It contains a fatty acid (in this case, cerebronic acid) which is linked with a base (usually sphingosine). This double molecule (termed a ceramide) is in turn attached to a sugar molecule, in this case, galactose) (Figure 1).

The term sulfatide is derived from the presence of a sulphuric acid ester group. This sulfate group is attached to the third carbon of the galactose. Lacking this sulfate group, the glycolipid molecule would be called a cerebroside, (ceramide monohexoside).

A CEREBROSIDE SULFURIC ACID ESTER
(SULFATIDE)

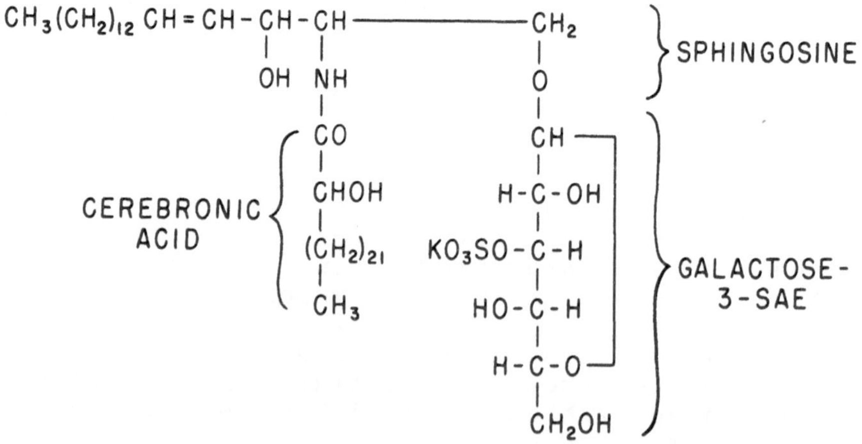

$$CH_3(CH_2)_{12} \ CH = CH - CH - CH \text{————————} CH_2 \quad \Big\} \text{ SPHINGOSINE}$$

$$
\begin{array}{c}
\text{CEREBRONIC} \\
\text{ACID}
\end{array}
\left\{
\begin{array}{l}
CO \\
CHOH \\
(CH_2)_{21} \\
CH_3
\end{array}
\right.
\qquad
\begin{array}{l}
CH \\
H-C-OH \\
KO_3SO-C-H \\
HO-C-H \\
H-C-O \\
CH_2OH
\end{array}
\left.
\begin{array}{l}
\\
\\
\\
\\
\\
\\
\end{array}
\right\}
\begin{array}{l}
\text{GALACTOSE-} \\
\text{3-SAE}
\end{array}
$$

Fig. 1. Structural formula of cerebroside sulfate, the prototype sulfatide. In the less common dihexose sulfatide, a glucose moiety is interposed, linked at one end with sphingosine and at the other end with galactose sulfate.

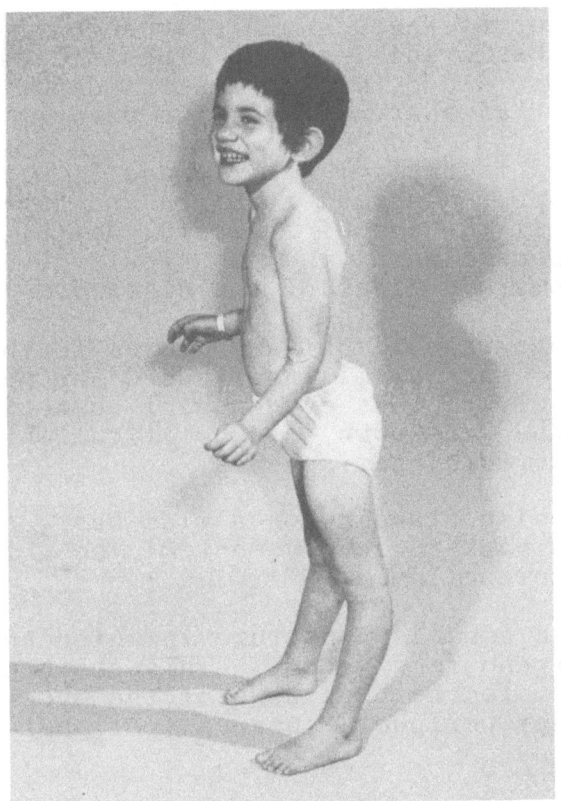

Fig. 2. An MLD patient, age five, after two and
one-half years of progressive symptoms. Note back-
ward bowing at the knees, the wide stance, and the
         smile, consistent with euphoria.

A DISEASE OF SULFATIDE CATABOLISM (MLD)

One disease of sulfatide metabolism is caused
by an abnormally slow breakdown of sulfatides.
This disease is also termed metachromatic leukodystrophy, or MLD (1-4). Leuko means white; dystrophy
means a wasting away or an attrition. Hence, leukodystrophy means a wasting away of white matter.
Metachromatic is a term used to describe a distinctive histochemical staining reaction of this white
matter. In this reaction the white matter turns

red to an unusual degree when placed in contact
with a blue basic aniline dye. This blue dye
changes color, hence, metachromatic--denoting a
color other than that of the stain used.

Figure 2 is that of a girl, age 5, with MLD.
The child appeared normal at birth and developed
normally until the age of 3 years. At this time,
her speech became nasal and she slowly developed
other symptoms and signs of the disease.

The photograph illustrates three features of
her disease. She is smiling (indeed she was smiling
inappropriately). This evidence of euphoria is referable to the breakdown of subcortical white matter in her cerebral hemispheres.

She is also standing on a wide base, a sign
referable in part to involvement of the white matter of her cerebellum.

She also assumes a posture in which her knees
curve back (genu recurvatum). This posture is related in part to the breakdown of the myelin
sheaths in her peripheral nerves (peripheral neuro-
pathy).

Briefly to summarize a good deal of accumulated evidence, one may say that: (1) histochemical
stains of the areas of abnormal white matter in MLD
show not only myelin breakdown, but also many discrete metachromatic clumps. These clumps indicate
a local excess of highly metachromatic lipids. (2)
chemical studies of this diseased white matter show
that sulfatides are increased, whereas all the
other typical myelin lipids are reduced. (3) tests
with pure lipids indicate that a sulfatide molecule
is highly metachromatic by virtue of its negatively-
charged sulfate group. Hence, the clumps of excess
metachromasia in MLD are attributable to a local
excess of sulfatides.

Included among the general questions which remain are the following:
(1) What is the mechanism of the increase in

sulfatides?
(2) How does the sulfatide increase affect
the glial cell?
(3) How can the increase in sulfatides be
correlated with the breakdown of myelin?

We may now be in a position to venture some
preliminary, oversimplified, answers to these questions (1). We may predict that the issues will
prove more complex in their final analysis.

WHAT ENZYME DEFICIENCY UNDERLIES THE INCREASE IN SULFATIDES IN MLD?

to help split off sulfate groups. The reaction
with respect to sulfatides may be summarized as
follows (7):

$$\text{Cerebroside-SO}_4 \xrightarrow{\substack{\text{Sulfatase A;} \\ \text{Cerebroside Sulfatase}}} \text{Cerebroside} + \text{SO}_4$$

It is reasonable to postulate that a sulfated lipid
might accumulate if sulfatase activity were deficient.

We have looked for the enzyme that normally
does split sulfatide. We have used a chromogenic
substrate (nitrocatechol sulfate). We find that
the activity of this enzyme (sulfatase A, cerebroside sulfatase) is markedly deficient in MLD (7,8).
Other enzymes are within normal limits. In particular, the activity of a different sulfatase,
sulfatase B, appears within the range of normal
in typical MLD.

HOW DOES THIS SULFATASE A DEFICIENCY AFFECT METABOLIC EVENTS INSIDE A GLIA CELL?

Glial cells are responsible for the formation
and the maintenance of the myelin sheath. Sulfatase A molecules are first made in relation to
the rough endoplasmic reticulum. They next pass
into the smooth reticulum and Golgi region. Sub-

sequently, sulfatase A passes into the digestive organelles of the lysosomal system (9). Normally, these digestive organelles appear to keep lipids (and other molecules) from reaching excessive levels. Hence, a lack of sulfatase A would impair the normal breakdown of sulfated lipids inside lysosomes. We would then expect to find that sulfatides spill out into the cell cytoplasm in excessive amounts. This accumulation of sulfatide inside cells of many organs is called a sulfatide lipidosis. Somehow this causes a breakdown of various membranes in which sulfatides are normal constituents. The question is how.

HOW DOES THE INCREASE OF SULFATIDES CAUSE AN ABNORMALITY OF THE MYELIN SHEATH?

In one of the early models of the myelin sheath proposed by Finean, we note that the axon is surrounded by many layers of lipids and proteins arranged like a jelly-roll (Figure 3) (10). To the degree that the molecules in these layers are present in normal amounts and normal proportions then one might expect these myelin lamellae to have a normal structural stability. Glycolipids are perhaps the most distinctive molecules of myelin. Sulfatides account for about 20% of the glycolipid molecules in the normal myelin sheath. However, sulfatide molecules may be much more important than their numbers indicate. To understand the basis for this reasoning, it may be recalled that the sulfate group confers a highly negative charge. There are cohesive forces which tend to bind this negative group with positive groups on adjacent molecules (11). Indeed, the strong metachromasia of isolated sulfatides (6), and of myelin sheaths, suggests that these unique glycolipids might function in part as polar keystones of the myelin sheath.

O'Brien and Sampson have isolated a myelin fraction from MLD patients (12). The "MLD myelin" contained an excess of sulfatides and a deficiency of cerebrosides. Sulfatides are increased to about three times the normal level. Correspondingly, cerebrosides are decreased to about one-third the

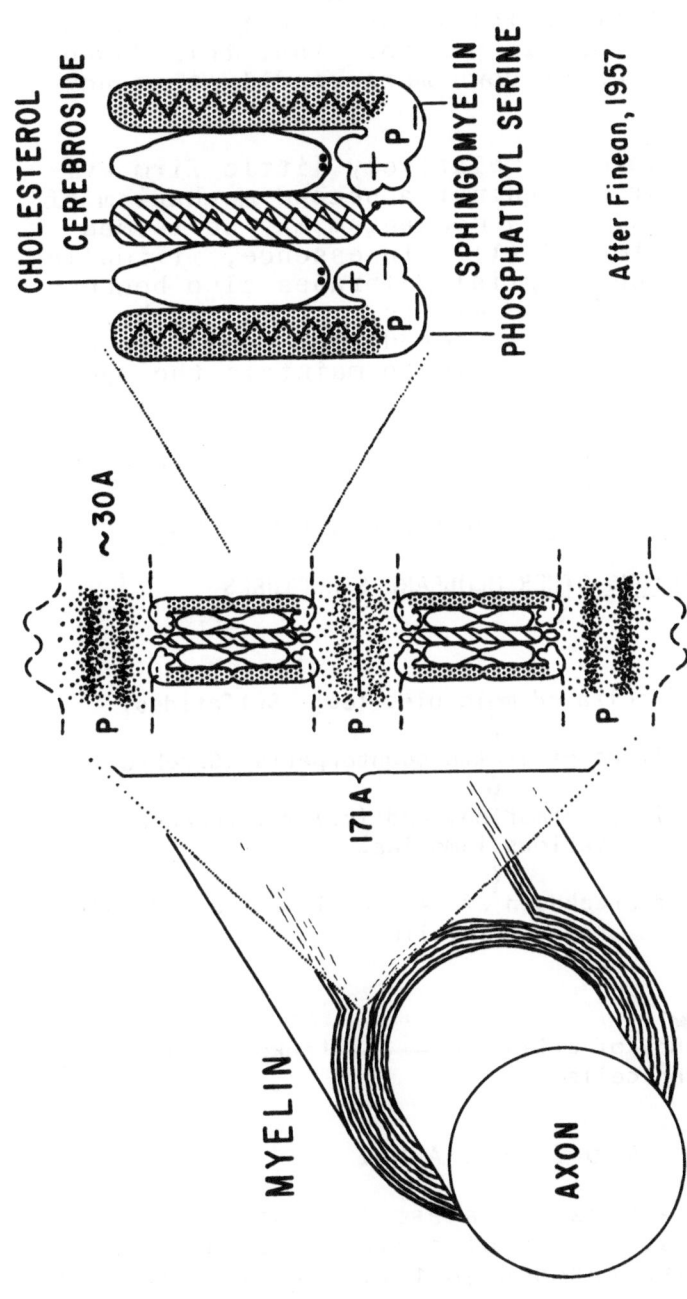

Fig. 3. An early schematic attempt to represent the ultrastructural arrangement of molecules in myelin of peripheral nerve(after Finean, 1957)(10). Note that the lipids may exist in certain architectural arrangements and in certain proportional relationships in myelin lamellae. Although not shown here, sulfatide molecules also occur in the bimolecular lipid leaflet which makes up each myelin layer. One sulfatide molecule is present for every three to eight cerebroside molecules in peripheral nerve (1).

normal level. Thus, MLD myelin has a rather remarkable inverse abnormal relationship between these
two glycolipids. It is possible that this "abnormal myelin" may break down more readily than normal myelin.

As yet, there is relatively little firm evidence either for or against a second mechanism of
myelin breakdown. This mechanism may be termed
"glial insufficiency" (1). In essence, it implies
that the function of glial membranes also becomes
impaired by a sulfatide excess. It may be hypothesized that these disordered glia would gradually
become metabolically unable to maintain the myelin

Myelin and glial mechanisms of leukodystrophy
are summarized in Table I:

PATHOGENIC SEQUENCES IN MYELIN LAMELLAE AND

IN VARIOUS OTHER MEMBRANE STRUCTURES

Excess of sulfated molecules (eg: Sulfatides)
↓
Deficit of their non-sulfated counterparts (Cerebrosides)
↓
Improper function, formation, and (or) maintenance of
various lamellae.
↓
Dissolution and breakdown of various lamellae, including
those of myelin

Insufficiency of Glia ⟶ Leukodystrophy.
and Schwann cells.

MLD IN THE ADULT

If one measures sulfatase A activity in the
rat, the highest levels of sulfatase A activity in
rat cerebrum are reached just at the peak of myelin-

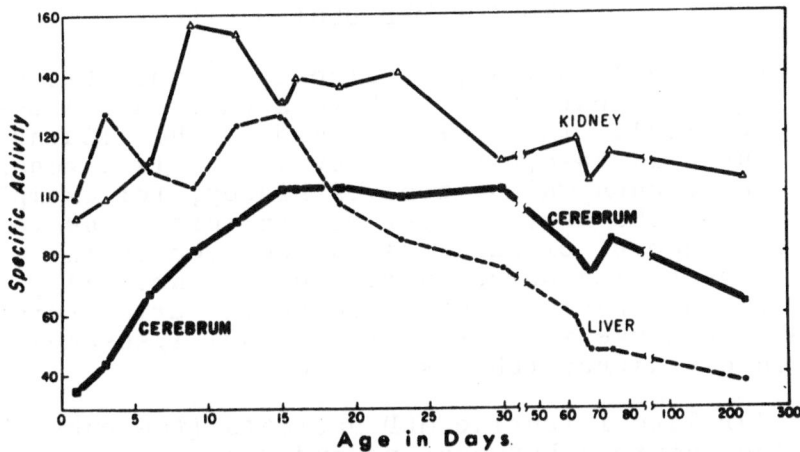

Fig. 4.   Changes in Sulfatase A with age in the
rat.   Note that the rat develops the highest levels
of sulfatase A activity in cerebrum between the age
of fifteen and nineteen days.   (At about this time,
the rate of sulfatide synthesis also reaches its
peak.)   Note, thereafter, that sulfatase A levels
are maintained throughout advanced age in normal
                    rat cerebrum.

ation (Figure 4).   Interestingly, rather high levels of sulfatase A are still found thereafter.
These levels persist into senescense.   In the human,
sulfatase A levels also persist into advanced age.
Perhaps the persistence of the enzyme reflects the
adults' metabolic need for the enzyme.

        It is of particular interest that a form of
MLD exists which may begin even in the twenties
and thirties (13).   We recently studied such a
patient, aged 62, who first developed signs of his
disease at about the age of 26.   The disease has
existed in clinically overt form for over 35 years.
The findings in this MLD patient are almost entirely restricted to a profound disorder of mentation
Despite his severe organic mental defect, he has
very few findings consistent with peripheral neuro-
pathy or cerebellar involvement.   He also has a
sulfatase A deficiency.

TESTS FOR MLD DURING LIFE

A recently-developed urine test indicates whether or not a patient does have a sulfatase A deficiency (14). The test is based on the finding that the sulfatase A deficiency of MLD is a generalized phenomenon which involves kidney, for example, as well as the nervous system. In this urine test, a normal amount of sulfatase A secreted in the urine gives a rich burgundy red color. Normally, some of the enzyme is excreted in the urine--presumably because it leaks out of renal lysosomes and enters kidney tubules.

Our late infantile MLD patients (the ones who develop symptoms between one and two years of age) have an almost complete absence of enzyme activity in their urine. Of special interest is the finding that the adult MLD patient still has a definite trace of enzyme activity (13). It may be, therefore, that the adult with MLD had his illness clinically delayed until the age of 30 because he was never completely lacking in sulfatase A, but was, instead, only moderately deficient in the enzyme. One may wonder how many adults with MLD exist in the back wards of our mental hospitals, still ambulatory, but slowly evolving their neurological signs.

The sulfatase A deficiency in MLD children also involves leukocytes (15). Adult heterozygote carriers of the disease do not appear to have lower than normal values.

ANTICIPATING FUTURE PROBLEMS OF CLASSIFICATION

There are a number of recent reports in which the information presented does not permit one to be certain that the diagnosis was, in fact, MLD (3,14, 16,17).

Confusion may increase if the term MLD "variant is now used to cover these indefinite borderline cases. The reason is that the term "MLD variant" has previously been used in a more specific

sense (3,18,19). Children with the MLD variant
have the sulfatide lipidosis plus a sulfated muco-
polysaccharidosis, plus a storage of gangliosides
in ganglion cells of cerebral cortex. The sulfated
mucopolysaccharide has certain resemblances to he-
paratin sulfate. These children have a distinctive
granulation abnormality in their leukocytes. They
have a deficiency not only of sulfatase A, but also
of sulfatase B, and sulfatase C (a tri-sulfatase
deficiency).

It is suggested that the term MLD variant be
reserved for this form of the disorder. In so doing, the essential features of the illness are
still regarded as prerequisites for the diagnosis.
Moreover, the additional features which are superimposed are documented chemically and enzymically.

GLOBOID LEUKODYSTROPHY (GLD; KRABBE'S DISEASE)

In MLD we are clearly dealing with a defect
of sulfatide breakdown. We turn next to a disorder
which appears to be associated with a defect in sulfatide formation. When a cerebroside precursor
comes in contact with a sulfate donor called "active sulfate" in the presence of the enzyme, cerebroside sulfotransferase, the sulfate group is
transferred to the cerebroside. The resulting
molecule is cerebroside sulfate (sulfatide). These
normal steps of sulfatide synthesis may be simplified and diagrammed as follows:

$$\text{cerebroside} + \text{PAP-SO}_4 \xrightarrow{\text{cerebroside sulfotransferase}} \text{cerebroside-SO}_4 + \text{PAP}$$

Recent studies suggest that globoid leukodystrophy
is associated with impaired conjugation of sulfate
to its cerebroside acceptor (20,21).

Figure 5 shows patient S.O. who was normal at
birth and essentially normal until the age of four
months. Gradually over the course of the next four
months she developed fretfulness, rigidity of the
arms and legs, optic atrophy, and extensor plantar

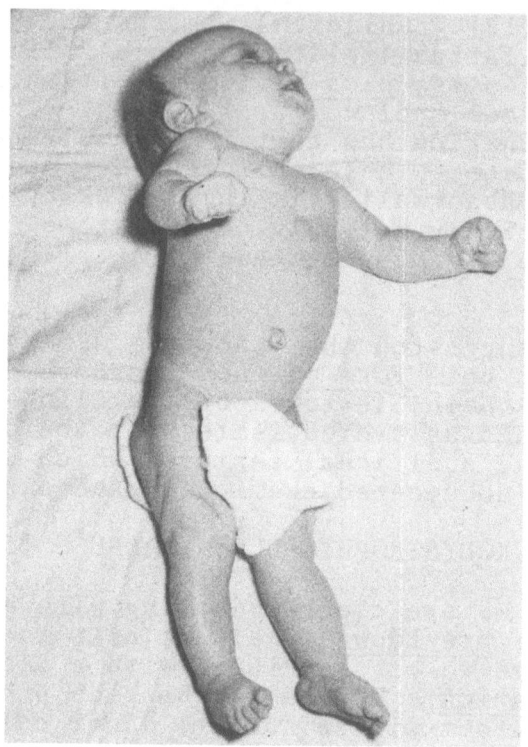

Fig. 5.   Early sustained decorticate posture in infantile Krabbe's disease.   Patient S. O. at five
months.

responses co-existing with ± deep tendon reflexes.
She died at twelve months of age. Histological
studies of white matter revealed widespread myelin
loss plus many large distended cells.   These cells
appeared stuffed with something.   Many had multiple
nuclei (5).   These multinuclear globular cells,
termed globoid bodies, are the hallmark of Krabbe's
disease   The sister, C. O., of the patient above
appeared more fretful than normal as early as two
months of age.   At three months, brief stiffenings
were noted.   By four months, her extremities were
unusually stiff.   At eleven months, she had a
slightly small head, profound dementia, bilateral

optic atrophy, rigidity of the arms and legs, and
bilateral extensor plantar responses, (but no deep
tendon reflexes). There was a slight response
to pin prick. She died at the age of one year
eleven months (Figure 6). Histological findings
were similar to those in her sister, but globoid
bodies were less frequent. Another older sister,
T. O., is normal.

If one extracts lipids from the abnormal white
matter in GLD and examines these by thin layer
chromatography, one finds that very few "myelin
lipids" still remain. Low levels of cerebrosides
are present, and low levels of sphingomyelins are
present. However, the greatest reduction is in
sulfatides (Figure 7). In the material we have
studied sulfatides are either markedly reduced or
completely absent. GLD thus affords a noteworthy

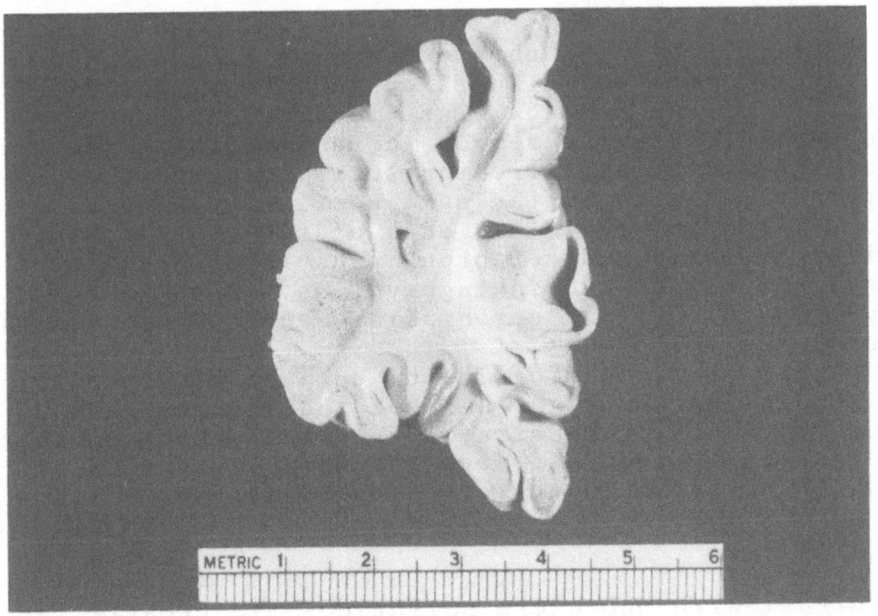

Fig. 6. Formalin-fixed frontal lobe of patient C.
O. Note that the area normally occupied by white
matter is shrunken and grey. (It lacks the white
appearance expected of normal myelin when it is
present.)

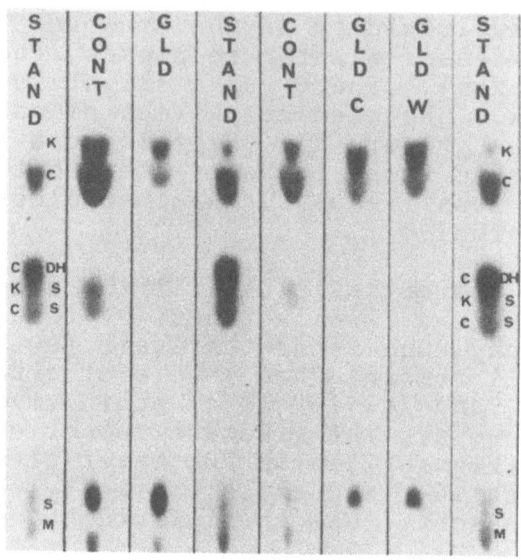

Fig. 7. Thin-layer chromatogram of sphingolipids
extracted from formalin-fixed white matter. At
left, from the top down, are the standard compounds: kerasin, cerebron, ceramide dihexoside,
kerasin sulfatide, cerebron sulfatide, and the two
sphingomyelins. In the next vertical channel, the
normal control child has each of the two cerebrosides, kerasin and cerebron, plus their counterpart sulfatides and sphingomyelins. The GLD child
(patient C. O.) has cerebrosides and sphingomyelins,
but lacks sulfatides.
     Sphingolipid profiles of dog white matter are
also seen in the right half of the figure. The
dogs with GLD (C = Cairn; W = Westie) have somewhat
  less sulfatide than does the normal dog (CONT.).
   Chlorox-benzidine method on Silica Gel G.

contrast with MLD where sulfatides are greatly increased.

     If one injects cerebrosides experimentally in-
to the rat, one finds that phagocytes develop many
nuclei when they are next to a very small concen-

tration of cerebroside (5). They also become PAS
positive. This evidence suggests that small local
concentrations of cerebrosides may stimulate a glo-
boid-like response.

The problem, then, is to find a primary enzymic
defect so positioned that it could cause: a) a
generalized reduction of sulfatides, b) a less profound, but still marked reduction of cerebrosides,
c) discrete pockets where cerebrosides are locally
increased. We have approached this problem by ana-
lysing white matter, cortical grey matter, and kidney for cerebroside sulfotransferase activity.

Three GLD patients, including C.O., have now
shown a marked deficiency of this enzyme system (20,
21). Other mechanisms exist which might also cause
an impairment in sulfate conjugation, and these are
now under investigation. Early in life a deficit
of sulfatide, of whatever cause, would seem to lead to
to structurally unstable myelin.

CANINE GLOBOID LEUKODYSTROPHY

A form of globoid leukodystrophy also affects
Westhighland Terriers and Cairn Terriers (5). Like
the human disease, canine GLD also appears to be
transmitted as an autosomal recessive. Four puppies
were affected in this pedigree of seventeen pups
(Figure 8) (22).

Histologically, the findings are similar to
those in the human, but globoid elements are less
frequent. There is also a similar but less striking reduction in sulfatides (Figure 7) (22). It
is too soon to say whether or not these animals
have the same enzyme defect as that noted in the
human. Clearly, however, one must search for and
preserve these special strains of animals which
seem to have a relevant genetically-determined enzyme deficiency. These special strains may provide
us with laboratory models uniquely suited for investigations of human disease.

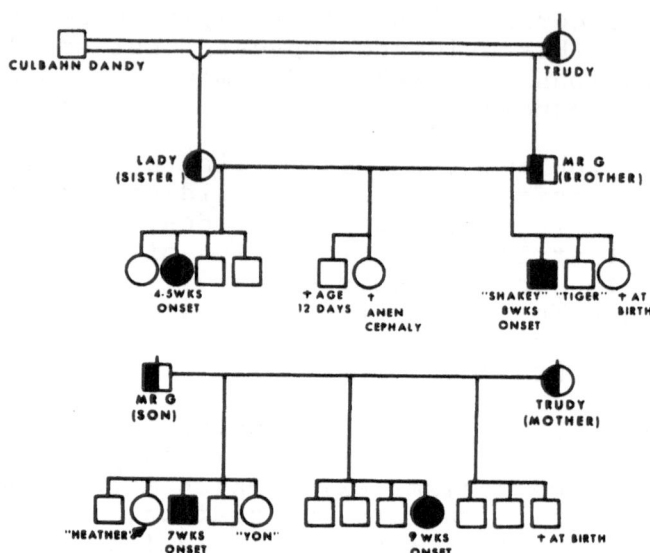

Fig. 8. Pedigree of Canine GLD in a Westhighland Terrier Strain. The mother (Trudy) had had two carrier puppies (Lady and Mr. G.). When inbred, these two clin cally-normal siblings have had puppies who developed GLD. Mother-son crossings (Trudy - Mr. G.) have also yielded GLD offspring.

SOMETHING OF THE FUTURE OF GLYCOLIPIDS

The purpose of our conference is to look into the future. From the foregoing examples, one may make certain predictions:

(1) There may ultimately emerge a separate enzyme deficiency associated with the synthesis, breakdown, or transformation of each major glycolipid.

(2) Diseases in which the enzyme deficiency is less marked may have a later onset and may run a more protracted course.

(3) There should be less confusion in classifying diseases when the characteristic enzyme

and substrate errors are regarded as essential
prerequisites for the diagnosis.

(4) Precise animal models of human disease
will play an increasingly important role in
investigations, both in clarifying the mechanisms of disease and in exploring newer methods of therapy.

REFERENCES

1. Austin, J., Some Mechanisms of Disease in Meta-
    chromatic Leukodystrophy, Chap. 47 in Shy, G.
    M., Goldensohn, E. and Appel, S., The Cellu-
    lar and Molecular Basis of Neurologic Disease,
    1969. In press.
2. Moser, H. and Lees, M., Sulfatide Lipidosis:
    Metachromatic Leukodystrophy, in Stanbury, J.,
    Wyngaarden, J. and Fredrickson, D., (ed.), The
    Metabolic Basis of Inherited Disease (2nd edi-
    tion), New York, McGraw-Hill, Inc., 539-564,
    1966.
3. Austin, J., Metachromatic Leukodystrophy, in
    Carter, C. (ed.), Medical Aspects of Mental
    Retardation, Springfield, Ill., Charles C.
    Thomas, Publisher, 768-812, 1965.
4. Kahlke, W., Metachromatic Leucodystrophy, in
    Schettler, G. (ed.), Lipids and Lipidoses,
    New York, Springer-Verlag, Inc., 310-331,
    1967.
5. Austin, J., Globoid (Krabbe) Leukodystrophy,
    in Minckler, J. and Neuberger, K. (ed.), Neuro-
    pathology, New York, McGraw-Hill, Blakiston,
    Chapter 119, 1968.
6. Austin, J., Observations in Metachromatic Leu-
    coencephalopathy, Tr. Am. Neurol. Assn., 83,
    149-152, 1958.
7. Mehl, E. and Jatzkewitz, H., Cerebroside 3-
    Sulfate as a Physiological Substrate of Aryl-
    sulfatase A, Biochimica Et Biophysica Acta,
    151, 619-627, 1968.
8. Austin, J., McAfee, D., Armstrong, D.,
    O'Rourke, M., Shearer, L. and Bachhawat, B.,
    Abnormal Sulphatase Activities in Two Human
    Diseases (Metachromatic Leucodystrophy and

Gargoylism), Biochem. J., 93, 15c-17c, 1964.

9. Novikoff, A., Lysosomes in Nerve Cells, in
    Hydén, H. (ed.), The Neuron, Amsterdam, Else-
    vier, 319-377, 1967.

10. Vandenheuvel, F., Study of Biological Struc-
    ture at the Molecular Level with Stereomodel
    Projections: I. The Lipids in the Myelin
    Sheath of Nerve, J. Amer. Oil Chem. Soc., 40,
    455-471, 1963.

11. Abramson, M. and Katzman, R., Ionic Interac-
    tion of Sulfatide with Choline Lipids, Science,
    161, 576, 1968.

12. O'Brien, J. and Sampson, E., Myelin Membranes:
    A Molecular Abnormality, Science, 150, 1613-
    14, 1965.

13. Austin, J., Armstrong, D., Fouch, S., Mitchell,
    C., Stumpf, D., Shearer, L. and Briner, O.,
    Metachromatic Leukodystrophy (MLD) VIII. MLD
    in Adults; Diagnosis and Pathogenesis, Arch.
    Neurol., 18, 225-240, 1968.

14. Austin, J., Armstrong, D., Shearer, L. and
    McAfee, D., Metachromatic Form of Diffuse
    Cerebral Sclerosis. VI. A Rapid Test for the
    Sulfatase A Deficiency in Metachromatic Leu-
    kodystrophy (MLD) Urine, Arch. Neurol., 14,
    259-269, 1966.

15. Percy, A. and Brady, R., Metachromatic Leuko-
    dystrophy; Diagnosis with Samples of Venous
    Blood, Science, 161, 594, 1968.

16. Bubis, J. and Adlesberg, L., Congenital Meta-
    chromatic Leukodystrophy, Report of a Case,
    Acta Neuropath., 6, 298-302, 1966.

17. Fogelson, M., Gonatas, N., Rorke, L. and Spiro,
    A., Oligodendroglial Lamellar Inclusions,
    Arch. Neurol., 19, 150-155, 1968.

18. Austin, J., Armstrong, D. and Shearer, L.,
    Metachromatic Form of Diffuse Cerebral Sclero-
    sis: V. The Nature and Significance of Low
    Sulfatase Activity; A Controlled Study of
    Brain, Liver and Kidney in Four Patients with
    Metachromatic Leukodystrophy (MLD), Arch.
    Neurol., 13, 593-614, 1965.

19. Bischel, M., Austin, J. and Kemeny, M., Meta-
    chromatic Leukodystrophy (MLD): VII. Elevated
    Sulfated Acid Polysaccharide Levels in Urine

and Postmortem Tissues, Arch. Neurol., 15, 13-28, 1966.

20. Bachhawat, B., Austin, J. and Armstrong, D., A cerebroside Sulphotransferase Deficiency in a Human Disorder of Myelin, Biochem. J., 104, 15c-17c, 1967.

21. Austin, J., Armstrong, D., Stumpf, D., Kretschmer, L., Mitchell, C., VanZee, B. and Bachhawat, B., Defective Sulfatide Synthesis in Krabbe's Disease (Globoid Leukodystrophy), Trans. Amer. Neurol. Assn., 175-179, 1967.

22. Austin, J., Armstrong, D. and Margolis, G., Studies of Globoid Leukodystrophy in Dogs, Neurol., 18, 300, 1968.

SOME BIOCHEMICAL FEATURES SHARED
BY DIFFERENT LIPIDOSIS SYNDROMES

DR. CROCKER: It is a pleasure to be able to
make a few remarks following those of Dr. Austin.
He has had a professional lifetime of enormous
contributions to the field of altered glycolipid
metabolism.

Since this has been an occasion for making
predictions of subsequent investigations, I thought
it might be of value to put on record my belief
that the usefulness of the one gene type of investigation in the lipidoses is leveling off. There
has been a trend in the tissue chemical studies
of the last few years to expose a multiplicity of
ultimate tissue handicaps that makes the univalent
concept of the lipidoses a little too simple. If
one looks at a diagram of the more common glycolipid lipidoses, and includes as well the mucopoly-
saccharidosis (Table I), one finds that a number
of common features reappear in different syndromes,
requiring explanations that have several phases.

It is now evident that glycolipid increases,
usually as ganglioside, are characteristic of the
cytoplasmic alterations in the gray matter neurones
of a broad group of inborn-error syndromes - including Hurler's disease, Norman-Landing disease,
the metachromatic leukodystrophy variant, lipogranulomatosis, and fucosidosis, as well as Tay-Sachs'
disease. Coordinately, viscera such as liver, and
the cultured fibroblasts from skin biopsies, in
such patients, commonly contain a mixture of glycolipids and mucopolysaccharides, in varying types
and ratios. The Hurler patients remain the prototype for mucopolysaccharide accumulation, and Tay-
Sachs' disease for gangliosides, while combinations
of storage appear to be occurring in I-cell disease
(Leroy's disease), Norman-Landing disease, lipomu-
copolysaccharidosis, and fucosidosis. Mild degrees
of increase in urinary mucopolysaccharides are
found intermittently in the pre-eminently "lipidosis" syndromes of I-cell disease, Norman-Landing
disease, the metachromatic leukodystrophy variant,

TABLE 1.

	Hurler Syndrome	I-Cell Disease (Leroy)	Norman-Landing Disease (GM-1 disease)	Fucosidosis	Lipogranulomatosis (Farber)	Metachromatic Leukodystrophy Variant (Austin)	Tay-Sachs Disease (GM-2 disease)
CEREBRAL GRAY MATTER GL:	↑		↑↑	↑	↑	↑	↑↑↑
LIVER CELLS (or cultured fibroblasts)	MPS ↑↑↑	MPS ↑ GL ↑	MPS ↑↑ GL ↑	MPS ↑ GL ↑	GL ↑	GL ↑	GL ↑(?)
URINE MPS:	‡	+1	0	0	+1	+1	0

Key: MPS – Mucopolysaccharides
GL -- Glycolipid or Ceramide

and lipogranulomatosis.

DR. AUSTIN: Dr. Crocker asked me several questions about ganglioside metabolism. In view of the limited time, let me summarize by saying that there is a non-specific increase in gangliosides in a number of storage diseases, and that in certain other diseases there is also a non-specific increase in certain gangliosides.

Dr. Suzuki found, for example, in Lafora's Myoclonus Epilepsy that there was an increase in a certain category of ganglioside in cortex (GA3) which, at least at this time, cannot be clearly related to the other moieties which are increased in this disease.

In what we call the Variant form of Metachromatic Leukodystrophy, we have scraped out distended, ballooned, ganglion cells. In these we have discerned an increased level of gangliosides - increased to the same degree as that found in the gargoylism. In this Variant form of MLD there is a marked increase in sulfated acid mucopolysaccharides in the urine. There is also evidence from Dr. Moser's laboratory that such patients may have an increase in cholesterol sulfate.

When we have analyzed organs from patients with the Variant form of MLD, we have found that they show not only a deficiency of sulfatase A, but also a deficiency of sulfatase B, and also a deficiency of sulfatase C. I do not know how these deficiencies can be related to a single enzyme defect, but I hope that the future may tell us.

B. THE GENETICS AND BIOCHEMISTRY OF SCHIZOPHRENIA

SCHIZOPHRENIC AND AFFECTIVE PSYCHOSIS: TWIN AND FAMILY STUDIES

James R. Stabenau

The relative role of genetic and environmental factors in the etiology of schizophrenia is not entirely known. This review will cover research efforts directed toward clarifying an issue. Kraepelin's early classification of the psychoses in 1899 according to descriptive and prognostic terms was into two major groups of psychoses: dementia praecox, dementing and irrecoverable insanity and the manic-depressive psychosis, a periodic recoverable psychosis (1). With this classification system efforts could begin toward determining etiologic factors. Bleuler conceived of a group of psychoses of no known organic origin which in 1911 he labeled as schizophrenia. In this group he placed dementia praecox, but also a group of recoverable psychoses, all of which shared a common set of primary symptoms including disturbance of association, affect and volition, and a varied number of secondary symptoms such as delusions and hallucinations. Thus schizophrenia

came to be considered primarily a disturbance of the
thought process which usually manifests itself between the ages of 15 and 45. Today the spectrum of
schizophrenic psychoses designated as schizophrenia
embraces 'true' schizophrenia, schizophreni-form
psychosis, process schizophrenia, and reactive schizophrenia. Manic-depressive psychosis has come to include acute periodic, self-limiting disturbances of
the affective state such as manic elation or retarded
depression, occurring mainly before the age of 50 and
without progressive residual personality disintegration before or after the psychotic episodes.

The report of the World Health Organization
Study Group on schizophrenia in 1957 stated: "This
clinical picture described by Kraepelin amplified and
given broader psychological significance by Bleuler
remains the most informative frame of reference whenever the criteria of illness are in question." (3)

A number of studies have attempted to assess the
reliability of the diagnostic classification of
patients. Schmidt and Fonda (1956) comparing diagnostic judgment by pairs of psychiatrists on 426
state hospital patients found agreement concerning a schizophrenic non-schizophrenic distinction
on 90% of the patients (4). Sandifer et al (1964)
report on a conference presentation of 91 patients
to ten diagnosticians, who achieved a 74% reliability figure for schizophrenic diagnosis, and
whose independent judgments were in agreement more
frequently in regard to schizophrenia than concerning any other diagnostic category (5). Babigian
et al (1965) demonstrated from a longitudinal case
register project of 1,215 patients that 70% of patients diagnosed as schizophrenic received that same
diagnosis on subsequent psychiatric contacts. They
concluded the category of schizophrenia has considerable diagnostic consistency (6). Astrup
(1959) isolated 70 cases of "pure" manic-depressive disorder. Seven to nineteen years after onset of that disorder, none had a schizophrenic
outcome. In contrast, 13 (50%) of a group of
26 cases diagnosed as schizophrenic affective
psychosis had schizophrenic symptomatology on

follow-up (7). Clark and Mallet (1963) did a three-
year follow-up study on 74 young adults diagnosed
as manic-depressive psychosis or reactive depression, and 76 who were diagnosed as schizophrenic.
Of the 76 patients initially diagnosed as schizophrenic, none was considered to have a depressive
disorder on readmission (8).

There seems to be sufficient certainty as to
placing psychotic individuals within the broad
classificatory categories of schizophrenia and
manic-depressive psychosis to justify research
attempts toward etiologic causation.

I will confine this review to schizophrenic
and manic-depressive psychoses and the study of
these disorders by means of twin and family studies.
Only the broadest operational classificatory definitions will be used in this report as each investigative group has defined the population of investigation along lines satisfactory to such categorization, but differing as to patterns of sub-classi-
fication. Recurrent depressive psychosis has been
included in the classification of the manic-
depressive psychoses for most investigations cited.
I have omitted involutional psychosis in order to
focus attention on the two major psychotic states
of schizophrenia and manic-depressive psychosis,
for which a greater quantity of comparative twin
and family study data are available.

Mendelian genetics (1865), the concept of the
genotype introduced by Johannsen in 1911, and
Galton's proposal in 1876 that the use of twins migh
provide a means of assessing the relative influence
of nature and nurture in human illness and behavior
led to the clasical twin method in the study of psychosis which began with Luxenburger's study in 1928
(9-11). In this method, one statistically compares
monozygotic (one egg) and dizygotic (two egg) pairs
in respect to their concordance for the psychosis
being studied. Significantly higher concordance
figures in a group of monozygotic twins has been
regarded as evidence in support of a genetic background for that condition. If the concordance is th
same for both monozygotic and dizygotic pairs, the

Table 1

TWIN STUDIES OF SCHIZOPHRENIA

Investigator	Year & Country	Concordance†						Sampling*	Diagnosis**	Zygosity***
		MZ		DZ SS		DZ OS				
		Pairs	%	Pairs	%	Pairs	%			
1. Luxemburger	(1928)Germany	13/17	76††	0/13	0	0/15	0	R,C,B.R.	A	S
2. Rosanoff	(1934)U.S.A.	28/41	68	10/53	19	5/48	10	R	H	S
3. Essen-Möller	(1941)Sweden	4/9	44	4/27	15			C,B.R.	A	S + F
4. Kallmann	(1946)U.S.A.	120/174	69	34/296	11	13/221	6	R,C	A	S
5. Slater	(1953)U.K.	28/41	68	11/61	18	2/54	4	R,C	A	S + F
6. Kallmann	(1956)U.S.A.	15/17	88	†††		†††		R,C	A	S + F
7. Mitsuda	(1957)Japan	6/12	50					?	?	?
8. Inouye	(1961)Japan	33/55	60††	2/11	18	0/6	0	R,C	A	F + B
9. Tienari	(1963)Finland	1/16	6	1/20	5			B,R,C (1920-29)	A	F + B
10. Kringlen	(1964)Norway	2/8	25	2/12	17			R,C	H	F + B
11. Harvald & Hauge	(1965)Denmark	2/7	29	2/31	6	2/29	7	B.R. C (1870-1910)	H	F + B
12. Gottesman & Shields	(1966)U.K.	10/24	42	3/33	9			C	H	F + B
13. Kringlen	(1967)Norway	21/55	38	9/90	10			B.R.C (1901-1930)	A	F + B

* R = resident population, C = consecutive admission, B.R. = birth register.

** H = hospital diagnosis, A = diagnosis established by author

*** S = similarity method, F = fingerprints, b = serologic blood typing

† uncorrected for period of risk

†† includes schizophrenic like psychoses; when cases listed as "schizophrenic-like" or "probably schizophrenic" are not counted MZ concordance figures are: Luxenberger, 10/17(59%); Inouye, 20/55(36%).

††† DZ SS and DZ OS pairs are separately reported.

condition is thought of as mainly determined by
environmental variables.

Before discussing the studies of psychosis in
twins it should be noted that there does not seem
to be significant difference in the admission rate
of twins with the diagnosis of schizophrenia as
compared to the general population (12-13). In
fact for functional psychoses (including schizophrenia, manic-depressive and reactive sub-groups)
in Norway, rates for twins are lower than for total
population (14).

Major twin studies of both schizophrenia and
affective psychoses have been conducted in eight
countries (Germany 1928; U.S.A. 1934, 1946, 1956;
Sweden 1941; United Kingdom 1953, 1966; Japan 1957,
1961; Finland 1963; Norway 1964; Denmark 1965).
Table 1 lists those studies by senior investigator,
country, and year of report. Numbers of monozygotic
and dizygotic same-sexed pairs, per cent concordance
for schizophrenia and the type of sampling, diagnostic and zygosity method is given (11,13,15,16,
17,18,19,20,21,22,23,24).

Very briefly, as research methodology improved,
i.e., zygosity for all twins was established by
comparison of blood types; diagnostic criteria were
rigorously employed in the personal examination of
all twins by the investigator; complete study of
large-scale demographic populations of twins born
during a fixed period of time and from a given geo-
graphic area, led to reduction in errors which
occurred previously due to the reliance on hospitalized populations of the more seriously ill twins
and from imprecise psychiatric and zygosity diagnoses. These methodologic improvements led to the
change of concordance figures for monozygotic twins
from well over 50% in studies done prior to 1961,
to well under 50% in subsequent well-controlled
studies. The concordance is still 2 to 4 times
higher for monozygotic twins as compared to dizygotic same-sexed twins. The assumption that environment is the same for monozygotic as dizygotic
twins has been challenged by the finding of higher

degrees of mutual interreliance, misidentification, and fewer periods of early sustained separation between monozygotic twins as compared to dizygotic twins (25). Thus twin studies of schizophrenia indicate that genetic influences are operative, however, the mode of inheritance and the nature of these genetically controlled factors are unclear.

Types and degrees of genetic influence for personality and somatic illness variables have been demonstrated in the comparison of monozygotic twins reared apart (26-28).

Since the concordance rate for schizophrenia in monozygotic twins is most likely below 50% the environmental etiological factors determining the emergence of schizophrenia become important to define. Detailed study of the monozygotic twin pairs who remained discordant over a substantial period of time has directed attention to some of the environmental components to schizophrenia. In a review of 100 cases of monozygotic twins discordant for schizophrenia the following was noted: Nine behavior traits were characteristic of the pre-illness schizophrenic twin: submissive; sensitive; serious-worrier; obedient-gentler; dependent; well behaved; quiet-shy; stubborn; and neurotic as a child; and six physical traits were also discriminating: having had a central nervous system illness as a child; any birth complications; neonatal asphyxia; weaker; shorter; and lighter at birth. The criterion employed was that the designated trait was present for the schizophrenic twin by a ratio of 2:1 as compared to the non-schizophrenic co-twin controls. Conversely six traits characterized the controls: more intelligent; better at school; the spokesman; outgoing-lively; the leader; and married while co-twin remained unmarried (29) (Figure 1). It is apparent from this list that in addition to the need to define the character of genetically controlled variables in the neonate and the environmental stresses associated later in life with the onset of schizophrenia, it has been clear that non-genetic prenatal, natal and early postnatal variables are of

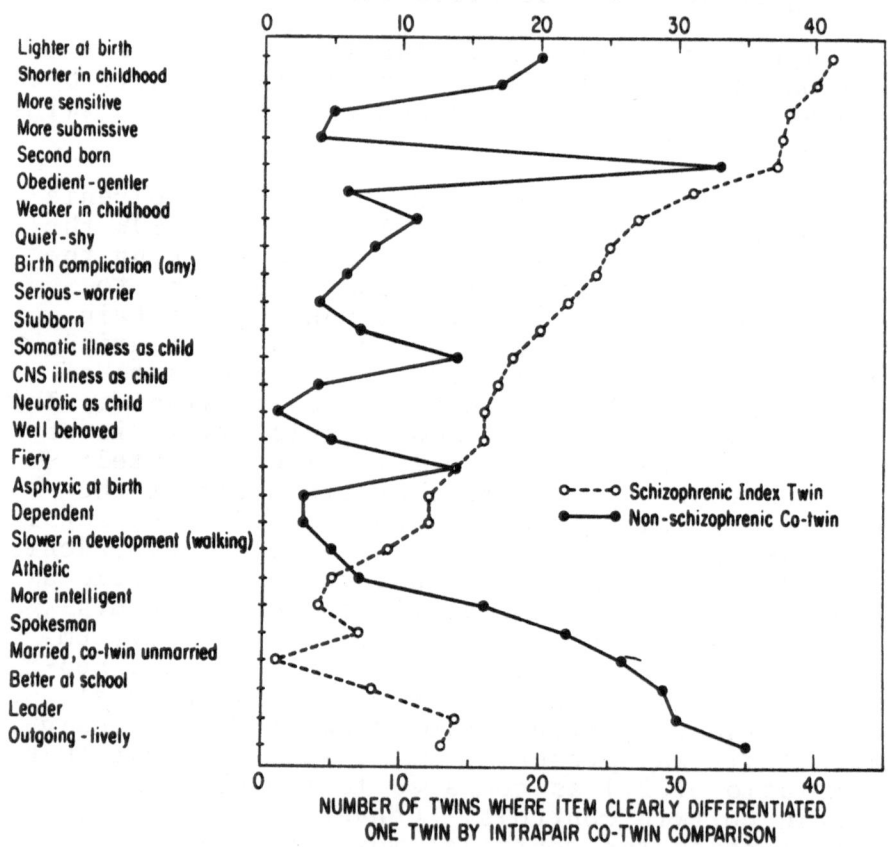

FIGURE 1

DIFFERENTIATING CHARACTERISTICS: MZ TWINS DISCORDANT FOR
SCHIZOPHRENIA (100 PAIRS)

additional importance and need detailed study (30-33).

Twin studies in manic-depressive psychosis are not as numerous as studies of schizophrenia, and with the exception of data from Da Fonseca have been conducted by the same investigators in the same studies as described in Table 1 and therefore the same issues of sampling, psychiatric diagnosis and zygosity diagnosis should be considered (11,14,18, 23,34, 35,36) (Table 2). The prevalence of manic-depressive psychosis is less than for schizophrenia in the general population. Prevalence figures have been reported as 0.4% for manic-depressive psychosis (37) and 1% for schizoprenia (38).

Fewer twin pairs with a diagnosis of manic-depressive psychosis were studied as compared to those with schizophrenia. This lends somewhat less statistical security to the concordance figures for the former group of data. In fact, more recent and methodologically more exacting demographic studies report on only 6 to 15 monozygotic pairs and 20 to 40 dizygotic pairs with manic-depressive psychosis diagnosis. However, a consistently higher concordance rate for monozygotic twins (33-96%) as compared to dizygotic twins (0-26%) has been found. There is no substantial difference in concordance between sexes or between the same sex and opposite sex dizygotic twins (14). These data suggest a stronger genetic component for the manic-depressive psychosis as compared to schizophrenia. Studies of large non-twin populations by Stenstedt in Sweden, Perris in Sweden and Angst in Switzerland have also supported the same conclusion (39-41).

The family studies in the major psychoses have included diagnostic studies of members of psychotic probands' families. These served to estimate the morbidity risk for same and different psychiatric diagnoses among close relatives, and in this sense measure the degree of genetic load for that psychiatric diagnosis. Böök gives morbidity risk figures of 7-15% for sibling of schizophrenics

Table 2

Concordance for manic-depressive psychosis in MZ and DZ twins†

Investigator	Year	Country	MZ		DZ	
			Number of pairs	Concordance* %	Number of pairs	Concordance* %
Luxenburger	1928	Germany	4	75	13	0
Rosanoff et al	1935	USA	23	70	67	16
Kallmann	1950	USA	23	96	52	26
Slater	1953	England	8	50	30	23
Da Fonseca	1959	England	21	75	39	39
Harvald & Hauge	1963	Denmark	15	60	40	5
Kringlen	1967	Norway	6	33	20	0

*Only the age-corrected figure has been given by Kallmann for the DZ. If the age-corrected figure had been given for the MZ, the figure would have been in excess of 100 per cent.

†From Kringlen, E: Heredity and Environment in the Functional Psychoses, Universitetsforlaget, Oslo, Norway, 1967, Table 28, p. 93.

7-16% for children of schizophrenics, and 5-10% for parents of schizophrenics (38). Data from four twin studies on morbidity risk in siblings is presented in Table 3. Kringlen concludes that findings of higher morbidity figures for dizygotic twins (though not statistically significant) are related to effects of the twinship itself, which is also considered in part to influence the consistently higher concordance figures among monozygotic twins (14).

Rainer states that morbidity figures for manic-depressive psychosis range from 10%-15% for parents, and for children (37). Stenstedt reports 11.4% for parents, 12.3% for siblings and 9.4% for children (39). Somewhat at variance with these large population studies, Kallmann's twin study of manic-depressive psychosis in twins and their siblings give the expectancy for manic-depressive psychosis of 23.4% for parents, 23.9% for siblings, 26.3% for dizygotic siblings and 95.7% for monozygotic co-twins (35). This can be seen to be substantially higher than comparative figures given for siblings and both types of twins with schizophrenic psychosis (Table 3). However, in spite of criticisms of Kallmann's diagnostic and sampling techniques, the data suggest, as have other data in this report, a more pronounced genetic component to manic-depressive psychosis.

From Table 4 it can be seen that a review of psychosis developing in monozygotic twins reared apart yields little information, as no pattern of either clear concordance or discordance for schizophrenia or other psychotic states is found. However, detailed history material from such pairs, if available, might provide important insight as to the effect of different environmental variables in the development of psychosis (13,14,19,22,27,34, 39,43,44).

The other type of "family studies" in the major psychoses is the investigative attempt not to demonstrate the genetic loading for an illness within a family or families, but to sort out the important

Table 3

Morbidity risk in siblings of schizophrenics, based on twin populations in per cent†

Studies compared			Relation to index-case*		
			MZ co-twins	DZ co-twins	Full sibs
Luxenburger	1935	Germany	54.0	14.0	11.8
Kallmann	1946	USA	69.0	10.3	10.2
			(85.8)	(14.7)	(14.3)
Slater	1953	England	68.3	11.3	4.6
			(76.3)	(14.4)	(5.4)
Kringlen	1967	Norway	25.4	8.1	5.2
			(38.2)	(12)	(6.8)

*The percentages in parentheses refer to age-corrected figures with respect to Kallmann's and Slater's studies. The figures in parentheses with respect to Kringlen's study refer to percentages based on personal investigation and employment of a wide concept of concordance.

†From Kringlen, E: Heredity and Environment in the Functional Psychoses, Universitetsforlaget, Oslo, Norway, 1967, Table 58, p. 140.

Table 4

MZ Twins with Psychosis Separated in Early Childhood and Brought Up Apart[†]

| Described by | Diagnosis of the twins | | Age at separation |
	Twin A	Twin B	
Kallmann (1938)	schizophrenia (cat.)	schizophrenia (cat.?)	shortly after birth
Craike and Slater (1945)	schizophrenia (par.)?'sensitiver Beziehungswahn'	react. psych. depr.par.(schiz.?)	nine months
Kallmann and Roth (1956)	schizophrenia	schizophrenia	?
Shields (1962)	schizophrenia (par.cat.)	schizophrenia	shortly after birth
Tienari (1963)	schizophrenia (heb.)	normal,introverted	three to six years
Kringlen(1964)	schizophrenia (cat.)	normal	one year ten months
Kringlen (1967)	schizophreniform (catatonic)	schizophreniform (paranoid-depr.)	shortly after birth
Kringlen (1967)	react.psych. (manic-depressive?)	normal	two years
Rosanoff (1935)	react. psych.	react. psych.	a few months
Stenstedt (1952)	manic-depressive?	manic-depressive?	one year

[†]From Kringlen, E: Heredity and Environment in the Functional Psychoses, Universitetsforlaget, Oslo, Norway, 1967, Table 60, p. 149.

environmental variables that may contribute to the
development of psychosis in one or more members of
a family by comparative study of the life history
of the family of a schizophrenic. Further study of
series of families of schizophrenics and families of
individuals with non-schizophrenic psychopathology
and non-psychiatrically disturbed families attempt
to define differences of possible environmental
etiologic nature.

Controlled investigation of the effects of
family are of only recent origin and in the main
have been conducted in the United States and the
United Kingdom. Most have involved families of
schizophrenics, only a few studies have considered
manic-depressive psychosis (45-48). The more numerous studies concerning the family relationship in
schizophrenia have attempted to investigate a number
of theoretical research variables. In a careful
and critical review Baxter has organized these
around 10 variables primarily of the clinical observational and retrospective recall type (49).
These may be evaluated as to the degree of specificity to schizophrenia which can be demonstrated.
The following are the conclusions Baxter reached
from his survey of the literature (49).

1. The sibling position of the patient (16
refs.). "From the review of the data available
from other pathological groups, the specificity of
birth order effects to schizophrenia would seem
doubtful," (an interactional effect, however, may
be present but as yet not demonstrated).

2. The loss of one or both parents(16 refs.).
"Comparisons have failed to find stable differences
in parental loss between schizophrenic and other
pathogenic groups and would appear to contradict
specificity of the effects to schizophrenia."

3. The presence of a distressed childhood
in the parents' backgrounds (3 refs.). "Evidence
concerning the nature of the parents' childhood
and early experience appears to be accumulating,
although at present it seems premature to speculate

on its significance."

4. Atypical mother-child relationship (41 refs.). "The question of the specificity of such deviant mother-child relationship also appears to exist when the issues of other pathologic, psychiatric and medical groups, and the effects of ethnic and social class variables are considered."

5. Atypical father-child relationship (27 refs.). "Both atypical mother-child and father-child relationships may be characteristic of the history of schizophrenics and, on the other hand, categorization of them neatly as the domineering and overpowering 'schizophrenogenic mother' and in the ineffective 'non-entity' father would appear to be an oversimplification. Especially when one differentiates schizophrenia into poor morbid (typical) and good premorbid (atypical) groups, it is in the families of the good premorbids that the tyrannical father is found more frequently."

6. Emotional immaturity in the parents (8 refs.). "It appears that families of schizophrenics tend to show more emotional immaturity than control families, although more stringent comparison with control families and with patients who have non-schizophrenic, psychiatric disturbance suggests that the relationship may be largely attributable to families of the poor premorbid patients."

7. Dominance disturbances between the parents (12 refs.). "While the existence of differences of overadequate and inadequate reciprocity are suggested by Bowen, there does not appear to be any systematic investigation of the interrelationships of parental role in family of schizophrenics."

8. Interpersonal conflict in the family (29 refs.). "High degrees of conflict within the families of schizophrenics appear quite consistently across numerous studies in the area. Further, these effects have proved consistent when parents are distinguished for levels of premorbid personality adequacy. Since at least sex and sub-cultural

differences in intrafamilial conflict have been
shown, careful attention must be paid to controls
used in study of the effects."

9. Family-centered pathological behavior
(17 refs.). "The fact that a great deal of emotional distress can be found in families of other
psychiatric groups, such as psychoneurotic patients,
would seem to argue against any high degree of
specificity of family disturbances to schizophrenia."

10. A special position in the sib-set for
the patient (12 refs.). "Evidence in this rather
critical area is certainly not conclusive, although
the weight of the observations, informal as they
are, and as subject to investigative biases as they
are, seem to support a general 'special family role'
concept." (However, this too may not be specific
to schizophrenia.)

Before dismissing any one of these variables
it must be kept in mind that an interactional effect
of one or more with each other may have important
etiologic consequences. The difficulties in research of any of these variables centers around
questions of whether the family variable is etio-
logically related to the schizophrenia in the child
or whether it is instead in response to the schizophrenia in the child. Families of children handi-
capped with cerebral palsy, mental retardation, and
phenylketonuria demonstrate that considerable effects upon family structure result in response to
the disturbed child (50-52).

A methodologic technique having research advantages over clinical observations and retrospective recall studies is that of direct observation
and recording of family interactions. There have
been a limited number of such studies since a prototype was first reported by Behrens and Goldfarb
in 1958 (53). Fontana concluded from an extensive
survey of 20 studies, published at the time of his
review (1966), which employed the direct recording
and systematic coding of family interactions, that

only four such studies met rigorous methodologic
criteria (54). Those four have utilized: a) only
schizophrenic families in an experimental group;
b) analyzed the data separately by sex; c) made
some statement indicating attention to reliability
of coding; d) specified adequate comparability of
control and experimental groups on demographic
variables; and e) included at least one hospitalized control group, Caputo, 1963; Farina et al,
1960; Fisher et al, 1959; Lerner, 1965 (55-58).
Table 5 provides a summary of demographic comparability of control in schizophrenic groups, sex of
subjects, hospital status of the schizophrenic
group, sub-division of the schizophrenic group,
type of control group, "blind" coding of data, reliability of coding, member interaction, and the
task characteristics (59-73).

It is clear from review of current studies of
families where pathology already exists that the
issue of its effect on the family is inextricably
enmeshed with any possible pre-psychotic causal
family interactional variables. Family interaction
study thus far has had the greatest research value,
not in establishing causal etiologic links between
family pathology and schizophrenia in an individual,
but in serving to narrow the field of such variables
and indicating which variables might be most profit-
ably employed in longitudinal prospective studies.

There have been only a few studies of manic-
depressive psychotics, their families and other
non-manic-depressive psychotic psychopathology
families (45-48). They have all been of the clinical observational and retrospective recall type.
The single, most relevant variable was in relating
a special relationship of striving for the family
in relation to the community and a special role for
the pre-psychotic manic-depressive within the family which served that purpose (45). Neither reports of family interactional studies; nor comparison studies between manic-depressive individuals
and their non-manic-depressive siblings within the
same family; nor studies of monozygotic twins discordant for manic-depressive psychosis were currently available, but might well be undertaken.

Table 5a

Methodological Summary of Family Interaction Studies†

Author	Demographic comparability of control and schizophrenic groups	Sex of subjects[a]	Hospital status of schizophrenic group	Subdivision of schizophrenic group
Baxter et al. (1962)	No control group	m & f	Acute	Poor premorbid[d]
Baxter & Arthur (1964)	No control group	m	Acute	Poor & good premorbid
Behrens & Goldfarb (1958)	Questionable	m & f	Ns	[e]
Caputo (1963)	Good	m	Chronic	—
Cheek (1964a)	Fair	m & f[b]	Convalescent	—
Cheek (1964b)	Fair	m & f[b]	Convalescent	—
Cheek (1965)	Fair	m & f[b]	Convalescent	—
Farina (1960	Fair	m	Acute	Poor & good premorbid
Farina & Dunham (1963)	No control group	m	Acute	Poor & good premorbid
Ferreira (1963)	Questionable	m & f	Ns	—

Table 5b

Methodological Summary of Family Interaction Studies[†]

Author	Demographic comparability of control and schizophrenic groups	Sex of subjects[a]	Hospital status of schizophrenic group	Subdivision of schizophrenic group
Ferreira & Winter(1965)	Good	m & f	Ns	—
Fisher et al.(1959)	Good	m	Ns	—
Haley(1962)	Fair	m & f	Ns	—
Haley(1964)	Questionable	m & f	Ns	[e]
Lennard et al.(1965)	Fair	m	Ns	[e]
Lerner(1965)	Good	m	Ns[c]	High & low genetic level[f]
McCord et al.(1962)	Good	m	Prehospitalization	[e]
Meyers & Goldfarb(1961)	Fair	m & f	Ns	Organic & nonorganic
Morris & Wynne(1965)	Questionable	m & f	Ns	—
Stabenau et al.(1965)	Good	m & f	Ns	—

Table 5c

Methodological Summary of Family Interaction Studies†

Author	Control Group	"Blind" Coding of data	Reliability of coding	Members in inter-action[h]	Task characteristics
Baxter et al.(1962)	—	Ns	Good	F,M,C	Joint interview
Baxter & Arthur(1964)	—	Ns	Good	F,M	Joint interview
Behrens & Goldfarb(1958)	Maladjusted persons	No	Fair	F(u)	Home interactions
Caputo(1963)	Normals[g]	Ns	Vs	F,M	RDT[j]
Cheek(1964a)	Normals	Ns	Vs	F,M,C	RDT
Cheek(1964b)	Normals	Ns	Vs	F,M,C	RDT
Cheek(1965)	Normals	Ns	Vs	F,M,C	RDT
Farina(1960)	Tubercular patients[g]	Ns	High	F,M	RDT
Farina & Dunham(1963)	—	Ns	High	F,M,C	RDT
Ferreira(1963)	Normals, maladjusted persons	Ns	Ns	F,M,C	RDT

Table 5d

Methodological Summary of Family Interaction Studies[†]

Author	Control Group	"Blind" Coding of data	Reliability of coding	Members in interaction[h]	Task characteristics
Ferreira & Winter(1965)	Normals, delinquents, maladjusted persons	Ns	Ns	F,M,C,	RDT
Fisher et al.(1959)	Neurotics,[g] normals,[g] normals	Yes	High	F,M	Family discussion, TAT story construction
Haley(1962)	Normals	Mt	High	F,M,C	Game
Haley(1964)	Normals	Mt	High	F,M,C	RDT, TAT story construction
Lennard et al.(1965)	Normals	Ns	Ns	F,M,C	Family discussion
Lerner(1965)	Normals[g]	Ns	High	F,M	RDT
McCord et al.(1962)	Maladjusted persons	Yes	Good	F,M,C	Home interactions
Meyers & Goldfarb(1961)	Normals	No	Ns	F(u)	Home interactions
Morris & Wynne(1965)	Neurotics[g]	Yes	Unobtained	F,M[j]	Joint therapy
Stabenau et al.(1965)	Delinquents, normals	Ns	Ns	F,M,C,S	RDT

Table 5e

Methodological Summary of Family Interaction Studies[†]

Note. -- Ns = not stated, Vs = vaguely stated, Mt - machine tabulated.

[a]m = male, f = female.

[b]Analyzed separately.

[c]Schizophrenic and control groups were equated on length of current hospitalization.

[d]Premorbidity rated according to the Phillips' (1953) scale.

[e]Included unspecified psychotics.

[f]According to Becker's (1956) system of Rorschach analysis.

[g]Hospitalized.

[h]F = father, M = mother, C = child, S = sib, F(u) = family (unspecified).

[i]Patient present but not a verbal contributor.

[j]Modification of the Revealed Difference Technique, after Strodtbeck (1951).

[†]From Fontana, A.: Psychol. Bull. 66: 219, 1966.

Studies of family relationship variables such as sibling order, early loss of parents, presence of disturbed childhood, have not provided any consistent body of data substantiating or refuting the importance of these variables in the establishment of manic-depressive psychosis (74-78).

Two other types of family studies which provide research strategies different from the family studies in which schizophrenics are compared to their own non-schizophrenic siblings, or non-schizophrenic monozygotic twin controls, and the studies of degree of concordance for schizophrenia among monozygotic and dizygotic twins are: the study of children of schizophrenic parents (high risk groups) and the study of schizophrenics who were raised by foster parents. Children of parents who have both been schizophrenic have been found to have a probability of developing schizophrenia of about 35% (79). Table 6 presents five studies demonstrating that the degree of psychopathology among all children of these schizophrenic couples is extensive (43, 80-83). The "environmental" effect of the pre-psychotic parent on the child only complicates drawing conclusions regarding the relative contribution of proto-schizophrenic genetic loading and issues of psychonoxious family state. A study by Sobel has shown that children of schizophrenic parents "developed clear-cut signs of emotional disturbance in infancy," when raised by their own parents (3 of 4), as compared to those raised by foster parents (0 of 4) (84).

In a well planned prospective study done in Scandinavia, 25 high risk children raised by their schizophrenic mothers were compared to 25 separated from their schizophrenic mothers and who were foster raised. Those raised by their schizophrenic mothers showed more signs of being asocial, withdrawn in general and unresponsive to social rewards (85, 86).

Two adoptive studies have provided data of a different nature. In the U.S.A. study by Heston, five of 47 individuals born to schizophrenic

Table 6

Psychopathology in the offspring of schizophrenic couples[†]

Study	Schizo-phrenia	Question-able schizo-phrenia	Other psycho-pathology	Normal	Total	Morbidity risk(%)
Kahn (1923)	7	2	5	3	17	
Schulz (1940)	13	5	22	20	60	35.0
Elsässer (1952)	12	3	12	32	59	
Lewis (1957)	4	0	3	20	27	
Kallmann (1938)	13	3	16	3	35	55.3 to
Total	49	13	58	78	198	68.1

†From Rosenthal, D: In The Origins of Schizophrenia, (ed) Romano, J.: Excerpta

Medica Foundation, Amsterdam, 1967, p. 23.

mothers, separated after three days and raised in
foundling homes or by family members other than the
mother or maternal relatives were schizophrenic at
the time of the study (mean age 36). None of 50
carefully matched control adoptives became schizophrenic (87). A recent ongoing Scandinavian study
of subjects given up for non-familial adoption at
an unreported early age, yielded three adoptive
subjects who were diagnosed as schizophrenic and
all three were from the 39 index cases (i.e., had
one biologic parent who was clearly schizophrenic).
Seven cases were called clear borderline schizophrenic and six of these were index cases. Only
one case diagnosed borderline schizophrenic appeared
in the control group (N=47) (88).

These are striking findings, however, it must
be kept in mind that for Heston's study all mothers
of schizophrenics were psychotic at the time of the
birth of the child (only 16% of the parents in the
Scandinavian study were psychotic prior to the birth
of the index case). It was not made clear in Heston's study how many subjects were raised by foundling homes with or without the information regarding
their "heritage." Further, the character of the
families where placement occurred was not described,
nor how many experimental subjects were raised by
the father and father's family. It is important to
know whether the index child grew up under the constant reminder that his mother was schizophrenic or
if he were free of this burden. Whether the Scandinavian adoptee families knew the maternal diagnosis in the 16% of mothers who had already been
psychotic at the time of placement, or the possible
relationship of knowledge of such diagnosis to the
three schizophrenic or seven borderline schizophrenic
subjects is not clear. Scott has shown that such a
"shadow of insanity" plays a significant "environmental" role in the evolvement of schizophrenia
within the families where individuals in previous
generations were known to be psychotic (89). In
the Scandinavian study the age of adoption or duration the child spent with the pre-schizophrenic
mother may be an equally important variable to
assess. No final conclusion regarding the relative

value of genes versus environment can be made for
samples in either study until a careful study of
the characteristics of the adoptive parents, their
home, family structure, and parent-child relationships has been made.

The majority of twin investigations in the
area of psychosis have concluded that simple Mendelian genetic expectancies for morbidity within
families of schizophrenics or manic-depressives
are not met. An alternative monogenic theory of
schizophrenia based on a differing degree of penetrance has been proposed by Slater (90). A polygenic model has been proposed by Gottesman and
Shields (used to account for data in their recent
and very well controlled study of twins and their
families (91)), which permits the conception of
schizophrenia as a "threshold character whose
phenotypic appearance would depend on both the
number of genes present and the amount of stress."
"The mode would predict the continual appearance
of segregants in the offspring of normal parents,
and increased risks of schizophrenia in loaded
families." The method converts incidences into
regression coefficients and the latter into estimates of heritability of liability. The polygenic
theory has been suggested by Kringlen as best
applying to his carefully controlled twin study
(14).

FUTURE RESEARCH:

Some issues for future research work have in
part already been considered in the preceding
material. Some variables related to personality
and social bonding in humans as well as animals
appear in part established during critical periods
of the early postnatal period, during which rapid
somatic and central nervous system growth and
development occurs (9,21,93). Study of neonates
(monozygotic, dizygotic twins and singletons) for
determination of discrete personality or behavioral
variables which are genetically controlled, or
which are primarily environmentally induced are
important. Such variables may be enduring types

of response dispositions, or, on the other hand, discrete, short-lived behavior variables whose full expression or extinction require a phase specific environmental interaction variable from the interpersonal, familial, or social environment, and which in turn become important in determining the appearance or extinctions of other subsequent behaviors in an epigenetic fashion.

Use of direct, reliably ratable behavior characteristics derived from the body of data currently available describing the pre-schizophrenic or pre-manic-depressive might be employed as predictive criteria in a large-scale study. Glueck and Glueck have reported remarkable success in the predictive identification of potential delinquents. 301 boys from a high delinquency rate area were evaluated at school entrance age of 5½. The Glueck prediction table encompassing five interpersonal family factors: affection by father for the boy; affection by mother for the boy; discipline of the boy by the father; supervision of the boy by mother; and family cohesiveness were used. Twenty-five of 33 boys were predicted as becoming delinquent (85%) while 9 of 25 where prediction was unsure became delinquent (32%). 236 or 97% of 243 predictions were correctly made toward non-delinquency (94).

Some potential predictive measures for schizophrenia are suggested from: a) the studies of Lane and Albee where the measures of IQ have been significantly lower for pre-schizophrenics as compared to their own non-schizophrenic co-siblings for both low and upper middle class families (95,96); b) the research of Wynne and Singer et al, where blind raters, using the Singer-Wynne Rorschach method were able to significantly differentiate the parents of schizophrenics from parents of neurotics and parents of normals, and also distinguish the parents of neurotics from parents of normals (97).

The ability to demonstrate the presence or absence of schizogenic features using the Singer-Wynne Rorschach technique in the adoptive parents

of children who become schizophrenic is important,
and is being planned by Rosenthal and Wynne. If
the Rorschach characteristics of the parents of
schizophrenics is in any way etiologic and not a
response to the raising of a pathogenic child, the
characteristics should conceivably be present at a
time when the pre-schizophrenic is of pre-school
age. Parents might be tested at such a time.
Examination of the child could be made for intellectual organization (IQ differences), neurologic
intactness (whether there are any of the soft CNS
signs shown to be important from the predictive
studies by Bender and Fish), as well as evaluating
the child for personality variables noted as differentiating early pre-illness characteristics for
schizophrenia such as shyness, dependency and
fearfulness (98-101). Any other important variables emerging from such well controlled family
interactional studies of schizophrenics such as
those being conducted by Mishler et al., and Reiss
and from the psychophysiologic predictive studies
of Mednick et al. could be added to a battery of
potential variables (85, 102, 103). A long-term
predictive study with well matched control subjects should then provide some leverage as to substantiating or refuting the effectiveness of these
variables as causally related to the development
of psychosis.

1. Twin studies of schizophrenia, especially
those of recent years where the methodologic errors
of earlier studies have been corrected, demonstrate
a higher degree of discordance than concordance for
schizophrenia among monozygotic twins. A consistently higher concordance rate for monozygotic
pairs as compared to dizygotic pairs was also demonstrated. These data suggest major environmental
components interacting with genetic determinants.

2. Twin studies of manic-depressive psychosis, with fewer recent studies which have been
able to control for methodologic errors of the
earlier studies, demonstrate the probability of a
greater degree of genetic control in manic-
depressive psychosis as compared to schizophrenia.

3. Genetic theories attempting to explain the morbidity risk for schizophrenia and manic-depressive psychosis in family members based on the Mendelian genetic model has not been adequate. A monogenic theory with reduced penetrance requires that the penetrance control relate to substantial environmental as well as other possible genetic factors. A polygenic theory places equally high importance on environmental factors for the phenotypic expression of schizophrenia.

4. Research for important environmental intrafamilial factors by methods involving primarily clinical observation and retrospective recall have been inconclusive due to major errors in the reliability and validity of these methods. Direct recorded observational family interactional studies have shown promise provided reliability and validity of coding is established, and the family variables studied can be established as being etiologic and not responsive to raising a pathologic child.

5. Future research should be directed at establishing which aspects of neonatal behavior and response tendencies are primarily genetically and which are primarily environmentally controlled. Direct interactional family studies may provide variables to be explored in future research conducted in the study of neonates and children and their families; and in predictive, long-term studies. The latter might utilize premorbid characteristics of individuals who became ill with schizophrenia and manic-depressive psychosis to delineate high risk groups for study and comparison with well matched control non-pathologic groups. By these means environmental variables of critical nature could be experimentally tested in a manner which could scientifically verify or refute their etiologic nature.

REFERENCES

1. Kraepelin, E., Psychiatrie, 6th Ed. J. A. Barth, Leipzig, 1899.

2. Bleuler, P. E., Dementia Praecox uder Die Grupbeder Schizophrenien, Deuticke, Leipzig, 1911.

3. World Health Organization Report of World Health Organization Study Group on Schizophrenia, Geneva, Amer. J. Psychiat., 116, 865-872, 1959.

4. Schmidt, H. and Fonda, C., The Reliability of Psychiatric Diagnosis:A New Look. J. Abnorm. Psychol., 52, 262-267, 1956.

5. Sandifer, M., Pettus, C.I. and Quade, D., A Study of Psychiatric Diagnosis, J. Nerv. Ment. Dis., 129, 350-356, 1964.

6. Babigian, H., Gardner, E., Miles, H. and Romano, J., Diagnostic Consistency and Change in a Follow up of 1215 Patients, Amer. J. Psychiat., 121, 895-901, 1965.

7. Astrup, C., Fossum, A. and Holmboe, F., A Follow up Study of 270 Patients with Acute Affective Psychoses, Acta Psychiat. Scand. Suppl. 135, 1959.

8. Clark, J. and Mallet, B., Follow up Study of Schizophrenia and Depression in Young Adults, Brit. J. Psychiat., 109, 491-499, 1963.

9. Johannsen, W., The Genotype Conception of Heredity, Amer. Naturalist, 45, 129-159, 1911.

10. Galton, F., The History of Twins as a Criterion of the Relative Powers of Nature and Nurture, Fraziers Magazine, 12, 566, 1876.

11. Luxenburger, H., Vorlaufiger bericht uber Psychiatrische Serien unter Suchungen an Zwillingen, Z. Ges. Neurol. Psychiat., 116, 297-326, 1928.

12. Rosenthal, D., Confusion of Identity and the Frequency of Schizophrenia in Twins, Arch. Gen. Psychiat., 3, 297, 1960.

13. Tienari, P., Psychiatric Illness in Identical Twins, Acta Psychiat. Scand., 39 (suppl 171), 1963.

14. Kringlen, E., Heredity and Environment in the Functional Psychosis, an Epidemiological -

Clinical Twin Study, Universitsforlaget, Oslo, Norway, 1967.

15. Rosanoff, A.J., Handy, L., Plesset, I. and Brush, S., The Etiology of So-Called Schizophrenic Psychoses with Special Reference to The Occurrence in Twins, Amer. J. Psychiat., 91, 247-286, 1934.

16. Essen-Möller, E., Psychiatrische Untersuchungen an einer Serie von Zwillingen, Acta Psychiat. (kbh.), Vol. 16, Suppl. 23, 1941.

17. Kallmann, F. J., The Genetic Theory of Schizophrenia: An Analysis of 691 Schizophrenic Twin Index Families, Amer. J. Psychiat., 103, 309-22, 1946.

18. Slater, E., Psychotic and Neurotic Illness in Twins, Med. Res. Counc. Spec. Rept. Ser. No. 278, London, Her Majesty's Stationery Office, 1953.

19. Kallmann, F. and Roth. B., Genetic Aspects of Preadolescent Schizophrenia, Amer. J. Psychiat., 112, 599-606, 1956.

20. Mitsuda, H., Klinische-erbbiologische Untersuchungder endogenen Psychosen, Acta Genet. (Basel), 7, 371-377, 1957.

21. Inouye, E., Similarity and Dissimilarity of Schizophrenia in Twins., Proc. Third World Congress Psychiat.,Montreal, 1, 524-30, Univ. of Toronto Press and McGill U. Press, 1961.

22. Kringlen, E., Schizophrenia in Male Monozygotic Twins, Acta Psychiat. Scand. (Kbl.), Suppl. 178, 1964.

23. Harvald, B, Hauge, M., Hereditary Factors Elucidated by Twin Studies. In: Neel, J.U., Shaw, M.W. and Schull, W.J. (Eds.), Genetics and the Epidemiology of Chronic Diseases., Wash. D.C., U.S. Dept. of Health, Ed. & Welfare, 61-76, 1965.

24. Gottesman, I.and Shields, J., Schizophrenia in Twins: 16 Years' Consecutive Admission to a Psychiatric Clinic, Brit. J. Psychiat., 112, 809-818, 1966.

25. Wilson, P.T., Study of Like-sexed Twins Health and Disease Records, Hum. Biol., 3, 270-281, 1931.

26. Juel-Nielsen, N., Individual and Environment, a Psychiatric-Psychological Investigation of Monozygotic Twins Reared Apart, Acta. Psychiat. Scand., 40 (Suppl. 183), 1964.

27. Shields, J., Monozygotic Twins Brought Up Apart and Brought Up Together, Oxford Univ. Press, London, 1962.

28. Newman, H., Freeman, F. and Holzinger, K., Twins: A Study of Heredity and Environm nt, Univ. of Chicago Press, Chicago, 1937.

29. Stabenau J. and Pollin, W., Early Characteristics of Monozygotic Twins Discordant for Schizophrenia, Arch. Gen. Psychiat., 17, 723-734, 1967.

30. Pollin, W., Stabenau, J. and Tupin, J., Family Studies with Identical Twins Discordant for Schizophrenia, Psychiatry, 28, 6 - 78, 1965.

31. Pollin, W., Stabenau, J., Mosher, L. and Tupin, J., Life History Differences in Identical Twins Discordant for Schizophrenia, Amer. J. Orthospsychiat., 36, 492-509, 1966.

32. Pollin, W. and Stabenau, J., Biological, Psychological and Historical Differences in a Series of Monozygotic Twins Discordant for Schizophrenia, Kety, S. and Rosenthal, D. (Eds.), The Transmission of Schizophrenia, 1967. In press.

33. Stabenau, J., Pollin, W. and Mosher, L., in collaboration with Frohman, C., Friedhoff, A. and Turner, W., A Study of Monozygotic Twins Discordant for Schizophrenia: Some Biologic Variables, Arch. Gen. Psychiat. In Press.

34. Rosanoff, A., Handy, L., Plesset, I. and Brush, S., The Etiology of Manic-Depressive Syndromes with Special Reference to Their Occurrence in Twins, Amer. J. Psychiat., 91, 725-762, 1935.

35. Kallmann, F., The Genetics of Psychoses. An Analysis of 1232 Twin Index Families. In: Congrés International de Psychiatric, Paris VI. Psychiatrie Sociale Rapports, Herman & Cie, Paris, 1950.

36. Da Fonseca, A., Analise Heredo-Clinca das Perturbacoes Afectivas (Estudio de 60 pares de gemeos, e sues conganguin neos) Impresna Portuguesa, Porto, 1959.

37. Rainer, J., Genetic Aspects of Depression, Canad. Psychiat. Ass. J., 11: Suppl. 29-33, 1966.

38. Böök, J. Genetical Etiology in Mental Illness. In: Causes of Mental Disorders: A Review of Epidemiological Knowledge, Milbank Memorial Fund, New York, 1959.

39. Stenstedt, A., A Study in Manic Depressive Psychosis, Acta Psychiat. Scand. Suppl., 79, 1952.

40. Perris, C., A Study of Bipolar (Manic-Depressive) Unipolar Recurrent Depressive Psychoses, I. Genetic Investigation, Acta Psychiat. Scand., 42, 15-44, 1966.

41. Angst, J., Zur Ätiologie und Nosologie Endogenen Depressiver Psychosen, Springer-Verlag, Berlin, 1966.

42. Luxenburger, H., Untersuchungen an Schizophrenen, Zwillingen und ihren Geschwistern, zur Prüfung der Realität von Manifestationsschwankungen, Z. Ges. Neurol. Psychiat., 154, 351-394, 1936.

43. Kallmann, F., The Genetics of Schizophrenia, J. J. Augustin, New York, 1938.

44. Craike, W. and Slater, E., Folie á deux in Uniovular Twins Reared Apart, Brain, 68, 213-221, 1945.

45. Cohen, M., Baker, G., Cohen, R., Fromm-Reichmann, F. and Weigert, E., An Intensive Study of Twelve Cases of Manic-Depressive Psychosis, Psychiatry, 17, 103-137, 1954.

46. Gibson, R., Comparison of the Family Background and Early Life Experience of the Manic-Depressive and Schizophrenic Patient. Final report on Office of Naval Research Contract (NOR-751(00), Washington, D.C., Washington School of Psychiatry.

47. Becker, J., Achievement Related Characteristics of Manic-Depressives, J. Abnorm. Soc. Psychol., 60, 334-339, 1960.

48. Spielberger, C., Parker, J. and Becker, J.,

Conformity and Achievement in Remitted
Manic-Depressive Patients, J. Nerv. Ment.
Dis., 137-162-172, 1963.

49. Baxter, J., Family Relationship Variables in
Schizophrenia, Acta Psychiat. Scand., 42,
361-391, 1966.

50. Schaffer, H., The Too-Cohesive Family: A
Form of Group Pathology, Int. J. Soc. Psychiat., 10, 266-275, 1964.

51. Siegel, G., Adult Verbal Behavior With
Retarded Children Labeled as "High" or "Low"
in Verbal Activity, Amer. J. Ment. Defic.,
68, 417, 1963.

52. Wood, A., Friedman, C. and Steisel, I.,
Psychosocial Factors in Phenylketonuria,
Amer. J. Orthopsychiat., 37, 671-679, 1967.

53. Behrens, M.L. and Goldfarb, W., A Study of
Patterns of Interaction of Families of
Schizophrenic Children in Residential Treatment, Amer. J. Orthopsychiat., 28, 300-312,
1958.

54. Fontana, A., Familial Etiology of Schizophrenia: Is A Scientific Methodology Possible?, Psychol. Bull., 66, 214-227, 1966.

55. Caputo, D., The Parents of the Schizophrenic,
Family Process, 2, 339-356, 1963.

56. Farina, A., Patterns of Role Dominance and
Conflict in Parents of Schizophrenic Patients,
J. Abnorm. Soc. Psychol., 61, 31-38, 1960.

57. Fisher, S., Boyd, I., Walker, D. and Sheer,
D., Parents of Schizophrenics, Neurotics, and
Normals, Arch. Gen. Psychiat., 1, 149-166,
1959.

58. Lerner, P.M., Resolution of Intrafamilial
Role Conflict in Families of Schizophrenic
Patients. I. Thought Disturbance, J. Nerv.
Men. Dis., 141, 342-351, 1965.

59. Baxter, J.C., Arthur, S.C., Flood, C.G. &
Hedgepeth, B., Conflict Patterns in the
Families of Schizophrenics, J. Nerv. Men.
Dis., 135, 419-424, 1962.

60. Baxter, J.C. and Arthur, S.C., Conflict in
Families of Schizophrenics as a Function of
Premorbid Adjustment and Social Class, Family
Process, 3, 273-279, 1964.

61. Cheek, F.E., The "Schizophrenogenic Mother" in Word and Deed, Family Process, 3, 155-177, 1964.

62. Cheek, F. E., A Serendipitous Finding: Sex roles and Schizophrenia, J. Abnor. Soc. Psychol., 69, 392-400, 1964.

63. Cheek, F. E., The Father of the Schizophrenic, Arch. Gen. Psychiat., 13, 336-345, 1965.

64. Farina, A. and Dunham, R. M., Measurement of Family Relationships and their Effects, Arch. Gen. Psychiat., 9, 64-73, 1963.

65. Ferreira, A. J., Decision-Making in Normal and Pathologic Families, Arch. Gen. Psychiat., 8, 68-73, 1963.

66. Ferreira, A.J. and Winter, W.D., Family Interaction and Decision Making, Arch. Gen. Psychiat., 13, 214-223, 1965.

67. Haley, J., Family Experiments: A New Type of Experimentation, Family Process, 1, 265-293, 1962.

68. Haley, J., Research on Family Patterns, An Instrument Measurement, Family Process, 3, 41-65, 1964.

69. Lennard, H., Beaulieu, M.R. and Embry, N.G., Interaction in Families With a Schizophrenic Child, Arch. Gen. Psychiat., 12, 166-183, 1965.

70. McCord, W., Porta, J. and McCord, J., The Familial Genesis of Psychoses, Psychiatry, 25, 60-71, 1962.

71. Meyers, D.F. and Goldfarb, W., Studies of Perplexity in Mothers of Schizophrenic Children, Amer. J. Orthopsychiat., 31, 551-564, 1961.

72. Morris, G.O. and Wynne, L.C., Schizophrenic Offspring and Parental Styles of Communication, Psychiatry 28, 19-44, 1965.

73. Stabenau, J.R., Tupin, J., Werner, M. and Pollin, W., A Comparative Study of Families of Schizophrenics, Delinquents, and Normals, Psychiatry, 28, 45-59, 1965.

74. Forrest, A., Frazer, R. and Priest, R., Environmental Factors in Depressive Illness, Brit. J. Psychiat., 111, 243-253, 1965.

75. Brown, F., Depression and Childhood Bereavement, J. Ment. Sci., 107, 754-777, 1961.

76. Monro. A., Some Familial and Social Factors in Depressive Illness, Brit. J. Psychiat., 112, 429-441, 1966.
77. Munro, A., Parental Deprivation in Depressive Patients, Brit. J. Psychiat., 112, 443-457, 1966.
78. Hopkinson, G. and Freed, G., Bereavement in Childhood and Depressive Psychosis, Brit. J. Psychiat., 112, 459-463, 1966.
79. Rosenthal, D., In: The Origins of Schizophrenia, Romano, J., Excerpta Medica, Amsterdam, 1967.
80. Kahn, E., Studien über Vererbung und Entstehung Geistiger Störungen, IV. Schizoid und Schizophrenie, im Erbung. Monogr. Gesamtgeb. Neurol. Psychiat., 36, 1, 1923.
81. Schulz, B., Kinder Schizophrener Elternpaare, Z. Ges. Neurol. Psychiat., 168, 332, 1940.
82. Elsässer, G., Die Nachkommen geisteskranker Elternpaare, G. Thieme, Stuttgart, 1952.
83. Lewis, A. J., The Offspring of Parents Both Mentally Ill, Acta Genet., Basel, 7, 349, 1957.
84. Sobel, D. E., Children of Schizophrenic Patients: Preliminary Observation on Early Development, Amer. J. Psychiat., 118, 12, 1961.
85. Mednick, S. and Schulsinger, F., A Pre-Schizophrenic Sample, Acta Psychiat. Scand., 40, (Suppl 180), 135-146, 1964.
86. Higgins, J., The Effects of Childrearing by Schizophrenic Mothers, J. Psychiat. Res., 4, 153-167, 1966.
87. Heston, L., Psychiatric Disorders in Foster Home Reared Children of Schizophrenic Mothers, Brit. J. Psychiat., 112, 819-825, 1966.
88. Rosenthal, D., Wender, P., Kety, S., Schulsinger, F., Welner, J. and Østergaard, L., In: The Transmission of Schizophrenia, Kety, S. and Rosenthal, D., Eds., Pergamon Press, Oxford. In press.
89. Scott, R. and Ashworth, P., The "Axis Value" and The Transfer of Psychosis, Bri . J. Med. Psychol, 38, 97-116, 1965.
90. Slater, E., The Monogenic Theory of Schizo-

phrenia, Acta Genet., 8, 50-56, 1958.

91. Gottesman, I. and Shields, J., A Polygenic
     Theory of Schizophrenia, Proc. Nat. Acad.
     Sci,, 58, 199-205, 1967.

92. Levine, S. and Mullins, Jr., R., Hormonal
     Influences on Brain Organization in Infant
     Rats, Science, 152-1585-92, 1966.

93. Eayrs, J., In: Brain-Thyroid Relationships,
     Ciba Foundation Study Group, 18, Little,
     Brown and Co., Boston, 1964.

94. Glueck, E., Identification of Potential Del-
     inquents at 2-3 Years of Age, Int. J. Soc.
     Psychiat., 12, 5-16, 1966.

95. Lane, E. and Albee, G., Early Childhood
     Intellectual Differences Between Schizophrenic
     Adults and Their Siblings, J. Abnorm. Soc.
     Psychiat., 68, 193-195, 1964.

96. Schaffner, A., Lane, E. and Albee, G., Intel-
     lectual Differences Between Suburban Pre-Schi-
     zophrenic Children and Their Siblings, J. Con-
     sult. Psychol., 31, 326-327, 1967.

97. Singer, M., Family Transactions and Schizo-
     phrenia: I. Recent Research Findings, In:
     Origins of Schizophrenia, J. Romano, Ed.,
     Excerpta Medica, Amsterdam, 1967.

98. Bender, L., Twenty Years of Clinical Research
     on Schizophrenic Children with Special Refer-
     ence to Those Under Six Years of Age, In:
     Emotional Problems of Early Childhood, Caplan,
     G. Ed., Basic Books, New York, 1955.

99. Fish, B., Shapiro, T., Halpern, F. and Wile,
     R., The Prediction of Schizophrenia in Infancy:
     A Ten Year Follow up Report of Neurological
     and Psychological Development, Amer. J. Psy-
     chiat., 121, 68, 1965.

100. Pollack, M., et al., Childhood Development
     Patterns of Hospitalized Adult Schizophrenic
     and Non-Schizophrenic Patients and Their
     Siblings, Amer. J. Orthopsychiat., 36, 510-
     517, 1966.

101. Fleming, P., Emotional Antecedents of Schizo-
     phrenia: Inner Experiences of Children and
     Adolescents Who Were Later Hospitalized for
     Schizophrenia, read before the Conference on
     Life History Research in Psychopathology,

New York, May 12, 1967.
102. Mishler, E. and Waxler, N., An Ppproach to the
     Experimental Study of Family Interaction and
     Schizophrenia, Arch. Gen. Psychiat., 15, 64,
     1966.
103. Reiss, D., Individual Thinking and Family
     Interaction. II. A Study of Pattern Recognition
     and Hypothesis Testing in Families of Normals,
     Character Disorders, and Schizophrenics, J.
     Psychiat. Res., 5, 193-211, 1967.

THE VARIABLES OF BIRTH ORDER AND FETAL NUTRITION

DR. GAULL: Dr. Stabenau gave an excellent summary of some of the problems involved in the consideration of the study of twins and taking into account some of the environmental factors that make interpretation difficult.

There is one general factor that troubles me about this sort of study that he didn't discuss, and I wonder if I could draw him out on this. This factor is the birth order. The infant that is second born is subjected to quite a different environment and has been shown to have a statistically lower chance of survival and increased morbidity. I wonder if it is possible to ever interpret twin studies?

DR. STABENAU: This has been of considerable interest to our group. The direction that we have taken is to get as much history as we can about the birth order and the type of delivery and any possible effects of the delivery process, especially the respiratory ability of the child. That is, whether there was anoxia.

It is known that second born twins have reduced oxygen tension and higher CO_2 capacities compared to the first born; that the neonatal morbidity is higher in the second born.

There is some evidence from the work of John Churchill that head positioning within the uterus and especially anatomical pressuring during delivery effects one twin as compared to the other. He feels this produces differences in the twin's EEG's which he has been able to predict from birth position during delivery.

Thus, there are a number of variables that do relate to second born position.

In addition, twins as a whole tend to be "premature." Their gestational age is about two weeks shorter than for singleton neonates. This in itself needs to be taken into consideration, especially for birth weight differences within a pair of monozygotic twins, as one will often weigh substantially less than the other. In our non-random NIMH twin series for 12 of the 16 pairs, the twin lower in weight at birth (mean 8%) was the one who subsequently became schizophrenic.

Now, in the population of the 100 MZ twin case reports that I presented, birth weight material was available in 69 pairs. For 61 of these pairs one twin was lighter than the other. 41 of those 61 were the twins who subsequently became schizophrenic. Thus the issue of maturity and relative immaturity at birth as it may indicate the relative development of the central nervous system becomes a very important issue that needs to be considered as one of the non-genetic constitutional differences in individuals which interact with the genetic predisposition and later environmental experience.

DR. GAULL: Surely the division of food that comes to the twins is not equal and therefore in the case of monozygotic twins there must be some differences in the nutrition. The intrauterine nutrition, it is clear, is becoming more and more important. One wonders how one can ever sort this out meaningfully from the genetic effects.

DR. STABENAU: The only way we have been able to pursue this is by paying as much attention to the stage of development of each of the twins at birth, the development as reported by the parents, the early intellectual differences, and other subtle or gross differences.

In twin pairs who share a common placenta, there is a high degree of anastomotic interconnection of the blood vessels. Competitive circulation does exist and it does lead to differences in growth and size which we feel need to be taken into account in terms of adult performance.

RELATIONSHIP BETWEEN BIRTH WEIGHT AND ADULT PROTEIN
BOUND IODINE

    DR. STABENAU: I wanted to add in connection
with this, that in a series of 23 pairs of monozygotic twins, many of whom were "normal", the twin
of lower birth weight had a significantly lower
protein bound iodine value as compared to his co-
twin and other twins of higher birth weight. The
relationship between birth weight and adult PBI
for 46 MZ twins and BEI (butanol-extractable iodine)
in 129 neonates is given in Figure 1. In a sub-
series of 12 of these individuals, this difference
was attributable to lower serum thyrobinding globulin. The possibility that the less mature neonate may have already established a difference in
protein synthesis is an important consideration.

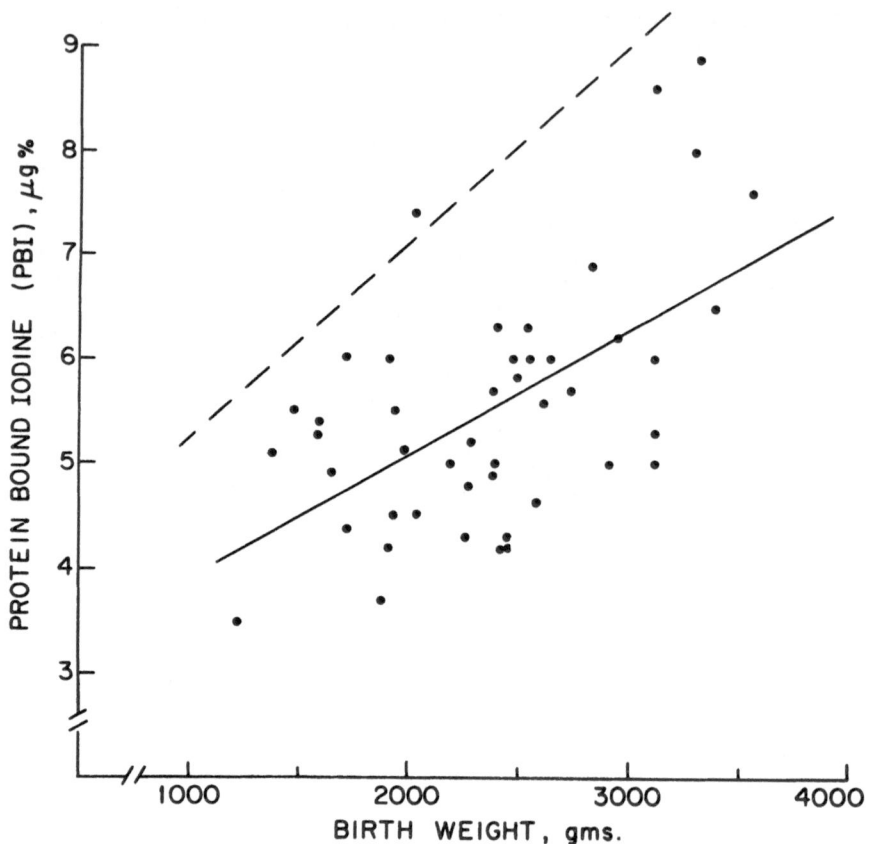

Fig. 1. Relationship between adult protein-bound
iodine (PBI) and weight at birth for 46 monozygotic twins. Solid line represents the least squares
linear fit. PBI (µg/100 ml) = 2.7424 + 1.1805 birth
weight (kg.) (N = 46). The correlation coefficient
is +0.58,p = <.001. The broken line represents the
least squares linear fit for 129 neonates studied by
Marks and Man. BEI (µg/100 ml = 3.3217 + 1.8963
birth weight (kg). The correlation coefficient is
+0.65,p = <001.
From Stabenau, J. and Pollin, W.: ADULT PROTEIN
BOUND IODINE AND MATURITY AT BIRTH IN MONOZYGOTIC
TWINS, J. Clin. Endocrin. and Metab., 28, 693-699,
1968, (Reprinted with the permission of the J.Clin.
Endocrin. and Metab.)

PRESENT DIRECTIONS IN THE INVESTIGATION OF

DIMETHOXYPHENETHYLAMINE

Arnold J. Friedhoff

Over the past ten years, tremendous advances have been made in the basic understanding of the biological substratum of mental processes. This has been reflected by increased understanding of the metabolism of neurohormones, the genetics of mental illness, the structure activity relationship of hallucinogens and the possible mechanism of action of psychopharmacological agents. Stemming from this new understanding have been a number of new techniques for the study of biochemical processes, and as a result of these techniques, several investigations of biochemical abnormalities related to schizophrenia have been undertaken which were not previously possible. All of these advances have given great forward momentum to the field of biological psychiatry, and it should be necessary to detail only a few to demonstrate this point.

It may have been the re-introduction of the drug reserpine as an anti-psychotic agent which initiated this revolution in psychiatry. In any event, this was one of the first in a series of developments which have resulted in the remarkable growth of basic research in psychiatry. Among these accomplishments has been the elucidation of the metabolic pathways of enzymes involved in the metabolism of catecholamine neurohormones. Extremely important basic research related to tryptamine and serotonin metabolism has also been carried out. A new science of psychopharmacology has been developed which has produced a vast number of important works. Among these must certainly be mentioned the dramatic improvement in the treatment of psychotic patients resulting from the use of new drugs. The clinical effects of hallucinogens have been very closely studied and it is now possible to design certain hallucinogenic molecules with some assurance about their potency and duration of action.

The observation that all effective anti-psychotic agents have the potential for producing Parkinsonian symptoms has stimulated several new avenues of research. The elaboration of a possible disturbance in dopamine metabolism in Parkinsonism has also illuminated a possible mechanism of action of the anti-psychotic agents themselves. In another area, a great advance in epidemiological and genetic studies in mental illness has been generated. Also the great strides in behavioral evaluation cannot be overlooked. The model for a good number of basic studies, carried out in many laboratories, was established by the interdisciplinary Laboratory of Clinical Sciences at the National Institute of Health. In this laboratory, the application of basic science techniques to the study of brain behavior relationships has reached a high level.

Perhaps more important than any single discovery has been a shift away from the breakthrough approach in the study of mental illness with the increasing maturity of the field of biological psychiatry. Frontal assaults on crucial disorders, such as schizophrenia, have been replaced by systematic and orderly basic studies of underlying biological systems. While the etiology and pathogenesis of schizophrenia remain largely uncertain, significant

progress has been made in the investigation of possible biochemical aberrations involved in this syndrome.

In our laboratory, we undertook, in 1961, to
study urine obtained from schizophrenic patients
and normal controls to determine whether there
were differences in the excretion of methylated
metabolites of dopamine and related compounds.
In our first group of 19 severely ill acute schizophrenic patients and 14 normal controls, we found
that a compound which we subsequently identified
as 3,4-dimethoxyphenethylamine (DMPEA) was present
in the urine of 15 schizophrenic patients and none
of the normal controls (Friedhoff and Van Winkle,
1962 a,b). Since these original studies we have
identified DMPEA in hundreds of urine samples from
schizophrenic patients. Urines from some patients
have been collected at intervals and reanalyzed,
and results are reproducible. In a recent tally
of 86 patients and 27 normal controls, assayed in
our laboratory by the same paper chromatographic
technique, we found that 67% of all patients excreted detectable amounts of DMPEA, as did 2 of 27
normals or about 8% of this group (Friedhoff, 1968).

Several studies have been carried out in other
laboratories which have confirmed our findings. In
the largest of these, a double blind study carried
out at the University of Liverpool, involving several hundred subjects, our original chromatographic
findings were confirmed (Ridges and Bourdillon,
1966; Bourdillon and Ridges, 1967). Another study
has been carried out by Predescu et al. (1968), in
which 221 patients and controls were studied. Approximately 45% of schizophrenic subjects were found
to excrete DMPEA, while only isolated subjects in
the other groups excreted this compound. Nordio
et al. in Italy have found that some children with
a diagnosis of childhood schizophrenia also excrete
this compound (1968). Several additional studies
have been partially confirmatory (Sen and McGeer,
1964; Takesada et al., 1963), while some investigators report either that a similar compound is excreted by non-schizophrenics (Vogel et al., 1967;

METABOLISM OF DMPEA

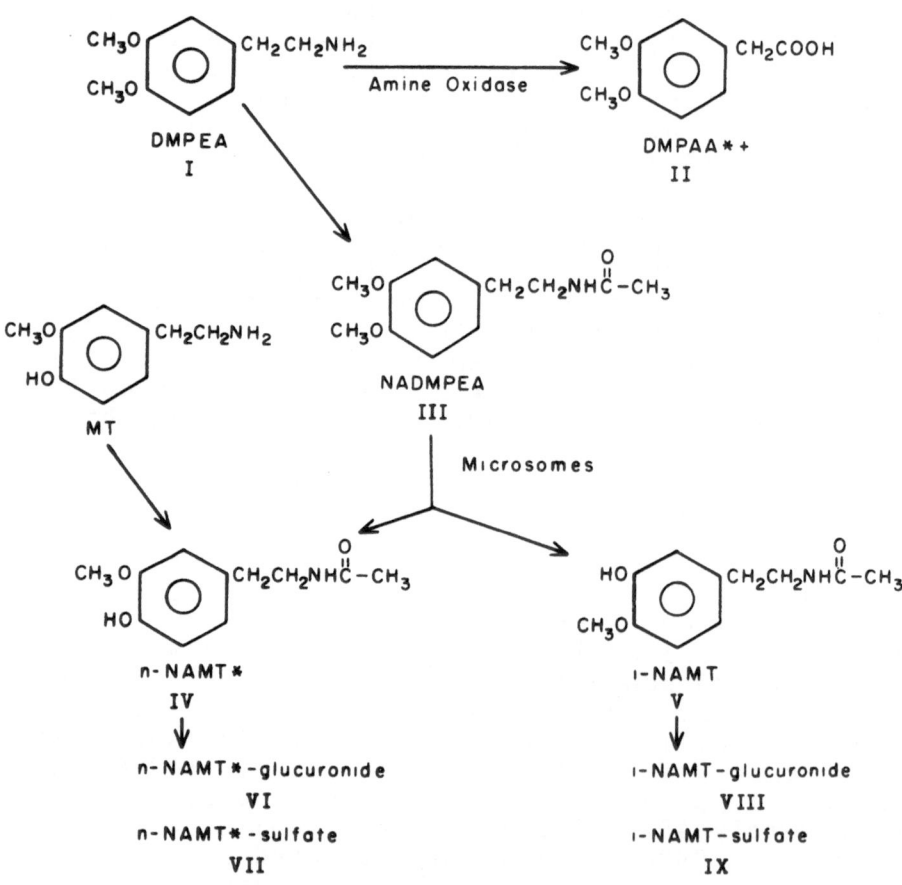

Fig. 1.

Takesada et al., 1963) or that DMPEA is not excreted at all (Perry et al., 1964; Faurbye and Pind,
(1964). However, that DMPEA is, in fact, excreted
has been demonstrated by convincing means in our
own laboratory (Friedhoff and Van Winkel, 1962 b;
1966) and by Creveling and Daly (1967) at the National Institute of Health, who used mass spectro-
metric means.

It sould be noted that in our laboratory we
used drug free, acutely ill schizophrenic patients
and a number of different groups of drug free controls. To our knowledge, there are no published
reports of non-confirmatory studies in which the
investigators have used a patient population and
techniques identical with ours; that is, in the
several non-confirmatory studies, either a different patient population was used--often chronic or
drug treated (Faurbye and Pind, 1964)--or different
techniques were used (Bell and Somerville, 1966;
Vogel et al., 1967). The failure of some investigators to use drug free populations raises serious
problems, since we have found that administration
of phenothiazines results in the disappearance of
DMPEA from the urine (Friedhoff, 1968).

Since our original identification of DMPEA,
we have also identified several degradation products of this compound. In Figure 1, it will be
seen that DMPEA is metabolized to 3,4-dimethoxy-
phenylacetic acid (DMPAA), which is the major metabolite of DMPEA (Friedhoff and Furiya, 1967), and
through a second pathway to N-acetyl-DMPEA (NADMPEA)
which is then demethylated to N-acetylmethoxytyra-
mine(NAMT). The first pathway is of interest because DMPEA has been identified in the urine of
schizophrenics, as well as normals. This identification has been made in our own laboratory (Friedhoff and Furiya, 1967) as well as that of Kuehl
(1966). From the finding that schizophrenics, as
well as normals, excrete DMPAA, while only schizophrenics excrete DMPEA, it would appear that: 1)
DMPEA is synthesized by both normals and schizophrenics, since its metabolite, DMPAA, is excreted
by both groups; 2) DMPEA is excreted by schizophre-

nics, but not by normals because its binding, transport or biosynthesis, and degradation are in some
way different in schizophrenics.

In order to investigate these possibilities,
we have administered labelled DMPEA to schizophrenics and normals, but have failed to find differences in the ability of these groups to metabolize
DMPEA. However, it should be borne in mind that
administration of this compound through the arm
vein results in metabolism primarily by peripheral
organs rather than brain. If disturbances in metabolism exist in the brain, they would probably not
be detectable by the techniques used. This fact
is of considerable interest because we have found
that demethylation of NADMPEA to NAMT does not take
place in the brain in vitro in any system that we
have been able to establish. This metabolic transformation does occur readily in the liver and,
in fact, can be localized to the liver microsomes.
We have now established, through circumstantial
evidence, that demethylation of NADMPEA does occur
in the brain in vivo. The possibility, therefore,
arises that there are different demethylating mechanisms in the brain from those found in the peripheral organs. The demethylation reaction serves
as an inactivation reaction since the acetylation
of DMPEA to NADMPEA activates this compound.
NADMPEA is about ten times as potent as DMPEA and
about five times as potent as mescaline in producing the hypokinetic rigid state in rats.

The importance of considering metabolism when
evaluating the effect of peripherally administered
compounds is further demonstrated by the findings
of Vacca et al. (1968). This group demonstrated
that DMPEA was more potent than mescaline in its
effect on neuronal transmission when injected into
the carotid artery, so as to avoid the large amounts
of monoamine oxidase (MAO) in the liver. On the
other hand, when DMPEA and mescaline are given into peripheral veins, mescaline is more potent because it is less readily metabolized by MAO.

We have also studied the biosynthesis of DMPEA

using dopamine as a precursor. We have found that schizophrenics are able to convert dopamine in low yield to DMPAA. Since DMPAA is a metabolite of DMPEA, it is presumed that DMPEA is an intermediate in this metabolic reaction. We are presently studying this reaction in normals.

Of considerable interest is the recent report of Proctor et al. (1968) that DMPEA, when incubated with plasma of schizophrenics and injected into mice aggregated in groups, results in the death of most of the mice. DMPEA incubated in a similar way with plasma from non-schizophrenic controls failed to cause this toxic effect. These findings are consistent with earlier reports of Bergen (1965) that DMPEA in extremely low dose, when incubated with serum from schizophrenics, can produce detectable behavioral effects in rats.

These findings, if confirmed, would raise interesting possibilities about the significance of DMPEA in schizophrenia. It may be that a number of seemingly unrelated studies represent different aspects of the same phenomenon. The metabolism and pharmacology of DMPEA will have to be further studied, as will its effects on neural transmission. It becomes increasingly clear that further basic studies of DMPEA and related compounds will have to be carried out before it can be determined whether this compound may have significance in schizophrenia.

REFERENCES

1.   Bergen, J.R., Possible relationship of a plasma factor to schizophrenia, Trans. N.Y. Acad. Sci., 28, 40-46, 1965.
2.   Bourdillon, R. E. and Ridges, A. P., 3,4-Dimethoxyphenylethylamine in schizophrenia? In: Amines and Schizophrenia, Himwich, Kety and Smythies (Eds.), Pergamon Press, Oxford, 1967.
3.   Creveling, C. R. and Daly, J. W., Identification of 3,4-dimethoxyphenylethylamine from schizophrenic urine by mass spectrometry, Nature, 216, 190-191, 1967.

4. Faurbye, A. and Pind, K., Investigation on the occurrence of the dopamine metabolite, 3,4-dimethoxyphenylethylamine, Acta Psychiat. Scand., 40, 240-243, 1964.

5. Friedhoff, A. J., Metabolism of dimethoxy-phenethylamine and its possible relationship to schizophrenia. In: Origins of Schizophrenia, Romano (Ed.), Excerpta Medica Internat'l Cong. Ser., 151, 27-34, 1968.

6. Friedhoff, A. J. and Furiya, K., 3,4-Dimethoxyphenylacetic acid in urine, Nature, 214, 1127-1128, 1967.

7. Friedhoff, A. J. and Van Winkle, E., Isolation of a compound from the urine of schizophrenics, Nature, 194, 897-898, 1962 a.

8. Friedhoff, A. J. and Van Winkle, E., New Developments in the investigation of the relationship of 3,4-dimethoxyphenethylamine to schizophrenia. In: Amines and Schizophrenia, Himwich, Kety & Smythies (Eds.), Pergamon Press, Oxford, 1966.

9. Friedhoff, A. J. and Van Winkle, E., The characteristics of an amine found in the urine of schizophrenic patients, J. Nerv. Ment. Dis., 135, 550-555, 1962 b.

10. Kuehl, F. A., Ormond, R. E. and Vandenheuvel, W. J. A., Occurrence of 3,4-dimethoxyphenylacetic acid in the urines of normal and schizophrenic individuals, Nature, 211, 606, 1966.

11. Nordio, S., Berio, A. and DiStefano, A., Pink spot and other amines in urine, Lancet, 2, 311-312, 1967.

12. Perry, T. L., Hansen, S. and Macintyre, L., Failure to detect 3,4-dimethoxyphenylethylamine in the urine of schizophrenics, Nature, 202, 519-520, 1964.

13. Predescu, V., Florescu, D. and Radulescu, C., Pink spot as a diagnostic test in schizophrenia, Nature, 217, 1150-1151, 1968.

14. Proctor, C. D., Cho, J. B., Ashley, L. G., Potts, J. L., Eaton, H. E., McGriff, J. E., Douglas, J. G. and Amoroso, C. P., Factors influencing a blood plasma-DMPEA test for schizophrenia, Fedn. Proc., 27, 219, 1968.

15. Ridges, A. P. and Bourdillon, R. E., Schizophrenia, transmethylation and pink spot, Proc.
Royal Soc. Med., 60, 555-560, 1966.
16. Sen, N. P. and McGeer, P. L., 4-Methoxyphenyl-
ethylamine and 3,4-dimethoxyphenylethylamine
in human urine, Biochem. Biophys. Res. Commun.,
14, 227, 1964.
17. Takesada, M., Kakimoto, Y., Sano, I. and Kane-
ko, Z., 3,4-Dimethoxyphenylethylamine and other
amines in the urine of schizophrenic patients,
Nature, 199, 203-204, 1963.
18. Vacca, L., Fujimori, M., Davis, S. H. and Mar-
razzi, A., Cerebral synaptic transmission and
behavioral effects of dimethoxyphenylethylamine: A potential psychotogen, Science, 160,
95-96, 1968.
19. Vogel, W. H., Ahlberg, C. D. and Horwitt, M.
K., Time study of the urinary excretion of
3,4-dimethoxyphenylethylamine and 3,4-dimeth-
oxyphenylacetic acid by schizophrenic and normal individuals, Internat. J. Psychiat., 3,
292-297, 1967.

RELATIONSHIP BETWEEN URINARY CONSTITUENTS

AND EXACERBATIONS OF BEHAVIOR

IN SCHIZOPHRENIC PATIENTS

Harold E. Himwich

The investigations on the urinary biochemistry of schizophrenic patients made at the Galesburg State Research Hospital represent the joint
efforts of many members of our staff (Brune and
Himwich, 1960; Brune and Himwich, 1962; Berlet
et al. 1964; Berlet et al, 1965; Spaide et al.
1967; Tanimukai et al In press). We chose patients with chronic schizophrenia and active
symptoms. 24-hour urine collections were analyzed for various indoleamines which are products
of tryptophan in the body. Our chief concern was
with tryptamine (See Fig. 1). In addition, 5-
hydroxy indoleacetic acid and 3-indoleacetic
acid were also determined. Behavior was observed
daily and later when the biochemical studies
were completed, the simultaneous biochemical and
behavioral results were compared.

For one study we chose eight chronic male
schizophrenic patients with active symptoms (Berlet et al. 1965). Psychotropic drugs were discontinued two to four months prior to the beginning of the observations and these drugs were
withheld throughout the course of this study.
The patients manifested symptoms such as hallucinations and delusions which they discussed with
the psychiatrists. Moreover, they could be man-

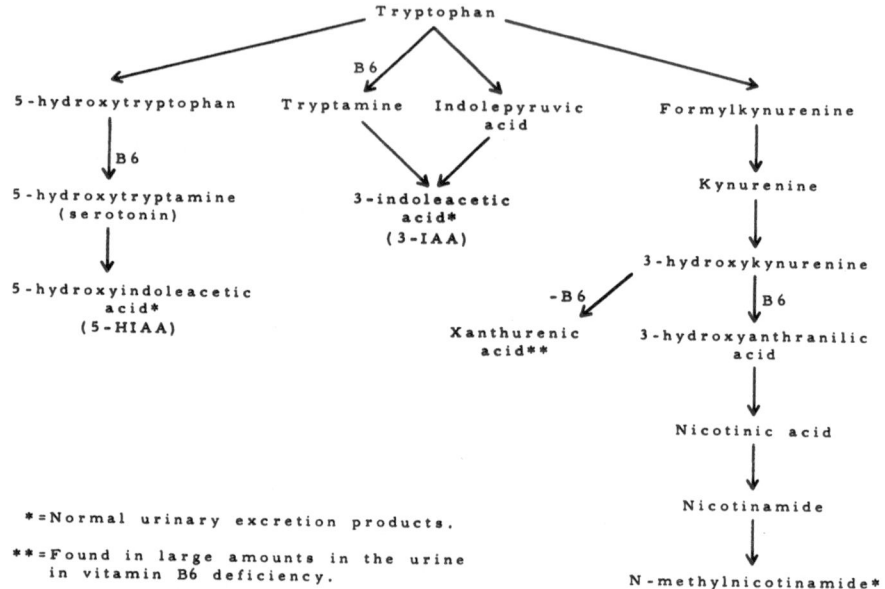

Fig. 1. Pathways for the metabolism of tryptophan.

aged without medication by our specially skilled
ward personnel. Weekly psychiatric examination
included cooperation and attention during the
interview, motor behavior, affectivity, thought
content, stream of thought, sensorium and ward
adjustment. Variations in these parameters were
described in terms of: not present (0), mild
(-1), moderate (-2), severe (-3), or extremely
severe (-4). Plus signs were used if behavior improved. The patients were maintained in the metabolic ward on a diet especially prepared in the
metabolic kitchen. All preformed indoleamines
were excluded from the diet.

In this study, five-day urine collections
were made and aliquots were used for biochemical
analyses. Additional 24-hour samples were collected at times of sudden behavioral changes in
order to compare biochemical alterations with
the more dynamic transitions of behavioral disturbances.

Exacerbations of schizophrenic behavior
and urinary indoles:

In two of the eight patients, wide behavioral
fluctuations developed and the indole excretions
were studied in detail in relation to chemical
determinations on daily urine collections obtained during the height of behavioral outbursts
(see Fig. 2). The chief result that emerged from
this study is that tryptamine increased in the
urine preceding the worsening of the schizophrenic

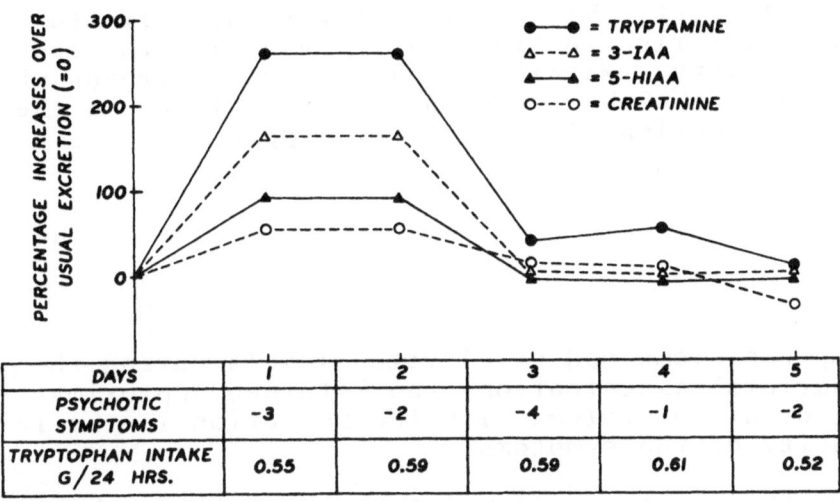

Fig. 2. Correlations between severity of symptoms
and levels of urinary constituents. On day one, an
aggravation of psychotic symptoms was first observed and thereafter 24-hour urinary samples were collected daily for the next 4 days. On percentage
changes in relation to usual excretion levels, an
increase in all three urinary indoles as well as
creatinine occurred. Tryptamine was by far the
most sensitive indicator. The most severe behavior worsening occurred on the third day, accompanied and followed by decreased concentrations of
all urinary constituents, with a return to the
patient's usual behavior.

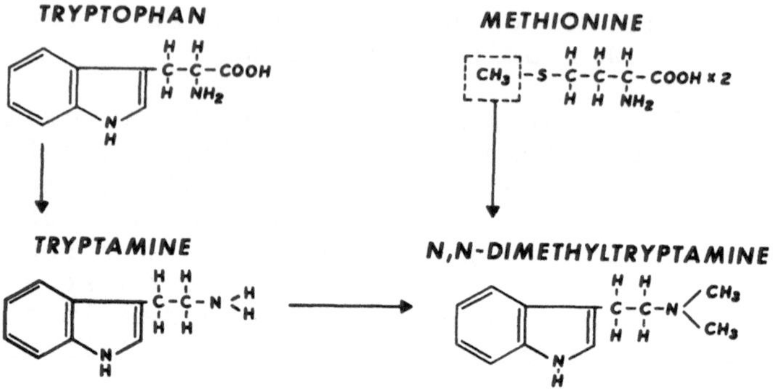

Fig. 3. Hypothetical mechanism for formation of psychotomimetic N:N-dimethyltryptamine. Tryptophan gives rise to tryptamine containing the indoleamine moiety. Methionine yields methyl groups which may combine with tryptamine thus resulting in the formation of N:N-dimethyltryptamine.

symptoms, remained at elevated levels during the behavioral exacerbation, and returned approximately to initial values with the reduction in the intensity of the symptoms.

This increase of urinary tryptamine probably reflects a similar elevation in the blood. It is probable, therefore, that these increases were the results of releases of free tryptamine in the body. Because the increases occurred only during behavioral worsening, it would seem that tryptamine or a tryptaminic compound acts like an endogenous metabolic factor in intensifying the psychotic symptoms (Berlet et al. 1964).

Pollin, Cardon and Kety (1961) performed brilliant experiments loading schizophrenic patients with methionine in addition to iproniazid, a MAO inhibitor. They noted marked exacerbations in the severity of the symptoms, findings which we confirmed (Brune and Himwich, 1962).

Taking into consideration the results that
methionine plus a MAO inhibitor evoked an aggravation of psychotic symptoms (Pollin et al,
1961), as well as our own, on the increased concentration of tryptamine in the urine in association with behavioral worsening, we suggested (see
Fig. 3) (Brune and Himwich, 1962) that a psychomimetic substance related to N:N-dimethyltrypta-
mine was formed in the body, the methylggroups
coming from methionine and the indole structure
from tryptophan. This suggestion is in accordance with the paper of Axelrod (1962), who has
shown that the mammalian body contains an enzyme
which can facilitate the formation of N:N-di-
methyltryptamines from unmethylated tryptamines
and that the latter occur naturally in the body.

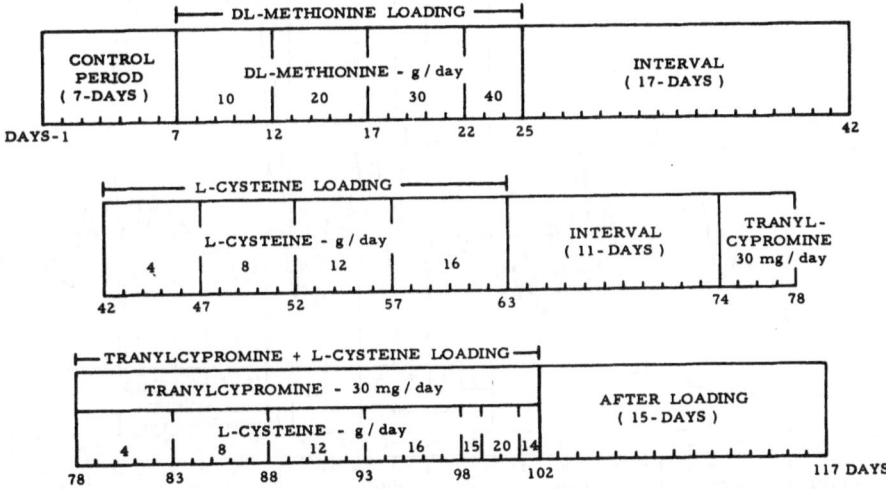

Fig. 4. Research design for patients receiving
cysteine with and without tranylcypromine. After
an interval of 17 days, thepatients were placed
on increasing doses of cysteine with increments of
4 gm each for four 5-day periods. After another
interval of 11 days, they received first tranylcypromine, 30 gm/day for four days, and then to the
tranylcypromine was added increasing doses of cystein for the next 24 days. The research design
was terminated with another interval free of drugs
for 15 days.

We regarded this hypothetical substance as a
mediator in the worsening of schizophrenic behavior.

In another series of experiments we gave
tranylcypromine (Parnate), a MAO inhibitor, and instead of methionine, substituted cystein (Spaide
et al. 1967) (Fig. 4).   The various phases of the
research design are as follows: placebo, cysteine,
placebo, tranylcypromine alone, tranylcypromine and
cysteine in increasing doses from 4 gm/day to 20
gm/day. We concluded with another placebo period.

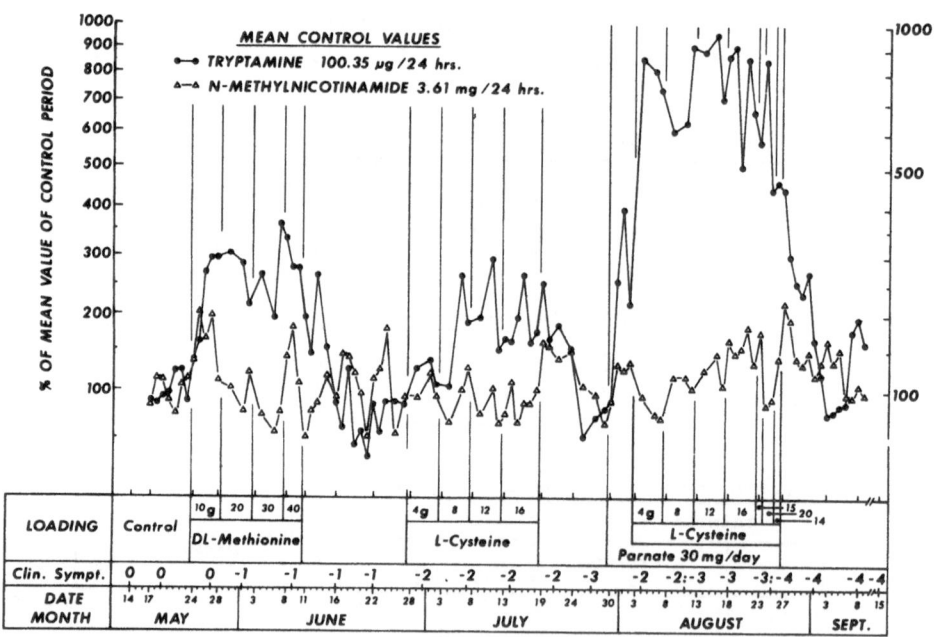

Fig. 5.   Relationships between alterations in behavior and changes in tryptamine and N-methylnicotinamide as expressed in terms of percentage differences from the controls. Immediately following
the combined treatment when tryptamine had been at
its highest levels, the patient's behavior became
gravely affected. Not shown on the slide is his
slow return to his usual behavioral levels. The
increases in N-methylnicotinamide were more moderate than those in tryptamine.

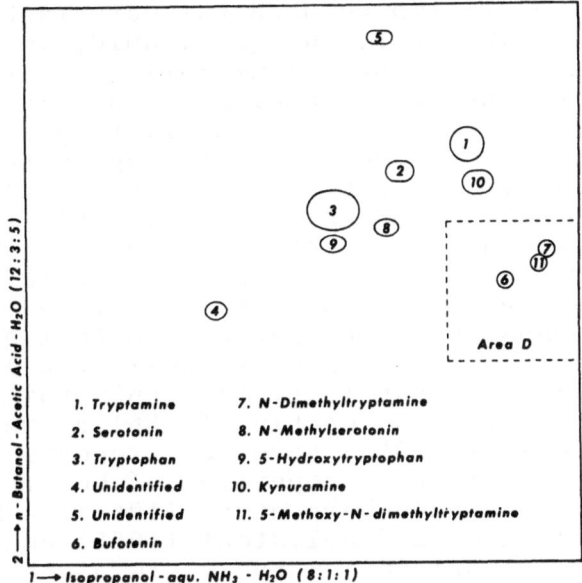

Fig. 6. Thin-layer chromatogram of indoleamines in the urine of a schizophrenic patient. Area D shows spots for bufotenin (6), N:N-dimethyltrypta-mine (7), 5-methoxy N:N-dimethyltryptamine (11).

Under the combined medication, tryptamine rose to high levels in the urine. We explained our results on the basis of cysteine-induced releases of tryptophan and methionine resulting from the breakdown of body proteins including those of muscle as indicated by elevations of creatinine. Behavior-ally, under the combined treatment, the exacerba-tions of schizophrenic symptoms in the four patients took place in accordance with their characteristic psychiatric profiles. The three paranoid patients became more paranoid and the catatonic patient more catatonic (Fig. 5). In this typical example you ob-serve that during a period of worsening of symptoms, with their ratings rising from -2 to -4, urinary tryptamine was greatly elevated. This patient form-ed new and bizarre delusions under the influence of the combined treatment.

In our next step we used chromatographic methods, paper, thin-layer, and gas-liquid, for further
urinary analyses on these four patients. After having subjected the urine samples to purification processes in order to eliminate interfering substances,
we performed the various chromatographic techniques
on the samples. You see in Figure 6 the results of
thin-layer analyses by which we were able to dis-
close, bufotenin, N:N-dimethyltryptamine, and 5-
methoxy N:N-dimethyltryptamine. We, therefore, have
some substantiation of our theory that the tryptaminic compounds, by combining with methyl groups,
were transformed from a non-psychomimetic agent into a psychomimetic one (Fig. 7) (Tanimukai et al.
In press). Fabing and Hawkins (1956) as well as
Turner and Merlis (1959) have made reports on the
possible psychomimetic properties of bufotenin;
Szara (1957) and Rosenberg et al. (1963) on N:N-
dimethyltryptamine and Holmstedt (1965) and Gessner and Page (1962) on 5-methoxy N:N-dimethyltryp-
tamine.

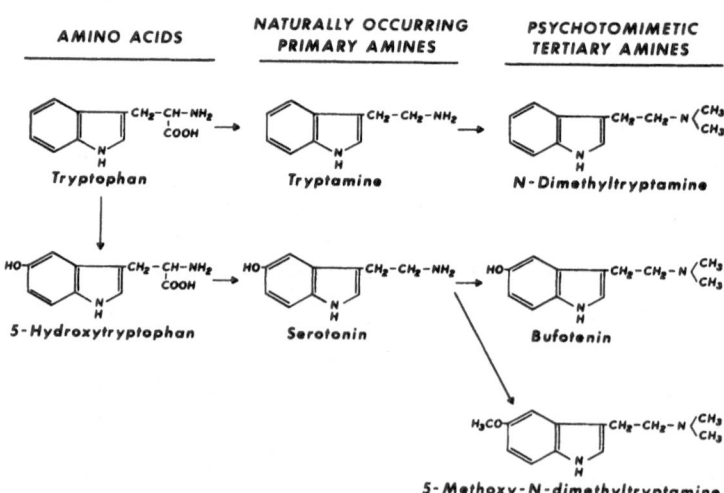

Fig. 7. Possible metabolic pathways for formation
of psychotomimetic indoleamines. These pathways
indicate the transformation of tryptamine to N:N-
dimethyltryptamine, serotonin to bufotenin, and 5-
methoxy N:N-dimethyltryptamine.

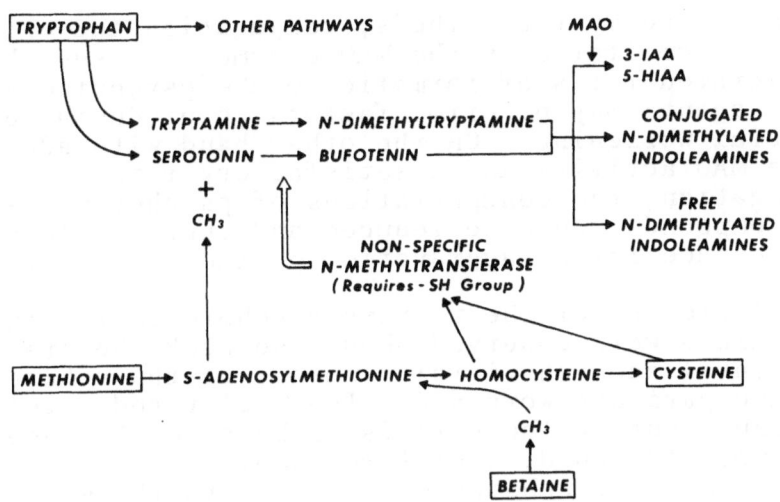

Fig. 8. Pathways for formation and detoxication of psychotomimetic indoleamines. Beginning on the left side of the diagram from the reader's viewpoint, tryptophan may give rise to tryptamine and serotonin. Methionine and betaine can yield methyl groups. Cysteine and homocysteine give rise to the sulfhydrhyl groups necessary for activation of non-specific N-methyltransferase. On the right side are presented the mechanisms for the detoxication of the free indoleamines, the MAO facilitated oxidation, and their conjugation. The remaining free N:N-dimethylated indoleamines have escaped the activities of the two detoxicating mechanisms.

But not all patients exhibit exacerbations of their symptoms in response to the combined treatment. In Figure 8 we are presenting some factors which may explain why some patients are reactors and others are not, at least from a biochemical viewpoint (Tanimukai et al. In press). Behavioral changes can be evoked only by the free tryptaminic compounds and not by the combined ones. The levels of the free compounds are determined by a dynamic balance between their rates of formation and detoxication. The latter process is

accomplished by two methods, conjugation and oxidation facilitated by the MAO enzyme. We see that
accelerated rates of formation of the psychotomimetic agents may be prime factors in producing behavioral worsening. On the other hand with adequate MAO activity and a satisfactory rate of
conjugation, the concentrations of psychotomimetic compounds would be reduced and there would be
less chance for an aggravation of the symptoms.

Increases of the urinary psychotomimetic indoleamines were observed about two weeks before
the mental and behavioral symptoms of the schizophrenic patients worsened. These elevated levels
continued during the periods of behavioral exacerbations. It would, therefore, seem that these
psychotomimetic substances may mediate the worsening of psychotic symptoms in schizophrenic patients, at least under our experimental conditions.

SUMMARY

From the results of experiments by Pollin,
Cardon and Kety, who treated schizophrenic patients with a MAO inhibitor drug in addition to
the amino acid, methionine, and observed an exacerbation of symptoms in schizophrenic patients,
and our findings that the behavioral worsening
was preceded and accompanied by rises of urinary
tryptamine, we suggested the hypothesis that
tryptamine combined with methyl groups, released
by methionine, to form a tryptaminic psychogenic
compound which acted as an endogenous metabolic
factor in the aggravation of the schizophrenic
symptoms. Some support for this hypothesis was
supplied by our observations indicating that bufotenin, N:N-dimethyltryptamine, and 5-methoxy N:N-
dimethyltryptamine accumulated in the urine of
schizophrenic patients before and during the
period of behavioral worsening. Our next step
will be to determine the responses of normal controls who will be subjected to a MAO inhibitor and
an amino acid, and to compare the new results with
those formerly obtained on schizophrenic patients.

REFERENCES

1. Brune, G. and Himwich, H. E., Effects of re-
    serpine on urinary tryptamine and indole-3-
    acetic acid excretion in mental deficiency
    schizophrenia and phenylpyruvic oligophre-
    nia, Acta of the International Meeting on
    the Techniques for the Study of Psychotropic
    Drugs, Bologna, 1-11, June 26-27, 1960.

2. Brune, G. G. and Himwich, H. E., Effects of
    methionine loading on the behavior of
    schizophrenic patients, J. Nerv. Ment. Dis.,
    134, 447-450, 1962.

3. Berlet, H. H., Bull, C., Himwich, H. E.,
    Kohl, H., Matsumoto, K., Pscheidt, G. R.,
    Spaide, J., Tourlentes, T. T. and Valverde,
    J. M., Endogenous metabolic factor in schizo-
    phrenic behavior, Science, 144, 311-313,
    1964.

4. Berlet, H., Spaide, J., Kohl, H., Bull, C.
    and Himwich, H. E., Effects of reduction of
    tryptophan and methionine intake on urinary
    indole compounds and schizophrenic behavior,
    J. Nerv. Ment. Dis., 140, 297-304, 1965.

5. Spaide, J., Tanimukai, H., Ginther, R.,
    Bueno, J. and Himwich, H. E., Schizophrenic
    behavior and urinary tryptophan metabolites
    associated with cysteine given with and
    without a monoamine oxidase inhibitor (tranyl-
    cypromine), Life Sciences, 6, 551-560, 1967.

6. Tanimukai, H., Ginther, R., Spaide, J.,
    Bueno, J. R. and Himwich, H. E., Psychoto-
    genic N:N-dimethylated indoleamines and be-
    havior in schizophrenic patients, Recent
    Advances in Biological Psychiatry, J. Wortis,
    (Ed.), In Press.

7. Fabing, H. D. and Hawkins, J. R., Intra-
    venous bufotenin injection in the human
    being, Science, 123, 886-887, 1956.

8. Turner, W. J. and Merlis, S., Effect of some
    indole-alkylamines on man, A.M.A. Arch.
    Neurol. Psychiat., 81, 121-129, 1959.

9. Szara, S., The comparison of the psychotic
    effect of tryptamine derivatives with the
    effects of mescaline and LSD-25 in self-ex-

periments, in Psychotropic Drugs, ed. S. Garattini and V. Ghetti (Amsterdam: Elsevier, 460-467, 1957.

10. Rosenberg, D. E., Isbell, H. and Miner, E. J., Comparison of a placebo, N-dimethyl-tryptamine, and 6-hydroxy-N-dimethyltrypta-mine in man, Psychopharmacologia, 4, 39-42, 1963.

11. Holmstedt, B., Tryptamine derivatives in epená, an intoxicating snuff used by some South American Indian tribes, Arch. int. Pharmacodyn. Ther., 156, 285-305, 1965.

12. Gessner, P. K. and Page, I. H., Behavioral effects of 5-methoxy-N:N-dimethyltryptamine, other tryptamines and LSD, Am. J. Physiol., 203, 167-172, 1962.

13. Pollin, W., Cardon, P. V. Jr. and Kety, S. S., Effects of amino acid feedings in schizophrenic patients treated with ipro-niazid, Science, 133, 104-105, 1961.

14. Axelrod, J, The enzymatic N-methylation of serotonin and other amines, J. Pharmacol. Exp. Ther., 138, 28-33, 1962.

PATHOGENETIC STUDIES OF SCHIZOPHRENIA:

FERRITIN AND FLUORESCEIN LABELING OF IgG

Robert G. Heath,

Austin Fitzjarrell, and Iris M. Krupp

Research in schizophrenia at Tulane during the past 19 years suggests that it is an immunologic disorder that affects primarily the septal region of the brain. Specifically, we postulate that the schizophrenic patient, probably because of genetically abnormal clones of cells, produces a unique abnormal immunoglobulin molecule that can enter cytoplasm of oligoglial cells of the septal region. The presence of this immunoglobulin at this specific site impairs neural transmission and thus causes pathologic activity in the septal region, which is basic to the psychotic state. This hypothesis has been substantiated by recent studies involving specific labeling of the unique molecule, taraxein.

BACKGROUND STUDIES

Previous studies that prompted this hypothesis have been extensively reported, and will only be briefly summarized here. In a series of 58 patients prepared with depth electrodes for long-term study and treatment, electroencephalograms (EEGs) recorded from the septal region of those who were display-

ing psychotic signs and symptoms consistently showed focal spiking or slow waves, or both (1,2,3).
Similar abnormal EEGs were recorded for monkeys
after intravenous injection of a protein serum fraction (taraxein) of acutely psychotic schizophrenic
patients (4,5). Injection of taraxein into non-psy-
chotic volunteer-subjects induced clinical symptoms
of schizophrenia, which persisted for about one
hour (6,7). Corresponding serum fractions from
patients with other diseases and from healthy control subjects, on the other hand, were inert.

In further testing of our hypothesis, we did
fluorescent antibody studies on tissues of brains
removed shortly after death (8). Presence of in
vivo globulin at focal brain sites in schizophrenic
patients who had shown psychotic signs just before
death, and its absence at these sites in control
subjects, suggested an intimate relation between a
human serum globulin and the site of abnormal
activity in the brain of psychotic schizophrenic
patients. Globulin was also detected at these
same sites in monkeys killed during peak response
to intravenous injections of taraxein (4,9).

Experiments involving papain digestion of the
schizophrenic IgG molecule (taraxein) indicated that
the antibody-binding fragment of the molecule was
responsible for the pathologic activity, and supported the hypothesis that psychosis-inducing schizophrenic IgG (taraxein) is antibody (10). This immunologic hypothesis of schizophrenia was further
substantiated by experiments in which antibody was
made against various tissues of monkey and human
brains by inoculating sheep with homogenates of
tissues-with-adjuvant and then obtaining sera at
the expected peak of antibody response (11,12). A
highly specific antibody produced against precise
antigenic sites of the septal region resembled
taraxein in its effects on behavior and brain function of monkeys (as judged by EEG and indirect
fluorescent antibody tests), as well as by immunoelectrophoretic patterns.

STUDIES WITH LABELED GAMMA G IMMUNOGLOBULINS

Materials and Methods

Serum fractions. Gamma G immunoglobulins (IgG) of acutely psychotic schizophrenic patients and control subjects, obtained by ion exchange chromatography over DEAE Sephadex A-50 (4,5), were first tested in the standard monkey assay for psychosis-inducing activity (4). Each of five schizophrenic IgG fractions had strong psychosis-inducing activity. When similarly tested in the monkey, four control serum fractions, three from healthy medical students and one from a patient with myasthenia gravis, were inert.

Immunoelectrophoretic analyses. Immunoelectrophoretic analyses (IEA) of all test serum fractions were performed before and after labeling on a special LKB apparatus. IEA of all unlabeled serum fractions against anti-human gamma globulin yielded IgG alone. Similar analyses of ferritin-labeled IgG molecules also showed only one band, but this band differed in mobility, the negative charge of the ferritin having caused it to move toward the positive pole (Fig. 1).

(In two instances, a small amount of unlabeled IgG was demonstrated along with the ferritin-labeled IgG.) Fluorescein against anti-human globulin does not, of course, show on IEA.

Total protein determinations. Total proteins of all fractions were read before labeling on a Beckman DU Spectrophotometer at a wave length of 280 millimicrons. The amount of protein in the starting material varied from 35 to 75 mg in 8 to 15 ml solution.

Monkey recipients of test materials. Eight Macaca rhesus monkeys received labeled gamma G immunoglobulins intravenously (Table 1): four received IgG of schizophrenic patients ($S_1 - S_4$), 2 received IgG of healthy medical students (C_1, C_2), one received IgG of a patient with myasthenia gravis

Table 1

RESULTS OF MICROSCOPY

IgG TAGGED WITH FERRITIN AND FLUORESCEIN ISOTHIOCYANATE

Experiment	Monkey No. (Type IgG Injected)	IFA	EEG	Clinical Response	Interval Inj to Death (min)	Microscopic Evidence of Labeling Material Electron (Ferritin)	Ultraviolet (Fluorescein)
Phase I: IgG tagged with Ferritin (F) + Fluorescein Isothiocyanate--brain not perfused							
I	S_1*	IgG + F	+	+	20	septal region	septal region
	C_1 †	IgG + F	-	-	20	none	none
II	S_2	IgG + F	no electrodes	+	35	septal region	septal region
Phase II: IgG tagged with Ferritin (F)--brain perfused							
III	C_2	IgG + F		-	45	none	
IV	S_3	IgG + F	no electrodes	+	45	septal region	
	C_4 (M. gravis)	IgG + F	no electrodes	-	45	none	
V	SC (1) S IgG (2) C IgG + F	IgG / IgG + F		+	35	septal region	
VI	S_4	IgG + F		+	15	septal region and cortex	
Ferritin alone injected--brain perfused							
III	C_3	-	no electrodes	-	45	none	

* Schizophrenic

† Control

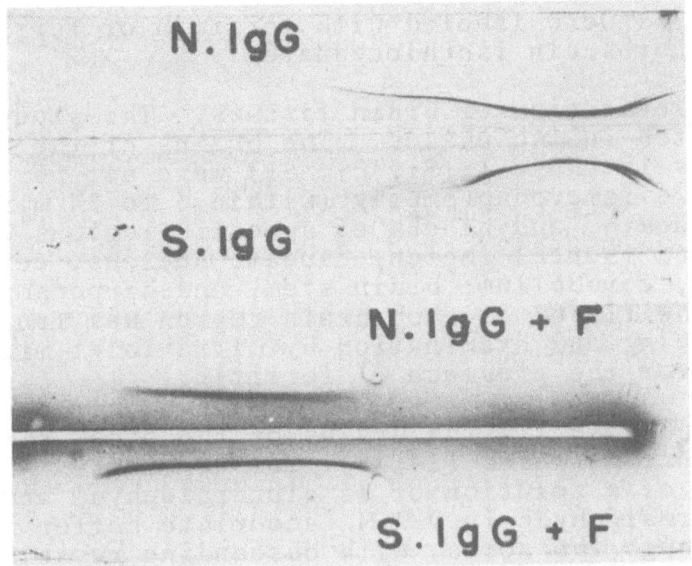

Fig. 1.   Immunoelectrophoretic patterns developed
to whole antihuman serum. Top: Schizophrenic IgG
(S IgG) (taraxein) and normal IgG (N IgG) before
labeling. Bottom: Schizophrenic IgG labeled with
ferritin (S IgG + F) and normal IgG labeled with
ferritin (N IgG + F).

(C_4), and one received unlabeled schizophrenic IgG
followed immediately by ferritin-labeled IgG from
a healthy medical student (SC). A ninth monkey
received ferritin alone (C_3).

Two (S_1 and C_1) of the nine monkeys were prepared with the standard array of cortical and subcortical electrodes for testing of taraxein activity (4). The EEGs permitted monitoring of physiologic brain activity, but it was possible that the
inevitable lesion from electrode implantation could
affect the blood-brain barrier. The other seven
monkeys used in the studies were therefore not
prepared with electrodes. A small polyethelene
cannula was introduced and fixed into the femoral
vein of each monkey three days before testing.

Labeling. By the method of Hsu (13), the IgG

molecules were labeled with ferritin or ferritin and fluorescein isothiocyanate.

Preparation of brain tissues. The study was conducted in two phases. The brains of the three monkeys in Phase I (S_1, C_1, S_2) were not perfused, but were removed promptly (within 6 to 14 minutes) after death, and blocks of specific regions were obtained (septal region, caudate nucleus, cerebral cortex, cerebellum, brain stem, and hippocampus). Half the tissue of each brain region was frozen for sectioning and examination by ultraviolet microscopy for the presence of ferritin.

In Phase II, the brains of the other six monkeys were perfused (14,15), for 40 minutes with 4 liters of a solution of 1% gluteraldehyde and 1% paraformaldehyde in 0.1 M cacodylate buffer delivered through the aorta, with descending vessels tied. This technic permitted better fixation for more accurate identification of the site of localization of the ferritin in the brain. Since this technic did not permit us to obtain frozen tissues for detection of fluorescein by ultraviolet microscopy, the IgGs used in Phase II were labeled with ferritin alone.

After perfusion, the cranium was removed, the dura opened, and the brain soaked in a 1 to 2% solution of gluteraldehyde and a 1 to 2% solution of paraformaldehyde in 0.1 M cacodylate buffer for 6 to 12 hours. Blocks of tissues (specified in Phase I) were then removed, and small 1-mm cubes were embedded in epon. Sections were cut on a Porter Blum II Ultramicrotome at 60-80 A and examined under a Siemens Elimiskop IA. The sections were either unstained or stained with an aqueous uranyl acetate.

RESULTS

EEGs and clinical responses. The EEG of the monkey that had been prepared with electrodes and that received labeled schizophrenic IgG (taraxein) (S) showed focal slow waves and spiking in septal

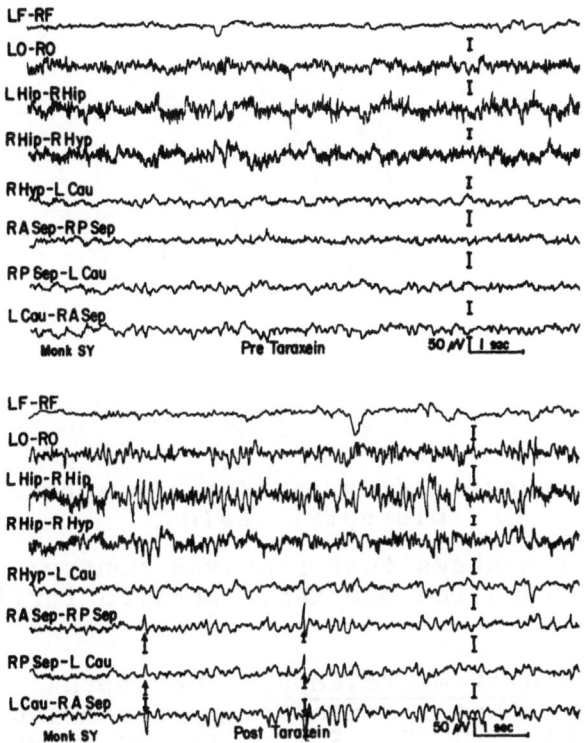

Fig. 2. Cortical and subcortical recordings from a monkey before and after intravenous injection of taraxein.

leads beginning four minutes after intravenous injection (Fig. 2). These physiologic abnormalities, and associated catatonic signs, increased in intensity for about 15 minutes after injection, and persisted at this peak thereafter until the monkey was killed. Neither EEG nor behavioral changes were evident in the monkey that received control IgG labeled with ferritin and fluorescein (C_1).

In the four monkeys that received schizophrenic IgG but had no electrodes (S_2, S_3, SC, S_4), moderate to pronounced catatonic signs appeared within 4 to 5 minutes after injection and persisted until the monkey was killed. No behavioral changes were noted in

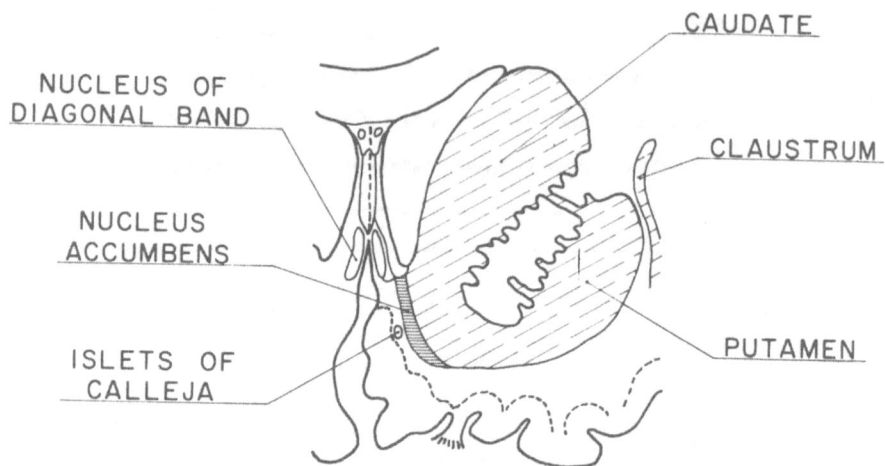

Fig. 3.   Schematic diagram of brain section through
          midseptal region.

the other 2 monkeys that received control injections
(C_2, C_4), or in the one that received ferritin alone
(C_3).

Ultraviolet microscopy. The septal tissues of
both monkeys that received fluorescein-labeled
schizophrenic IgG (S_1, S_2) contained brightly flu-
orescing cells in the septal region, but tissues of
their other brain regions showed no fluorescence.
Nor did septal or other brain tissues of the monkey
that received fluorescein-labeled normal IgG (C_1)
show fluorescence.

Electron microscopy. Electron microscopic examination showed ferritin molecules localized in
the center of the septal region (16) of all 4 monkeys that received ferritin-labeled schizophrenic
IgG (S_1 - S_4), as well as of the monkey that received unlabeled schizophrenic IgG (taraxein) immediately
followed by ferritin-labeled normal IgG (SC). The
ferritin could be seen in the vicinity of the nucleus
accumbens septi, islets of Calleja, and nucleus of
the diagonal band (Fig. 3). No ferritin was detected in more caudal reaches, near the septal
nuclei proper, in its more rostral aspect, or in
the very ventral cells of the orbital cortex. In
blocks of septal tissues of all of these monkeys

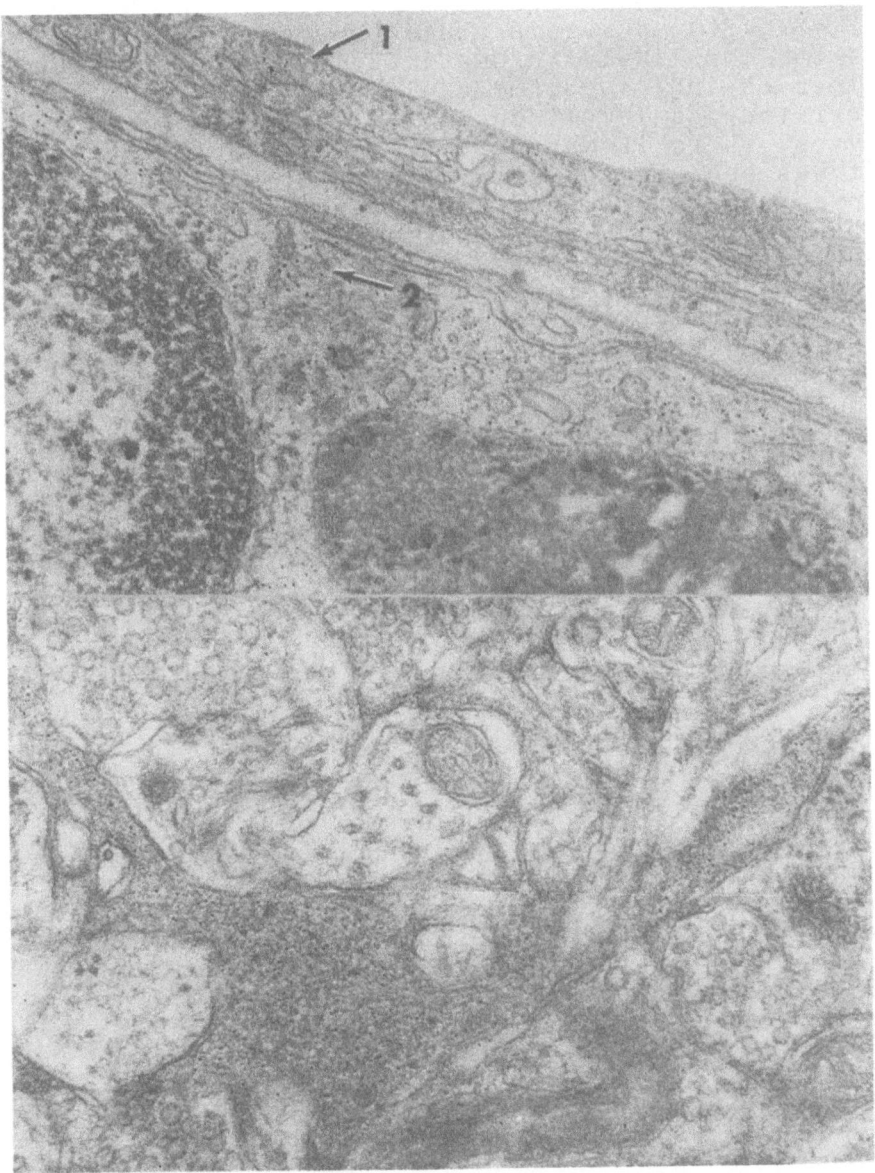

Fig. 4. Electromicrographs showing ferritin-labeled
taraxein in oligoglial cells of the septal region.
Top: Ferritin (dark specks) in capillary wall (arrow 1), and distributed through the cytoplasm of an
oligoglial cell (arrow 2). Bottom: Ferritin-labeled taraxein (dark specks) in glial process intertwined with axones and near synapses. (Magnification 70,000; 2% uranyl acetate stain.)

(S_1-S_4, SC), ferritin could be seen in the walls of the capillaries, sometimes in clumps and sometimes as individual molecules (Fig. 4, top). Most ferritin was localized in the cytoplasm of oligocytes, where it was uniformly distributed (Fig. 5). It was seen in the oligoglial processes near synapses and intertwined with neural cells, axones, and dendrites (Fig. 4, bottom). The ferritin was never seen at synapses, in intercellular spaces, on neural cells, or in any glial cells other than the oligodendrocytes. On small myelinated axons, it seemed at first to be inside as well as outside the myelin, but close examination showed it to be enclosed by at least one thread of a loop of myelin, and thus localized in a process of the oligoglial cell (Fig. 6). Distribution of the ferritin within the cytoplams of oligoglial cells of the septal region was almost identical for the monkey killed 15 minutes after injection and the one killed 45 minutes after intravenous injection.

Ferritin was not detected on brain tissues of the monkey that received ferritin alone (C_3), of the two that received normal-labeled IgG (C_1 and C_2), or of the one that received ferritin-labeled myasthenia gravis IgG (C_4).

Discussion

The schizophrenic IgG differs from other IgG in that it moves easily from the blood stream into the septal region of the brain. Furthermore, the schizophrenic IgG moves rapidly and precisely, lodging within one specific type of cell of the

---

Fig. 5. Oligoglial cell showing cell body and processes in relation to neuronal processes. Ferritin is not visible at this magnification. (Magnification 18,000; 2% uranyl acetate and lead citrate stains.)

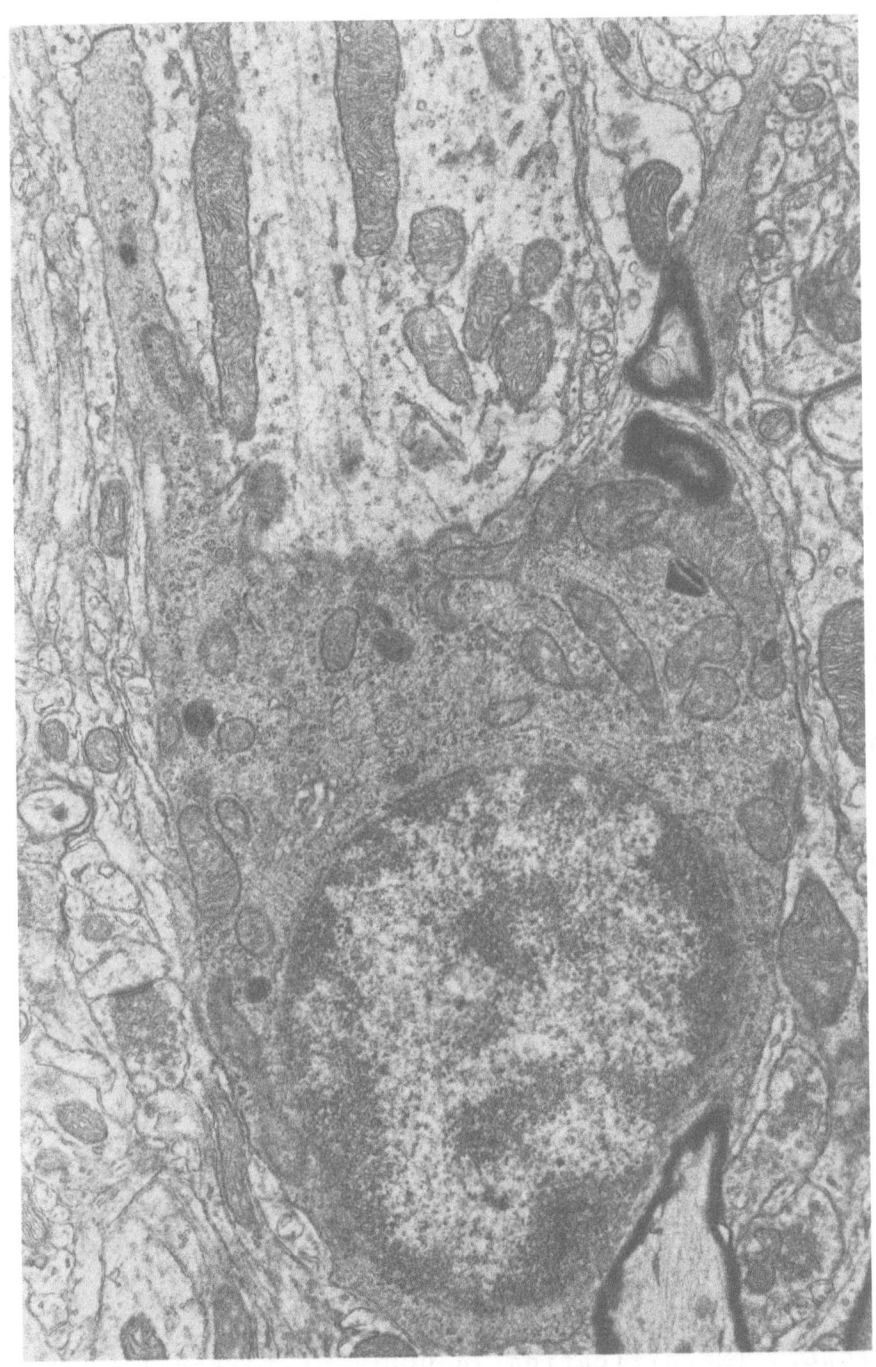

septal region, simultaneously with onset of behavioral changes and abnormal EEG changes at that site.
We can only speculate on the mechanism by which
schizophrenic IgG affects neural cells to produce
catatonia and abnormal EEGs in monkeys similar to
those recorded for psychotic schizophrenic patients.

Since the ferritin-labeled schizophrenic IgG
localized in almost identical sites whether the
monkey was killed at 15 or 45 minutes after intravenous injection, the change in neural cell activity
was presumably not caused by direct entrance of
the schizophrenic IgG into the nerve cell. Instead,
the change would seem to be caused by the action
of the IgG in the adjacent oligoglial cells, since
it stays at that site. At this point we cannot explain how the schizophrenic IgG arrived at the site
within a few minutes after its injection. The capillary walls and the cell membranes were structurally intact. There was no ferritin in intercellular spaces, and there was none seen in the end
feet or, in fact, anywhere in the astrocytes, which
are considered by some as a possible system for
passage of materials from the blood stream into
neural cells. Pinocytosis, a process by which in-
trathecally introduced ferritin can enter brain
tissues (14), was not observed in the brains of
these monkeys. Nor had sufficient time elapsed for
development of an inflammatory or phagocytic process; in fact, even the brain tissues of monkeys
that received repeated taraxein injections over
extended periods have never shown structural changes
of the septal or other region of the brain.

The entrance into the brain of ferritin-labeled control IgG, given intravenously immediately after
injection of unlabeled, active schizophrenic IgG

Fig. 6. Ferritin-labeled taraxein (dark specks)
in propinquity to myelin sheath of myelinated
axone. (Magnification 74,000; unstained.)

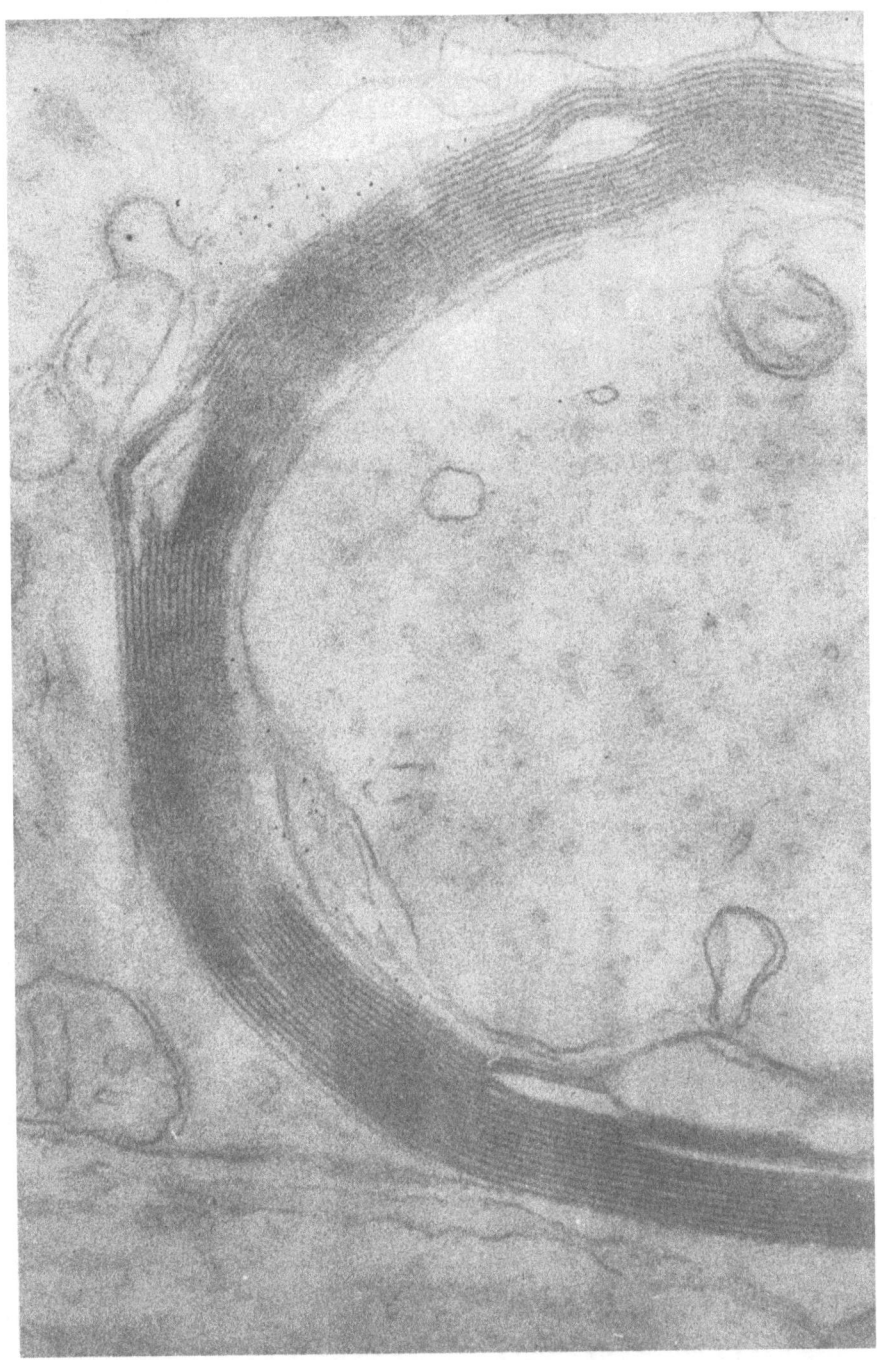

(taraxein), suggests that taraxein alters permeability, not only of the blood-brain barrier, but of specific cell membranes at this particular brain site. Earlier, less definitive data (17,18), had suggested that taraxein acted by altering the blood-brain barrier, and the present experiments add support to this speculation.

If the taraxein studies have a role in the future course of the brain sciences, it may be in the identification of the antigen (glial material?) for taraxein. Such identification could open the way for precise diagnostic tests of schizophrenia and, ultimately, for specific treatment of the disease. Because of the ability of taraxein to move into glial cells and thereby affect adjacent neurons, these studies may also help explain the neural cell-glial cell interrelation.

REFERENCES

1. Heath, R.G., and the Department of Psychiatry and Neurology, Tulane University, Studies in Schizophrenia, Cambridge, Harvard University Press, 1954.
2. Heath, R.G., Developments toward new physiologic treatments in psychiatry. J. Neuropsychiat., 5, 318, 1964.
3. Heath, R.G., and Gallant, D.M., Activity of the Human Brain During Emotional Thought. In: The Role of Pleasure in Behavior, ed. by Heath, R.G. New York, Hoeber Medical Division, Harper & Row, 83-106, 1964.
4. Heath, R.G., Krupp, I.M., Byers, L.W., and Liljekvist, J.I., Schizophrenia as an immunologic disorder. II. Effects of serum protein fractions on brain function. Arch. Gen. Psychiat., 16, 10, 1967.
5. Heath, R.G., Taraxein (Schizophrenic IgG): Recent fractionation methods. In preparation.
6. Heath, R.G., Martens, S., Leach, B.E., Cohen, M., and Angel, C., Effect on behavior in humans with the administration of taraxein., Amer. J. Psychiat., 114, 14, 1957.

7. Heath, R.G., Schizophrenia: Biochemical and physiologic aberrations. Int.J. Neuropsychiat. 2, 597, 1966.

8. Heath, R. G. and Krupp, I. M., Schizophrenia as an immunologic disorder: I. Demonstration of antibrain globulins by fluorescent antibody techniques, Arch. Gen. Psychiat., 16:1, 1967.

9. Heath, R. G., Schizophrenia: Studies of Pathogenesis, in Biological and Clinical Aspects of the Central Nervous System, Basle, Switzerland, Buchdruckerei, Kohlhepp AG, Neuallschwil, 189-214, October, 1967.

10. Williams, C. A., Liljekvist, J., Heath, R. G. In preparation.

11. Heath, R. G., Krupp, I. M., Byers, L. W. and Liljekvist, J. I., Schizophrenia as an immunologic disorder. III. Effects of antimonkey and antihuman brain antibody on brain function, Arch. Gen. Psychiat., 16:24, 1967.

12. Heath, R. G. and Krupp, I. M., Catatonia induced in monkeys by antibrain antibody, Amer. J. Psychiat., 123:1499, 1967.

13. Hsu, K. C., Protocol for labeling of globulin with fluorescein and/or ferritin, Dept. of Microbiology, College of Physicians and Surgeons, Columbia University, New York.

14. Brightman, M. W., The distribution within the brain of ferritin injected into cerebrospinal fluid compartments. I. Ependymal distribution, J. Cell Biol., 26:99, 1965.

15. Palay, S. L., Harvard Medical School, Personal communication.

16. Heath, R. G., Definition of the Septal Region, in Heath, R. G. and the Department of Psychiatry and Neurology, Tulane University, Studies in Schizophrenia, Cambridge, Harvard University Press, 2-5, 1954.

17. Heath, R. G., Leach, B. E. and Cohen, M., Relation of Psychotic Behavior and Abnormal Substances in Serum, in The Effect of Pharmacological Agents on the Nervous System, Vol. 37, Proc. Ass. Res. Nerv. & Ment. Dis., Baltimore, Williams & Wilkins Co., 397-411, 1959.

18. Melander, B. and Martens, S., Experimental studies on taraxein and LSD, Acta Psychiat. 35 Neurol. Scand., Suppl. 136, 34:344, 1959.

ORGANIC APPROACHES TO SCHIZOPHRENIA

John R. Bergen

and

Hudson Hoagland

Great strides have been made in many areas of
science in recent decades but progress in the behavioral sciences has been slow by comparison. The
mechanisms involved in the normal operations of the
brain are not yet understood and, as a consequence,
few facts to explain the function of the abnormal
mind have been established. The relative neglect
of research in neuropsychiatry cannot be due to the
unimportance of the problem because the problem is
very great. Perhaps the apparent disinclination to
investigate mental disease is related to the special
problems inherent in psychiatric research. Advances
in medical research have generally resulted from
experimentation on animals. While animal studies
have shed valuable light on some behavioral problems, this approach has limited applications to man
because humans possess the ability to develop unique
problems within the systems governing behavior, e.g.
neuroses and psychoses. Aberrations in behavior can
be induced experimentally but it is highly question-
able if they correspond to the states seen in the
mentally ill.

Research in mental illness would be facilitated considerably if the illnesses could be diagnosed unequivocally on the basis of established,
objective criteria. However, since the causes of
many mental illnesses are unknown, etiological

classification systems cannot be constructed and
in their stead, classification systems using descriptive criteria have been developed. These
systems, which were justified, perhaps, on the
limited number of cases studied, have restricted
usefulness and none of them has been universally
adopted. With respect to schizophrenia, "the disease is characterized by a specific type of thinking, feeling and relationship to the external world
which appears nowhere else in this particular
fashion" (Bleuler), and so the diagnosis must
necessarily be made from the subjective evaluation
of symptoms by psychiatrists. It is not uncommon
that uniform agreement is not reached concerning
the exact nature of the patient's disease because
different psychiatrists subjectively attribute
different levels of importance to the signs and
symptoms displayed by the patient. There is lack
of agreement whether schizophrenia is a single
disorder or a collection of different disorders
grouped together by a common symptomatology; should
the disease be referred to as "schizophrenia" or
"the schizophrenias"?

Although great strides have been made from
the times of our ancestors when mental illness was
considered to be an invasion of the body by demons,
we have travelled over only a short part of the
path that will lead to a true understanding of the
etiological factors responsible for mental derange-
ments. I believe that for the most part we have
passed through the period when the mind, or psyche,
was treated by psychiatrists as if entirely divorced
from the physical processes of the body and have
entered a more realistic period in which it has
long been recognized by many that mind and body
relate directly to each other. The more biologically oriented psychiatrists thus have adopted a
holistic view which has led not only to the rejec-
tion of purely psychodynamic concepts in mental
illness but to the development of psychosomatic
concepts in medicine.

We, and others, believe that normal mental
activity, including behavior, represents the manifestation of specific patterns of neuronal activity

operating in a manner which gives high priority to
the processing of messages of importance for proper
maintenance of the individual and which suppresses
or rejects these messages which are unimportant.
From the operation of these selective mechanisms,
rational, directed behavior results and, conversely,
a pathological alteration of these relationships
leads to abnormal mental activity and abnormal
behavior. It is a working hypothesis with biochemists and physiologists that abnormal chemical reactions in cells cause the changes in mental activity which lead to pathological behavior. Support
for the belief that psychological and biochemical
events are inextricably bound together comes from
important contributions made in the past which
have shown that certain disorders of the mind can
be traced to specific biochemical abnormalities.
Thus, the severe psychosis characteristic of pell-
agra has virtually disappeared following discovery
that it is related to vitamin B lack and the subsequent use of vitamins as preventative agents.
Feeblemindedness accompanying phenylketonuria (PKU)
is now understood to be genetically determined in
terms of the lack of a specific enzyme system. It
is now successfully treated by exclusion from the
diet of the amino acid, phenylalanine, that cannot
be metabolized normally due to the enzyme deficit.
Other mental disorders are likewise becoming understood in terms of enzyme or hormone deficits. There
is good reason to believe that progress in the study
of brain chemistry will, in time, yield insights
into other major disorders, such as schizophrenia
and ultimately make available rational chemothera-
peutic approaches.

The basic data on which any theory of genetical
contributions to the causation of schizophrenia
would have to be based were reviewed at a recent
meeting by Slater (1). He stated that for the
general population, the data from different countries agreed rather well for an expectancy of somewhat below 1% (0.80 - 0.85%). Figures for blood
relatives of schizophrenic patients were significantly higher with a higher expectancy in first
degree relatives than for more remote relatives.
Attention was called to the correlation reported

by Kallman in 1938 that the risk of schizophrenia
for the children of schizophrenic patients is about
twice as great (18.5%) for hebephrenic and catatonic
schizophrenics as for paranoid and simple schizophrenics (9.8%). Slater reported that the combined
observations of a number of workers on the risk of
schizophrenia borne by the child of parents, both
of whom have had the illness, is about 36%, a much
higher incidence than when only one parent is affected; a figure which is compatible with genetical
theory though not a theory of simple dominance or
recessivity.

Concordance rates for dizygotic twins were
reported to approximate the expectancy for other
children of a schizophrenic parent but for monozygotic twins the rate was considerably greater.
In older studies the concordance rates for monozygotic twins has been reported to be 65% but more
modern researches show a figure considerably lower,
approximately 25%. The difference between the older
and newer studies probably reflects differences in
case material. The older studies contained patients
suffering from severe illnesses and in many instances were hospitalized invalids. The newer studies
still show that the monozygotic twin of a schizophrenic proband stands a much higher risk of developing schizophrenia than a dizygotic same sexed
twin. Slater concluded that the difference is
what one would expect on a genetic basis but that
genetics would not necessarily be the only hypothesis capable of explaining it.

Whatever the predisposing factors may be, it
seems probable that when overt symptoms of the
disease appear, they must be referable to derange-
ments at the molecular level which contribute to
the cause of the disorder. Considerable interest
in the possibility of a biochemical change associated with the plasma proteins was stimulated by
the reports of Heath and his group (2,3) of a
unique globulin which they isolated from the blood
of schizophrenic patients and which they named
taraxein. Dr. Heath has already reviewed his work
with taraxein at these meetings and has presented
evidence that the factor in question is contained

in the γ G immunoglobulin fraction of the plasma proteins.

Our work has also been centered primarily in the area of the plasma proteins and some of the earlier studies were presented at the previous International Conference of the Manfred Sakel Foundation (4). We reported evidence for a factor contained in human plasma which may be related to schizophrenia. Comparison of the effects produced by plasma extracts from schizophrenic, non-schizophrenic psychotic and normal persons has shown that significantly greater changes were produced in several test situations by the extracts from schizophrenic patients compared with those from non-schizophrenic persons. One test involved measurement of the deterioration of trained rat rope climbing performance following intraperitoneal injections of the protein extracts (5) and another concerned the evaluation of changes due to the intravenous administration of plasma extracts on the primary potential evoked photically and recorded from the visual cortex of unanesthetized rabbits (6). By the former test it was shown that the biologically active component is present in only a small portion of the total plasma proteins. The factor is contained in the euglobulins and is readily associated with a subgroup, the β lipoproteins. When separated from the β lipoproteins by chromatographic means, the active component has the characteristics of an α-2 globulin (7). Indirect evidence is suggestive that the biological activity resides in a dialyzable moiety which is associated firmly with the protein (8). Active fractions have been electrodialyzed but this procedure has given no evidence for dissociation of a small molecule (7). The active protein extract is extremely labile, but biological activity may be retained by prompt preparation following collection of blood, by maintenance of low levels of a reducing agent throughout the fractionation procedure (e.g., 4 mM ascorbic acid or 5 mM β-mercaptoethanol) and by storage of the final product under hydrogen.

Other groups who have investigated the plasma proteins for an abnormality referable to schizo-

phrenic patients and which may be related to the
disease have collected information which points to
a group of proteins possessing similar characteristics (Table 1). The tests used are, for the
most part, different between laboratories, as are
the methods for the extraction of the biologically
active principle. The protein isolated is thought
by most to belong to the globulins. Chromatographic separation of the globulins has shown that the
factor is contained in a borderline group between
the α and β types. An exception is the γ globulin
proposed by the New Orleans group although they
had previously reported that the active factor
appeared to be associated with the α-β and possibly
with the α-2 globulins (9). Lability is a common
characteristic, which in many cases has been controlled by the use of reducing agents. Some investigators claim that the globulin is a lipoprotein and estimates of the lipid content range as
high as 80%.

Because many of the investigations appear to
involve the same type of protein the question frequently arises whether different laboratories are
working with essentially the same factor or with
dissimilar factors. Comparative tests of plasma
factors prepared by two of these groups were
undertaken when the Lafayette Clinic invited investigators from the Worcester Foundation for Experimental Biology (WFEB) to come to Detroit to
collaborate in an experiment. Details of the
experimental procedures employed have been published elsewhere (16) and only some of the pertinent
information will be summarized here.

For this experiment, each group prepared fractions by its own separation procedure, then measured the activity of the fractions with its accustomed assay procedures, and coded the fractions in
anticipation of the double-blind experimental design. Upon arrival of the Worcester Foundation
group in Detroit, each group gave to the other its
fractions from schizophrenic and control subjects.
Each then assayed the other group's fractions using
its own assay method and predicted on the basis of
the assay results whether the fractions were of

Table 1

Characteristics of Blood Protein Factors in Schizophrenia

Investigator	Test	Protein Type	Lipid in Protein	Molecular Weight	Lability	Protection	Small Molecule	Ref.
Bergen and Pennell	Rat rope climbing	α_2	+	High	++++	Vit. C, H_2 β-mercapto-ethanol	+	8
Ehrensvaard	Organic acid oxidation	α or β	-	High	++++	Vit. C or glutathione	+	10
Frohman	L/P ratio	α_2	+	400,000 ± 50,000	++++	Vit. C, H_2	?	11
Heath	Monkey and human behavior	?	-	?	++	-	?	12
Sanders	Rat rope climbing	α_2 or β	+	?	++++	Vit. C	?	13
Vartanian and Krasnov	L/P ratio	α or β	+	?	++++	Vit. C	?	14
Walaas	Rat diaphragm glucose uptake	α or β	+	High	++++	Glutathione	-	15

schizophrenic patient or of control subject origin.

Of eight Lafayette Clinic samples tested by the Worcester Foundation using the rat rope climbing test five samples were predicted to have been prepared from control donors and three from schizophrenic patients. When the code was revealed, three schizophrenic and four control subjects had been correctly predicted (Table 2). The single schizophrenic patient sample found to be inactive by the Worcester group was also found to be inactive according to the Lafayette Clinic's tests. Although prediction was correct in seven of eight samples, agreement between groups was found in all of the eight samples.

Of sixteen samples prepared by the Worcester Foundation, thirteen predictions in agreement were made by the Lafayette Clinic using their lactate

Table 2

Assay by WFEB of Lafayette Clinic Samples

Sample No.	Lafayette Clinic Rating	WFEB Test Rat Rope Climb
1	S	S
2	S	S
3	C	C
4	S	S
5	C	C
6	C*	C
7	C	C
8	C	C

Agreement between groups 8/8

C = control   S = Schizophrenic patient

*Sample from schizophrenic patient whose plasma extract previously yielded values in "control" range by Lafayette Clinic Tests.

Table 3

Assay by Lafayette Clinic of WFEB Samples.

Sample No.	WFEB Rating	Lafayette Clinic Test	
		L/P ratio	Glutamic Acid
1	C	C	C
2	S	S	S
3	S	S	S
4	C	S	C
5	S	S	S
6	S	S	S
7	C	S	S
8	C	C	C
9	S	S	S
10	C*	C	C
11	C	C	C
12	C	C	C
13	S	S	S
14	C*	S	S
15	S	S	S
16	S	S	S
	Agreement between groups	13/16	14/16

C = Control   S = Schizophrenic patient

*Samples from schizophrenic patients but biological activity in

"control" range.

to pyruvate ratio method and fourteen by using the glutamic acid uptake method (Table 3).

These results indicate that the two groups in great probability are studying the same plasma protein factor. The direct relationship of the factor to schizophrenic illness remains to be proved, however.

Attempts to extract and identify an active principle by the application of a variety of techniques to biologically active plasma extracts has not yet been fruitful. However, results from other of our experiments show that the factor

affecting trained rat behavior has characteristics
similar to those of the biogenic amines of the
substituted phenylethylamine series.

A hypothesis was put forward in 1952 by Osmond
and Smythies (17) and Harley-Mason (18) that schizophrenia might be associated with an abnormal
methylation of catecholamines with the production
of a substance such as 3,4 - dimethoxyphenylethan-
olamine which might have psychotogenic properties
like its close chemical relative mescaline. Many
of the known psychotogenic drugs possess methoxy
or N-methyl groups in their chemical structure.
Following reports of the isolation of the trans-
methylating enzymes, catechol O-methyl transferase
and N-methyltransferase (19,20) subsequent research
has established the importance of these enzymes in
biogenic amine metabolism. The experiment by
Pollin, Cardon and Kety (21) in which large doses
of l-methionine, (a methyl group donor), were administered to chronic schizophrenic patients receiving iproniazid resulted in exacerbation of
schizophrenic symptoms and brief intensification
of the psychosis in four of the nine patients so
treated. In a similar experiment by Brune and
Himwich (22) an increase in the severity of the
disease was observed in seven of nine male schizophrenic patients treated with isocarboxazid and
DL-methionine.

In 1962 Friedhoff and Van Winkle produced
their first report concerning the isolation of an
abnormal methylated catecholamine metabolite from
the urine of fifteen of nineteen schizophrenic
patients tested and its absence in fourteen non-
schizophrenic subjects (23). Since then many workers have contributed to the literature on this
subject and have referred to the amine as 3,4-
dimethoxyphenylethylamine (DMPEA) according to its
identification by Friedhoff and Van Winkle or by
its more euphonious name "the pink spot." A review
of these papers has been published by Ridges and
Bourdillon (24). The most important recent finding
has been confirmation of the identity of DMPEA by
infra-red and mass spectrometry (25).

Our interest in investigating some of the pharmacological and biochemical properties of DMPEA was stimulated because its chemical structure falls within the class of compounds that may be associated with the plasma factor according to our evidence.

In our hands DMPEA produces changes in rat rope climbing behavior similar to that seen with plasma protein fractions and the time course for the performance changes (time to climb the rope) is likewise similar (Fig. 1). Another phenylethylamine, mescaline, has a similar time course of action as DMPEA but differs in the response produced when administered daily to the rats. Relative refractoriness to mescaline (5 mg/kg) develops rapidly over several days but has not been noted for DMPEA administered at a constant dose (50 mg/kg) over five days. LSD (400 µg/kg), a known psychotomimetic, has a shorter course of action and, in addition, produces a characteristic ptosis which is not seen with other agents. Injections of adrenochrome, a derivative of adrenaline thought by some to be involved in schizophrenia, produces a decrement in trained rat performance but the development of impaired function is considerably slower than for the other agents and is accompanied by hind leg extension, a phenomenon which is not seen following DMPEA or plasma globulin extract administration.

Tracer labeled DMPEA (10 µc 1-C^{14} DMPEA contained in 100 mg of drug/kg) has been injected I.P. and appreciable plasma uptake was noted within five minutes as judged from measurement of radioactivity. Total counts in plasma and brain reached a maximum at 20-25 min. or approximately at the same time when rope climbing performance changes were most noticeable. In another experiment in which liver uptake as well as plasma and brain uptake were measured at five minute intervals, maximal counts were recorded for liver five minutes post injection and no significantly higher values were recorded in succeeding test periods even though the plasma levels continued to climb (Fig. 2). This would indicate rapid concentration and saturation by the liver with drug shortly after injection and may help to explain the rapid inactivation of DMPEA.

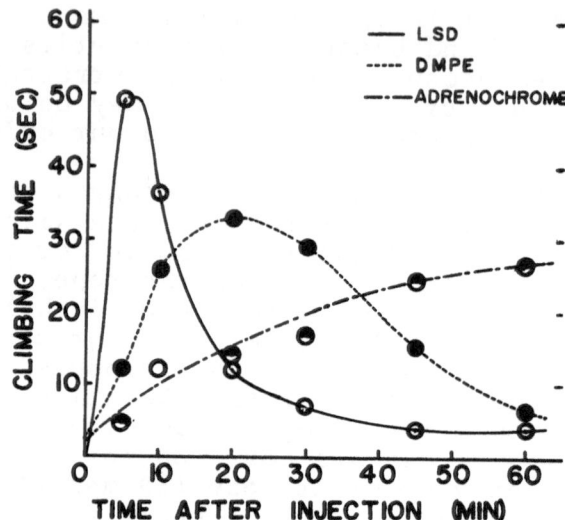

Fig. 1. Mean climbing time for 5 rats at intervals
after I.P. injection of each of the following -
400 μg LSD/kg, 50 mg dimethoxyphenylethylamine/kg
and 100 mg adrenochrome/kg.

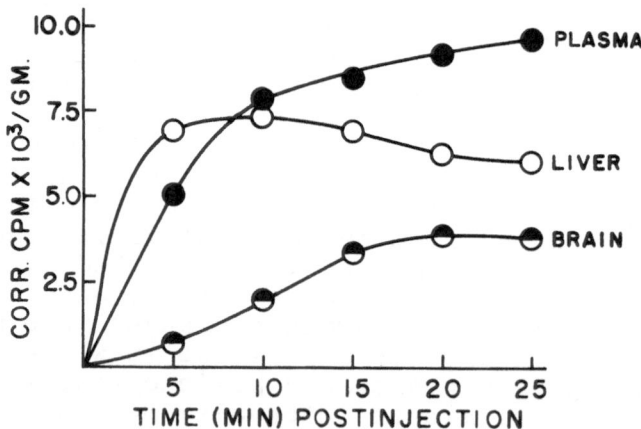

Fig. 2. Plots of radioactivity measurements
(counts x 10^3/min corrected for background and
quenching) in plasma, brain and liver of rats
following I.P. injection of 10 μc $1-C^{14}$ DMPEA
contained in 100 mg of carrier DMPEA/kg. Each
point represents average of 2 counts on duplicate
samples from 2 rats i.e. 8 determinations per point.

Protection of DMPEA from metabolic inactivation
mechanisms does, in fact, lead to potentiation of
behavioral effects in trained rats. If the rats are
pretreated with a monoamine oxidase inhibitor (MAOI)
such as isocarboxazid (5 mg/kg) the effect of I.P.
administered DMPEA is potentiated approximately
seven times as compared with non-pretreated rats.
Another method investigated for protection against
rapid inactivation was acetylation of the amine
group in DMPEA. Similar potentiation as with MAOI
pretreatment was observed except that the N-acetyl
derivative appeared to enter the brain sooner and
to have a shorter duration of action. Maximal biological effects in trained rats occurred five minutes
post injection and were parallel to the observed
brain uptake of tracer labeled N-acetyl DMPEA. By
twenty minutes post injection considerable improvement in climbing performance had occurred and radioactivity measurements of brain and blood showed substantially fewer counts than at five minutes.

Because of the large amounts needed by I.P. injection to produce behavioral effects in rats (50 -
100 mg/kg), DMPEA does not appear to be a potent
agent. As already stated, this appears to be due to
the rapid metabolism of DMPEA to the corresponding
acid which is biologically inactive. Koella, of the
Worcester Foundation has shown however, that injections
of DMPEA directly into the fourth ventricle of the
cat brain are effective in producing EEG changes in
quantities too small to be effective when administered by the arterial route to the brain. He found
that amounts as small as 50 - 100 nanograms increased the synchrony of recordings from the supra-
sylvian gyrus. Intra-arterial injections were effective only in doses approximately 1000 times
greater in altering the EEG recorded from the supra-
sylvian gyrus and on potentials recorded from the
visual cortex following electrical stimulation of
the chiasma. These experiments show that DMPEA is
a potent agent capable of altering neural mechanisms
if delivered close to the site of action, thus
avoiding peripheral inactivating mechanisms. The
catecholamines, epinephrine and norepinephrine,
tested under these conditions and at the same dosage
proved to be inneffective.

The capacity for DMPEA to be transported in a form whereby it is protected from inactivating mechanisms has been demonstrated by _in vitro_ experiments. Samples of normal human plasma were incubated at 37° C for thirty minutes with tracer labeled DMPEA (40 μg DMPEA/ml containing ca. 4000 cpm) then dialyzed in the cold room against large volumes of saline changed at frequent intervals. The counts recovered from plasma, corrected for background and quenching, showed that more than 90 % of the drug had been removed from the plasma in 24 hours. After 85 hours of dialysis the amount remaining amounted to approximately 2.5 μg/ml plasma. Over the first 24 hours equilibration occurred slowly and this could indicate release of weakly bound drug from secondary binding sites. What appears to be a primary site of binding is contained in the proteins which constitute our biologically active protein factor. Sanders (26) also has reported binding of DMPEA to the same group of plasma proteins. We have investigated the uptake of DMPEA by red cells under conditions similar to those used for plasma and based on the counts recovered at the end of the dialysis period, blood cells contained approximately 65 μg/ml in contrast to the 2.5 μg bound per ml of plasma.

Smythies and his group conducted an exhaustive study of certain of the methoxylated phenylalkylamine series using behavioral tests with rats, based on the Sidman Avoidance Schedule and the Bovet-Gatti profiles (27). He found that with the phenylethylamine series the 3,4,5 (mescaline), 2,3,4,5 and 2,3,4,5,6 methoxylated derivatives were active and all other derivatives tested were inactive. A property of these active compounds is that they are not metabolized by monoamine oxidase in contrast to the mono- and dimethoxy derivatives. From his investigation of the phenylisopropylamine series (amphetamines), in which the amine group is protected by an adjacent methyl group, quite different results were obtained (28). In the dimethoxy series, the 3,4-methoxy compound was approximately twice as active as an equivalent amount of mescaline and in the monomethoxy series, p-methoxy amphetamine was found to be the most potent drug tested, aside

from LSD. From his experiments Smythies concludes
that a characteristic of the essential hallucinogenic
molecule of this series would involve 4-methoxyl-
ation and that the psychosis found in amphetamine
psychosis may be associated with the production of
4-methoxyamphetamine. He draws attention to the
report by Boulton et al., (29) that one pink spot
excreted by schizophrenics is paratyramine and by
Jenner et al., (30) that a manic depressive patient
excreted large amounts of paratyramine metabolites.
These reports are of interest because his data suggest that the production of 0-methyl paratyramine
produced in the synapse (separated from MAO) would
have long lasting effects on behavior. Therefore,
he entertains the hypothesis that schizophrenia
may be associated with an aberration of 0-methyl-
ation whereby 4-methoxy instead of the normal 3-
methoxy compound is produced in the synaptic cleft
away from the site of MAO activity in the mitochondria of the axon terminals. Thus these compounds
given by mouth or by injection without MAOI would
appear inactive because they would be broken down
rapidly by MAO. Produced locally in the synapse or
given with MAOI, they might well produce a potent
disruption of behavior (31). Certainly our data
do not contradict such a view.

Concerning the future of the brain sciences in
the area of mental health I quote from Hudson Hoagland. "The behavior of each of us is determined
by our past history, by our genes on the one hand
and by the modification of our brain chemistry and
physiology by environmental factors in the course
of a lifetime of conditioning. Behavior both normal and pathological is a product of the correlated
action of molecular events in our brain, and functional disorders in psychiatry are names given to
behavior patterns, the molecular basis of which
has so far eluded analysis. I expect that in the
coming decades one of the most spectacular advances
of science will be in the understanding of brain
mechanisms in relation to our psychosocial behavior.
Research in the broad field of neuropsychiatric

problems is aimed at helping us live more effective
emotionally balanced lives. There are few fields
of action with more promise for man than basic
research in the processes underlying mental health."

REFERENCES

1. Slater, E., In: Biological Research in Schizo-
    phrenia, Academy of Medical Sciences of the U.
    S.S.R., Institute of Psychiatry, 244-249, 1967.
2. Heath, R.G., Martens, S., Leach, B.E., Cohen,
    M. and Angel, C., Am. J. Psychiat., 114, 683,
    1958.
3. Heath, R.G., Martens, S., Leach, B.E., Cohen,
    M. and Feigley, C., Am. J. Psychiat., 114, 683,
    1958.
4. Bergen, J.R., Pennell, R.B., Saravis, C.A.,
    Freeman, H. and Hoagland, H., In: Biological
    Treatment of Mental Illness, M. Rinkel (ed.),
    L.C. Page and Co., New York, 18, 323, 1966.
5. Bergen, J.R., Gray, F.W., Pennell, R.B., Free-
    man, H. and Hoagland, H., Arch. Gen. Psychiat.,
    12, 80, 1965.
6. Bergen, J.R., Czicman, J.S. and Koella, W.P.,
    J. Neuropsychiat., 4, 219, 1963.
7. Pennell, R.B., Pawlus, C., Saravis, C.A. and
    Scrimshaw, G., Trans. N.Y. Acad. Sci., 28, 47,
    1965.
8. Bergen, J.R., Pennell, R.B., Saravis, C.A.,
    and Hoagland, H., In: Serological Fractions
    in Schizophrenia, R.G. Heath, (ed.), Harper
    and Row, Pub. Inc., 23, 1963.
9. Leach, B.E., Byers, L.W. and Heath, R.G., In:
    Serological Fractions in Schizophrenia, R.G.
    Heath, (ed.), Harper and Row, Pub. Inc., 7,
    1963.
10. Ehrensvaard, G., In: Biological treatment of
    mental illness., M. Rinkel, (ed.), L.C. Page
    Co., New York, 20, 355, 1966.
11. Frohman, C.E., Goodman, M. Backett, Latham, P.
    G.S., Snef, R. and Gottlieb, J.S., Ann. N.Y.
    Acad. Sci., 96, 438, 1962.
12. Heath, R.G. and Drupp, I., Arch. Gen. Psychiat.,
    16, 1, 1967.
13. Sanders, B.E., Smith, E.V.A., Flataker, L.,

and Winters, C.A., Ann. N.Y. Acad. Sci., 96, 448, 1962.

14. Karsnov, A.I., Zh. Nevropat Psikhiat., Korsakov, 65, 1206, 1965.
15. Haalvadsen, R., Lingaerde, O. and Walaas, O., Confin. Neurol, 18, 270, 1958.
16. Bergen, J.R., Frohman, C.E., Mittag, T.W., Arthur, R.E., Warner, K.A., Grinspoon, L. and Freeman, H., Am.J. Psychiat., 18, 471, 1968.
17. Osmond, H. and Smythies, J., J. Ment. Sci., 98, 309, 1952.
18. Harley-Mason, J., J. Ment. Sci., 98, 313, 1952.
19. Axelrod, J. and Tomchick, R., J. Biol. Chem. 233, 702, 1958.
20. Axelrod, J., Science, 134, 343, 1961.
21. Pollin, W., Cardon, P.V. Jr. and Kety, S.S., Science, 133, 195, 1961.
22. Brune, G.G. and Himwich, H.E., J. Nerv. and Ment. Dis., 134, 447, 1962.
23. Friedhoff, A.J. and Van Winkle, E., Nature, 194, 897, 1962.
24. Ridges, A.P. and Bourdillon, R.E., Proc. Roy. Soc. Med., 60, 555, 1967.
25. Creveling, C.R. and Daly, J.W., Nature, 216, 190, 1967.
26. Sanders, B.E., Pandey, R.S. and Leitch, W.E., Fed. Proc., 26, 833, 1967.
27. Smythies, J.R., Johnston, V.S., Bradley, R.J., Benington, F., Morin, R.D. and Clark, L.C., Psychopharmacologia, 10, 379, 1967.
28. Smythies, J.R., Johnston, V.S., Bradley, R.J., Benington, F., Morin, R.D. and Clark, L.C., Nature, 216, 128, 1967.
29. Boulton, A.A., Pollitt, R.J. and Majer, J.R., Nature, 215, 132, 1967.
30. Jenner, F.A., Gjessing, L.R., Cox, J.R., Davies-Jones, A., Hullin, R.P. and Hauna, S.M., Brit. J. Psychiat., 113, 895, 1967.
31. Smythies, J., In: Biological Research in Schizophrenia, Academy of Medicine of the U. S.S.R., Institute of Psychiatry, 215-218, 1967.

C. THE BIOCHEMISTRY OF EMOTION

THE BIOCHEMISTRY OF AFFECTIVE DISORDERS

Alec Coppen

In this paper I want to summarise what
seems to me to be the most important aspects of
the biochemistry of depression and mania. To do
so in the brief time allocated for this paper will
be difficult for this field is one of the most
rapidly expanding areas of psychiatry. My summary,
will therefore, have to be selective and dogmatic
and I will be unable to do justice to the various
controversies that surround nearly every aspect
of the subject. I hope to show in this talk that
the biochemical changes are of primary aetiological
importance but I should like to emphasise that, as
a psychiatrist, I recognize that other factors -
social and psychological are also of great importance. I think one of our tasks in the future will
be to show how these environmental events can interact with the biochemistry of the nervous system
to produce the syndromes of depression and mania.
In other words I think the affective disorders can
be described as a truly psychosomatic disorder.

A great deal of research has been centered on
the biogenic amines - catecholamines and 5-hydroxy-
tryptamine (5-HT). There is indirect evidence
that increasing or reducing the brain amines will
alleviate or produce depression. The anti-depres-
sive drugs act by increasing the amount of free

amine present. The monoamine oxidase inhibitors
(MAOI) do this by decreasing the rate of metabolism of the amines--the imipramine type of drug
acts by interfering with the rebinding of the released amine and thus increasing the proportion
of unbound amine present. These drugs act on both
the catecholamines and 5-HT, and the question
arises, is it the increase in the catecholamines
or 5-HT or both that is responsible for the antidepressant effect? We (Coppen, Shaw and Farrel,
1963) investigated this problem by comparing the
antidepressant effect of a MAOI with a MAOI and
tryptophan. We argued that if increasing 5-HT
by use of a MAOI had an antidepressant effect then
feeding tryptophan, the precursor of 5-HT, should
potentiate the antidepressant effect of the drug.
We found that tryptophan did indeed have a very
marked potentiating effect. Subsequently we found
that tryptophan, used alone, had an antidepressant
effect comparable to electro-convulsive therapy
(Coppen, Shaw and Maggs, 1967).

We carried out a similar experiment using tyrosine, the precursor of the catecholamines but
could not detect any antidepressant effect (Coppen and Shaw, unpublished observation). Other investigators have tested intravenous dopa and again
failed to detect any antidepressant effect.

Reserpine is another drug which has indirectly
implicated the biogenic amines in the causation of
depression. Reserpine will deplete amine stores,
and there is now good evidence (Bunney and Davis,
1965) that in up to 15% of patients treated with
reserpine a depressive illness develops that is
often indistinguishable from an endogenous depression. This indirect evidence that a decrease in
brain monoamines may be a cause of depression has
prompted studies of these amines in depressed patients.

Investigation of the catecholamines has led
to conflicting reports, and Bergsman (1959) in one
of the most extensive studies reported that the
urinary excretion of both adrenaline and noradre-

naline was normal in depression. Schildkraut,
Green, Gordon and Durell (1966) reported a lower
excretion of normethanephrine in patients when they
were depressed than when they had recovered.

Turning now to investigations on the indoleamines there is considerable evidence of an abnormality. We found (Coppen, Shaw, Malleson, Eccleston
and Gundy, 1965) that the urinary excretion of
tryptamine was about half normal in a group of depressed patients, although it returned to normal
after clinical recovery. Two studies, one from
Edinburgh (Ashcroft, Crawford, Eccleston, Sharman,
MacDougal, Stanton, and Binns, 1966) and one from
Gothenburg (Dencker, Malm, Ross, Werdenius, 1966)
have both reported that 5-hydroxyindoleacetic acid,
a metabolite of 5-HT, is considerably reduced in
depression. This metabolite has also been found
to be one third lower than normal in the brains
of depressed suicides (Bunney, Davis, Shaw and Coppen, to be published). Hind-brain 5-HT has also
been found to be significantly lower in subjects
who committed suicide while depressed (Shaw, Camps
and Eccleston, 1967).

The evidence to date therefore points to an
abnormality of indoleamine metabolism in patients
suffering from depression and the therapeutic results obtained with tryptophan suggest that this
deficiency may be causally related to depression.
There is little data about the role of the biogenic
amines in mania.

Electrolytes have been another area of interest
to many investigators. Initially balance studies
were made on patients suffering from rapidly alter-
nating mania and depression and most investigators
reported changes in water and salt excretion with
change in clinical state (see review by Gibbons,
1963). More recently, isotope-dilution techniques
enabled estimates to be made of the distribution
of electrolytes--that is the relative concentrations
of sodium and potassium in cells and in the extracellular space. The most consistent finding was
an increase in the amount of "residual sodium"

that is, mainly intracellular sodium together with a small but unknown amount of exchangeable bone sodium). Residual sodium returned to normal after recovery (Coppen and Shaw, 1963). Low concentrations of intracellular potassium were found in depressed patients but this did not improve with clinical recovery. These changes measure body electrolytes as a whole and it is not known if such changes occur in the C.N.S. If comparable events did occur in the brain, cerebral excitability would be much affected by the reduction of resting and action potentials. In addition to these changes in electrolyte distribution a reduction in extracellular water in depression has been observed in investigations using either the thiocyanate space (Altschule and Tillotson, 1949) or the volume of distribution of radioactive bromine (Coppen and Shaw, 1963). In mania, distribution studies have revealed similar but greater changes in the distribution of sodium compared with those found in depression; the average residual sodium in mania is approximately twice normal.

Are these changes primary or secondary? At the moment it is difficult to say because we lack the means of manipulating the distribution of sodium back to normal. However, consideration of the therapeutic action of lithium in mania suggests that electrolytes may have a central role in the affective disorders as this ion has a unique influence on sodium transport across biological membranes (Keynes and Swan, 1959) and increases both extracellular and intracellular water (Coppen and Shaw, 1967). However, it is possible that lithium exerts influence in some other way by affecting the metabolism of the monoamines (Schildkraut, Scanberg, Kopin, 1966).

The last area of research I should like to touch on is endocrinological studies. These have been limited, for most part to the thyroid hormone and to hydrocortisone and its metabolites. Investigations on the thyroid have shown no abnormalities or change in affective disorders. Because of the familiar affective changes in Cushing's syndrome or during the administration of corticotrophin or hydrocortisone, adrenocortical activity in depression has been repeatedly examined using various indices, including plasma-hydrocortisone levels, urinary excretion of 17-hydroxycorticosteroids and hydrocortisone secretion rates. Here, however, cause and effect are hard to separate as it is known that the normal individual responds readily to everyday stressful situations. Much of this work has been admirably summarised by Sachar (1967) and my own view is that although modest increases in hydrocortisone secretion have been found by many investigators, they did not bear a very close correlation with clinical state and that the changes are probably secondary.

I hope, in this brief summary to have presented the salient points in the biochemistry of affective disorders. Much has been achieved but it is clear we are only at the beginning in understanding the condition. We need to know much more about the nature of the changes in amine metabolism, what amines are altered and indeed if the pattern can vary from patient to patient. Secondly, we need to know more about the factors responsible for the changes in metabolism. It is eminently feasible that these changes can be initiated or sustained by hormonal changes. The endocrinology of affective disorders is still largely an unexplored field. What is the role of sex hormones or aldosterone or anterior or posterior pituitary hormones? How are the changes in electrolytes related to the other biochemical changes: and again what produces this quite marked shift in sodium distribution?

The changes in monoamines and electrolytes may be but part of a very complex change in affective disorders but it is clear that the understanding already gained offers the possibility of a rational treatment and prevention of depression and mania.

REFERENCES

1. Altschule, M.D. and Tillotson, K.J., Amer. J.. Psychiat., 105, 829, 1949.
2. Ashcroft, G.W., Crawford, T.B.B., Eccleston, D., Sharman, D.F., MacDougal, E.J., Stanton, J.B., Binns, J.K., Lancet, ii, 1049, 1966.
3. Bergsman, A., Acta psychiat. neurol. Scand., suppl. 33, 133, 1959.
4. Bunney, W.E., and Davis, J.M., Arch. gen. Psychiat. (Chic.), 13, 483, 1965.
5. Coppen, A., and Shaw, D., Brit. Med. J., ii, 1439, 1963.
6. Coppen, A., and Shaw, D., Lancet, ii, 805, 1967.
7. Coppen, A., and Shaw, D., and Farrell, J.P., Lancet, i, 79, 1963.
8. Coppen, A., Shaw, D. and Maggs, R., Lancet, ii, 1178, 1967.
9. Coppen, A., Shaw, D., Malleson, A., Eccleston, E. and Gundy, G., Brit. J. Psychiat., iii, 993, 1965.
10. Dencker, S.J., Malm, U., Roos, B.E., Werdenius, B., J. Neurochem., 13, 1545, 1966.
11. Gibbons, J.L., Postgrad. Med. J., 39, 19, 1963.
12. Keynes, R.D., and Swan, R.C., J. Physiol., Lond. 147, 626, 1959.
13. Sachar, E., Arch. gen. Psychiat., (Chic.), 17, 544, 1967.
14. Schildkraut, J.J., Green, R., Gordon, E.K., Durell, J., Amer. J. Psychiat., 123, 690, 1966.
15. Schildkraut, J.J., Scanberg, S.M., and Kopin, I.J., Life. Sci., 5, 1479, 1966.
16. Shaw, D., Camps, F.E., Eccleston, E.G., Brit.J. Psychiat., 113, 1407, 1967.

HORMONE IMBALANCE IN DEPRESSIVE STATES

David J. McClure and Robert A. Cleghorn

The topic of endocrine influences on brain and behaviour is a large and complex one and justice cannot be done to it in this short presentation. Since this International Conference is called "The Future of the Brain Sciences", we thought it would be appropriate to present something which we feel is new and has a future in psychiatry. For this reason we will confine ourselves to examining the reported hormonal disturbances and our own biochemical findings in affective disorders with a view to supporting our hypothesis:

(a) that part at least of the hormonal disturbances in psychotic depressive illness is an imbalance between the adrenocorticosteroids and the catecholamines whereby there is a relative or absolute deficiency in the steroids and norepinephrine of the brain and (b) that clinical recovery is induced when appropriate medication is administered to correct this biochemical disturbance.

Human behaviour is a complex phenomenon which is as yet poorly understood. Knowledge and understanding of it is basic to our function as psychiatrists and we welcome information from any source which helps to clarify it for us. The growing body of knowledge in psychoendocrinology lends hope that more careful exploration of psychological states in

relation to endocrine function in man may provide
a better understanding of the role of brain mechanisms in behaviour (Buchan, 1779). The more one
pries into the bewildering complex of interactions
between mind and bodily functions, the less one
feels able to name precise mechanisms responsible
for any one disturbance (Cleghorn, 1956). In the
thirty odd years since Selye (1952) introduced his
concept of the General Adaptation Syndrome, the
field of the emotions and endocrines has become
more clear. This has been due in great part to
the development of more specific and sensitive
biochemical methods for the estimation of hormones
in the body fluids.

At present there is a large body of evidence
indicating an interplay between the endocrines and
psychological dysfunction. The mental aberrations
occurring in Addison's Disease and Cushing's Disease and the changes arising from the therapeutic
administration of adrenocorticotrophic hormone
(ACTH) and the cortisone-like compounds demonstrate
that insofar as abnormal human behaviour is concerned, the steroids secreted by the adrenal cortex are most important (Rome and Braceland, 1952;
Cleghorn, 1952). The adrenal cortical steroids
have been shown from animal experiments to participate in the metabolism of the central nervous
system (Woodbury, 1958). It is, however, from
clinical observations that we find clear evidence
of its importance to mental functions. That adrenocortical deficiency led to psychological de-
rangements was noted by Addison in 1855 and later
by Engel and Margolin (1941) and confirmed by the
findings of Cleghorn (1951) and Stoll (1953). The
increased steroid secretion occurring in Cushing's
Syndrome also carries with it a high incidence of
psychiatric symptoms (Maclay et al, 1938; Wittkower and Cleghorn, 1954). The primary clinical
symptoms reported in each of these conditions are
the same, namely depression with retardation, agitation, anxiety, paranoid ideation which has led
Bleuler (1951) and Reiss (1955) to remark that
there was no characteristic difference in the psy-
chopathological changes seen in hypo- or hypercor-

ticalism. That is to say, the type of abnormal
behaviour observed in one is also described in the
other. The expression of the disturbed behaviour
seems to depend on the previous mental pattern of
the individual rather than on the level of circulating steroids. Reiss maintains that since many
endocrinological disturbances do not show any psychopathology that one must therefore postulate a
constitutional or personality predilection for
those who do. It is difficult then from the
studies so far, to deduce what the action of the
steroids may be but Kaplan (1956) has pointed out
that there should be a tolerance for ambiguity
when facts are insufficient to provide adequate
explanatory hypotheses.

The first to study corticosteroids using
modern biochemical methods in depressed patients
were Board et al.(1956, 1957). They found that
severely depressed patients, especially those
with intense suffering, retardation and inability
to cry, had elevated plasma 17-hydroxycorticoster-
oid levels. These elevated levels subsided markedly after two weeks in hospital. However, Sachar
(1967A, 1967B) has pointed out that this study was
conducted during the hospital admission period and
that the initial steroid effect may probably have
been due to the admission experience itself. In
1962, Gibbons and McHugh measured plasma cortisol
at weekly intervals in depressed patients and
found a positive correlation between the steroid
levels and the severity of the illness. During
treatment there was a gradual lowering of the high
initial corticosteroid levels and this coincided
with clinical improvement in the depression. One
patient with cyclothymic mood swings had high plasma cortisol during these depressive epidoses which
returned to normal levels during periods of mild
elation or euthymia. Gibbons (1964) reported a
combined study of plasma cortisol and cortisol
secretion rate in 15 cases of depression. Ten of
the patients had increased cortisol secretion
rates which decreased after clinical improvement.
Those with the most severe depression had the greatest increase in secretion rate. A high correlation

was also found between the plasma cortisol level
and the cortisol secretion rate. However, a small
number of patients failed to show this change in
cortisol secretion and the author states that significant depression can occur in some subjects without any increase in their adrenocortical activity.

Bunney et al (1965) reported longitudinal
studies of urinary 17-hydroxycorticosteroids and
continuous behavioural ratings in 17 depressed
patients. The authors attempted to correlate psychological fluctuations during illness with fluctuations in urinary 17-OHCS excretion. Days of
maximum emotional disturbance were compared with
days of minimum disturbance. It was noted by the
investigators that the days of maximum emotional
disturbance were accompanied by almost as much
anxiety as depression. The overall correlations
for the group were not significant, but in a subgroup of 12 patients they were. The results supported an earlier paper of Gibbons (1960) that
anxiety during the depressive illness was associated with elevated urinary 17-OHCS levels but the
severity and course of the illness itself was not.
In addition about 1/3 of the patients had normal
or subnormal urinary 17-OHCS which suggested to
the authors the possibility of a depressive subgroup with low 17-OHCS excretion. In a different
paper, Bunney et al.(1965) studied crisis periods
in patients with psychotic depression. During
these episodes of turmoil, which were characterized
by great anxiety, there was substantial elevation of
the urinary 17-hydroxycorticosteroids.

The findings of the studies thus far reviewed
are in general agreement that there is marked elevations of plasma cortisol and urinary 17-OHCS
levels in depressed patients characterized by great
stress, features of anxiety, agitation and retardation. In these studies there is a positive correlation between the degree of adrenal cortical overactivity and independent ratings of depression
for a large proportion of the patients studied.

However, other investigators using different
measures of adrenocortical activity have advanced

different interpretations of pituitary adrenocortical function in severe depressions. Kurland. (1964A, 1964B) studied urinary 17-hydroxycorticosteroids and 17-ketogenic steroid excretion for two 24 hour periods before treatment and two 24 hour periods after recovery in 10 depressed subjects. He reported elevations in the 17-ketogenic steroids which correlated well with the severity of the depression while the 17-hydroxycorticosteroids were elevated only in the initial phase of the depression and there was no correlation with the severity of the illness. The author hypothesized an increased corticosteroid turnover rate in the liver during depression which resulted in a faster metabolism of cortisol and cortisone through their di-hydro- and tetrahydro-derivatives to cortol and cortolone. In the urine the parent compounds were detected as hydroxycorticosteroids and ketogenic steroids but cortol and cortolone were detected as ketogenic steroids only. He felt that the correlation between the clinical state and 17-ketogenic steroids could be explained by postulating a faster turnover of corticosteroids in depression with a return to normal with recovery since this would mean a slower production of cortol and cortolone. In support of this interpretation, Kurland quotes Persky (1957, 1962) who found an increased rate of cortisol breakdown in anxious patients. This suggested that despite an increased rate of production, if there was an even greater rate of cortisol breakdown, it could result in a relative deficiency occurring in depression.

More recently Kurland (1965) successfully treated a heterogeneous group of 12 patients, who were previously unresponsive to antidepressant medication, with an artificial glucocorticoid, prednisone. This adds further supporting evidence for the possibility of a relative deficiency of cortisol in depression.

Jakobson in 1966 reported an increase in the excretion of compound S in depressed patients. This precursor of cortisol has a low glucocorticoid activity. He suggested that this increased ex-

cretion could result from an alteration of normal
steroid metabolism similar to that caused by the
administration of blocking agents such as metopirone.
These alterations in steroid metabolism are known
to occur in conditions such as congenital adrenal
hyperplasia and in some other forms of adrenocortical hyperplasia. This finding also supports the
hypothesis of Kurland.

In a recent publication Sachar (1967) pointed
out that urinary 17-hydroxycorticosteroid excretion
dropped slightly following recovery, but the corticosteroid changes were not commensurate with and
did not correlate with the severity or course of
the depressive illness. The urinary 17-hydroxycor-
ticosteroid response to hospital admission far exceeded the change from illness to recovery. Seven
of the twenty patients he studied had lower urinary
17-OHCS levels during illness than after recovery.
In 10 patients plasma 17-OHCS levels were also
measured and they dropped slightly but not significantly following recovery. His findings suggested
little effect of depressive illness per se on adrenal cortical function. He pointed out also that
this was at variance with some reports on the literature, but that many studies had failed to take
into account the important control issues such as
the adrenal cortisol response to hospital admission
and the interference of certain medications.

An apparent paradox has therefore arisen with
regard to the endocrine findings in depressive illness. Some investigators postulate a state of hyper-
adrenocorticism while others postulate a state of
relative adrenocortical hypofunction in severe
depression.

Our knowledge of the adrenocortical response
in mania is equally paradoxical. Rizzo et al. (1964)
found low urinary glucocorticoids in a manic depressive female patient during her hyperactive episodes, with the levels returning to normal after
clinical recovery. Gibbons and McHugh (1962) reported a case with high plasma cortisol levels
during phases of depression and low levels during

phases of mild elation. Psychiatrists have suggested that mania is a defence against the pain of depression (Lewin, 1950) and in 1965 Bunney et al. attempted to find biochemical confirmation of this psychological theory. They simultaneously studied the behaviour and urinary 17-hydroxycorticosteroids in a female patient with regular 48 hour manic-depressive cycles over a two year period. The patient's mania was characterized by intense denial of her illness and was accompanied by low 17-hydroxycorticosteroid levels. Her days of depression were associated with feelings of suffering and pain and her 17-hydroxycorticosteroid levels were high. Bunney and his associates felt that these findings support the concept that the mechanism of denial is associated with low levels of 17-hydroxycorticosteroids.

In contrast to these findings, Schwartz et al. (1966) studied a 48 year old female manic-depressive patient through a complete 42 day cycle. They found that both manic and depressed phases were associated with increased urinary excretion of 17-hydroxycorticosteroids which suggested to them that the manic state is associated with peak elevations in corticoid excretion.

Bliss et al. (1956) reported a four month study of a schizophrenic patient who showed agitated and hypomanic behaviour. On admission to hospital, the patient was hyperactive and had high plasma 17-hydroxycorticosteroid levels. These levels remained high, but dropped later when the patient settled down.

These high levels reported in the last two papers during the hypomanic behaviour of patients are in contrast to the previous studies mentioned.

Let me now turn to a consideration of the catecholamines. The function of norepinephrine, dopamine and serotonin in the brain has interested investigators for over a decade. This was triggered by the observation that these substances could be influenced by psychopharmacological agents. Be-

cause of this interest, we have in the last few
years gathered important new information about the
cerebral amines and their functions (Sourkes, 1967).
Pharmacological treatments for depression so far
suggest that the tricyclic compounds and the monoamine oxidase inhibitors increase the functionally
free norepinephrine in the brain and that the anti-
hypertensive drugs, reserpine and methyldopa, decrease free cerebral norepinephrine and cause clinical depression (Klerman, 1963; Schildkraut, 1964;
Bunney and Davis, 1965).

Rosenblatt et al. in 1959 suggested that changes
in brain norepinephrine may be involved in depression. Their hypothesis was that the depressive
state might be associated with a relative decrease
of norepinephrine or serotonin, although they have
hitherto lacked direct evidence in its support.
Strom-Olsen and Weil-Malherbe (1958) found that
patients with manic-depressive illness had larger
urinary excretion of norepinephrine and epinephrine
during the manic phase than during the depressed
phase. Kopin and Gordon (1962, 1963) have shown
in animals that normetanephrine excretion may reflect noradrenergic activity. Schildkraut et al.
(1965, 1966) observed a rise in normetanephrine
excretion during clinical improvement in depressed
patients treated with imipramine. In the retarded
patient, normetanephrine was lower before treatment
than after treatment. Increased excretion of normetanephrine was observed in one patient during a
hypomanic phase. The implication is that if the
Kopin - Gordon hypothesis applies in humans, clinical improvement from depression is associated with
increased noradrenergic activity and that noradrenergic activity is decreased in retarded depressions
and increased in mania (Schildkraut, 1965; Kety,
1966; Schildkraut and Kety, 1967). Although caution
must be exercised in interpreting these results
for a wide variety of reasons (Karki, 1956) and
especially since only a small fraction of the urinary norepinephrine or normetanephrine estimated
comes from the brain, nevertheless it is possible
that catecholamine metabolism in the periphery may
in fact reflect changes in brain.

An Edinburgh group in 1960 (Ashcroft and Sharman) found that 5-hydroxy-indoles were reduced in concentration in the cerebrospinal fluid of depressed patients and the more severe the depression the lower the titre. These findings have since been confirmed by Dencker of Sweden (1966).

It has appeared to us for some time now that the discussed two hormone systems were both disturbed in psychotic depression. So let us consider some of the evidence that these two groups of hormones interact with each other.

There is evidence that adrenalectomized dogs maintained on cortical extract responded to pressor influences in a substantially normal way (Armstrong et al. 1938) and those in which the extract was withdrawn had a diminished response (Cleghorn et al. 1938). Conclusion drawn from these experiments at the time was that the adrenal cortex was concerned with the capacity of the animal to respond to sympathomimetic substances (Remington et al. 1941; Cleghorn et al. 1950). There is also evidence that epinephrine or nerve stimulation in adrenalectomized animals increases the demand for cortical hormone and that sympathectomized and adrenalectomized dogs need less cortical hormones for maintenance.

We know that cortisol, norepinephrine, dopamine and serotonin are found in the brains of humans. Cortisol has been found in higher concentrations in brain tissue than in blood (400 ug/100 mg. brain and 15 ug/100 ml. of blood: Touchstone, 1966). In rats isotopically labeled cortisol, when injected intravenously, enters brain tissue. There it has a prolonged half life in contrast to its short half life in the periferal circulation. Peterson (1963) has also shown that rat brain tissue accumulates cortisol even when there is only a very small increase in cortisol in the blood and that it accumulates cortisol with every increase in blood concentration. Eik-Nes (1965) demonstrated that dogs accumulated isotopically labeled cortisol when injected intravenously and that high levels were

maintained in the hypothalamus for 24 hours. It
would appear then from these animal experiments that
the major source of brain cortisol comes from the
adrenal cortex. Fawcett and Bunney (1967) point out
that if similar changes in the brain concentration
of cortisol occur in humans, small changes in the
plasma cortisol concentration might significantly
change the central adrenal steroid concentration.

There is no direct evidence as yet that cortisol affects the activity of the biogenic amines in
the brain; it does, however, affect the amines in
other organs and indirect methods strongly suggest
that it also does in the brain. Wurtman and Axelrod (1965) demonstrated in animals that adrenal
steroids enhance the activity of the enzyme which
converts norepinephrine into epinephrine in the
adrenal medulla. Ramey and Goldstein (1957) in
discussing the relationship between the cortical
steroids and epinephrines put forward the suggestion
that the steroids are "permissive" for the action
of the epinephrines and that in their absence, the
threshold for epinephrine and norepinephrine action
is raised. They also point out that corticosteroids
may play a role in the transmission of impulses in
the central nervous system by the potentiation of
neurohormonal effects. The neurohormones would include the epinephrines, serotonin and acetylcholine
and the steroids have been shown to affect the reactions of all of these substances (Barnard, 1953;
Halberg, 1954; Porter, 1954).

From clinical studies there is also some indirect evidence of a close relationship between
the action of norepinephrine and the cortical steroids. Kurland and Freedberg (1951) demonstrated
in normotensive subjects that there was a striking
potentiation of the blood pressure response to norepinephrine within 24 hours after the administration
of ACTH or cortisone and that the potentiating effect disappeared within 36 hours after the omission
of these agents. They felt their data were consistent with the hypothesis that ACTH increases
the pressor response to norepinephrine by the release of C-11-oxysteroids from the adrenal cortex

and that cortisone alone was equally effective in increasing sensitivity to norepinephrine. Prange et al. (1967) studied the systolic blood pressure response of depressed patients to infused norepinephrine, feeling that some depressed patients would be refractory to this hormone. They found in all 38 cases that norepinephrine pressor response was clearly related to clinical improvement. Gellhorn and Loofbourrow (1963) reviewing the work in the field of blood pressure response to infused norepinephrine concluded that it was a measure of hypothalamic parasympathetic reactivity; the smaller the response the greater the parasympathetic reactivity. Taking these findings into consideration, Prange and his associates concluded that improved depressed patients had moved towards sympathetic dominance compared to unimproved patients.

A case reported by Cumming and Kort (1956) may be an example of the relationship between adrenal cortical deficiency and depression. They found an Addisonian patient with an agitated psychotic depression which was refractory to 12 electroconvulsive treatments. However, after he was placed on hydrocortisone, the patient made a dramatic recovery, responding to two E.C.T.'s and maintaining his improvement. Electroconvulsive therapy which is an old, established and effective treatment for psychotic depression may owe its beneficial effects to an alteration in catecholamine metabolism since elevation in plasma and urinary norepinephrine and epinephrine has been reported after E.C.T. has been given without an anaesthetic and muscle relaxant, i.e., unmodified E.C.T. (Weil-Malherbe, 1955; Cochran and Marbach, 1962; Holmberg, 1963). Whether or not in this instance of Cumming and Kort the steroids facilitated a favourable response to E.C.T. cannot be decided but such unique instances sometimes provide unexpected results.

May we now turn to our own present research project which is designed to investigate the possibility of an imbalance between the adrenocorticosteroids and the catecholamine hormone systems in psychotic depression. This is a preliminary report

on the findings so far. I should start by pointing
out how this research programme came about. In an
earlier study of depression we plotted the diurnal
variation in plasma cortisol levels and found the
curve (Graph I) was similar to that of the normal
curve, except that the amplitude was different
(McClure 1966A, 1966B). In the normal person the
lowest levels of plasma cortisol are found between
10:00 p.m. and 2:00 a.m. then there is a gradual
elevation until 4:00 a.m. followed by a steeper
rise until maximum values are reached between 6:00
a.m. and 8:00 a.m.; the values then fall steadily
until noon and more slowly until 10:00 p.m. when
they remain approximately the same until 2:00 a.m.
when the cycle starts anew (Pincus, 1943; Bliss,
1953).

We then found that the diurnal variation in
the plasma cortisol levels changed during treatment
with imipramine. The amplitude of the patients'
graphs fell steeply in the first 14 days and then
more gradually after 28 days so that they returned
to a normal diurnal curve (Graph II).

The next step was to find out what happened
to the clinical picture when the cortisol levels
were reduced at the beginning of the illness. Small
physiological doses of dexamethasone (0.75 mgm.)
were given at midnight to suppress ACTH secretion
from the pituitary gland just before its diurnal
secretion began (Nichols et al. 1965). Complete
suppression does not occur so it is a reasonably
safe method of lowering the plasma cortisol levels
with the minimum risk of the patient having an
Addisonian crisis when the medication is withdrawn.
Since in the beginning it was not known what effect
this medication would have on the depressive picture,
the duration of time in which patients were kept on
it varied. The first four psychotically depressed
patients (Mayer-Gross, 1960; Murphy, 1964) we
studied on this regime gave very interesting results. Patients were placed on dexamethasone 0.75
mgm. given at midnight and during this time plasma
cortisol levels were estimated together with 17-hy-
droxycorticosteroids in the urine over a 24 hour

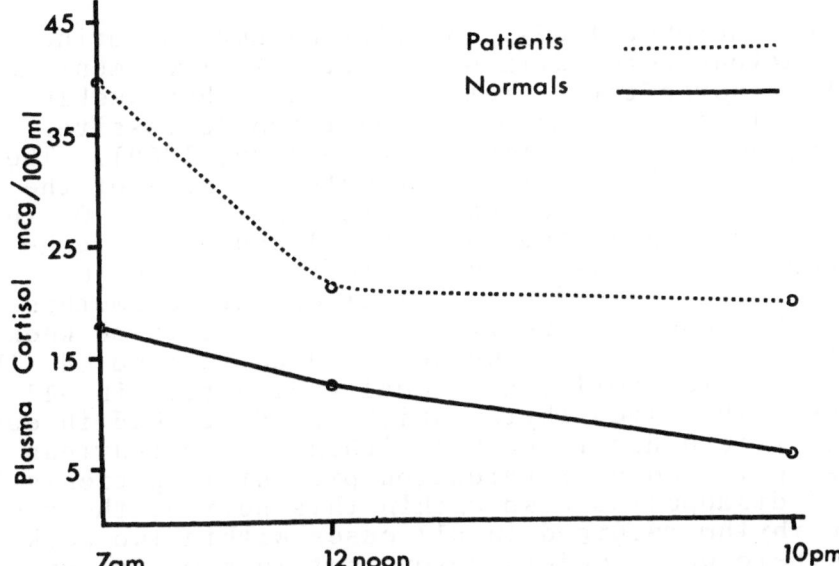

Graph I. The graph shows the diurnal variation in the plasma cortisol concentrations of normal and depressed subjects.

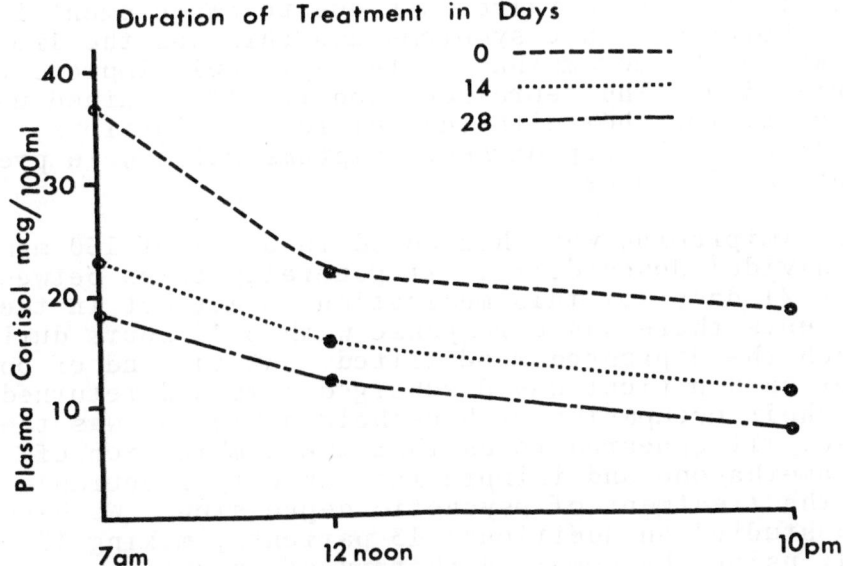

Graph II. The graph shows the change in the diurnal variation of the plasma cortisol levels during treatment.

period (Murphy, 1967). Similar plasma and urine
estimations were carried out prior to treatment and
weekly thereafter. Pretreatment and then weekly
clinical assessments on the Hamilton Depressive
Rating Scale were performed (Hamilton, 1960). The
cortisol levels did fall markedly in three of the
cases and to a lesser extent in case no. 3. Of the
first 4 patients studied, cases 1 and 2 were continued on the medication for four weeks, but by
this time a limited clinical effect of dexamethasone
could be observed, so it was reduced to three weeks
in patient no. 3 and two weeks in patient no. 4. The
results were similar and equally effective in all
cases. The sleep rhythm which was disturbed in each
subject returned to normal within ten to fourteen
days; psychomotor retardation present in patients 1
and 3 disappeared also within this period; the appe-
tite rhythm returned in all cases within two weeks
and there was a major improvement in somatic com-
plaints. The Hamilton Rating Scale showed that the
depression improved to a limited extent during this
phase of dexamethasone treatment but reached a score
below which it could not be reduced further. The
fall in the scale values was due to improvement in
the above-mentioned symptoms and this was the limit
to which the dexamethasone therapy could improve the
depression. The depressed mood itself remained unchanged, together with the suicidal ruminations,
guilt and self reproaches, symptoms which were present in all cases.

Imipramine was then added in doses of 150 mgm.
in divided doses daily. It generally takes between
10 - 21 days for this medication to act but in these
patients there was a response within 72 hours during
which the depressed mood lifted. At the end of one
week each patient was discharged home and returned
to their occupation or household duties as was the
case. It appeared to us that the combination of
dexamethasone and imimpramine was very favourable
in the treatment of psychotic depression. We have
now studied an additional 13 patients, making 17 in
all, using the combined therapy of an antidepressant
medication and dexamethasone. The method we now use
is to place the patient on dexamethasone 0.75 mgm.

PATIENT NO. 1

TABLE "A"

PHASE	MEDICATION	HAMILTON RATING SCALE	TIME	PLASMA		URINE	
				FULLERS	FLORISIL	FULLERS	FLORISIL
I	Placebo	74	7:45 a.m. 11:45 a.m. 10:00 p.m.	16.0 10.9 3.7	17.7 6.7 1.0	35.1	50.4
II	Dexamethasone	51.5*	7:45 a.m. 11:45 a.m. 10:00 p.m.	0.075** 0. 0.9	0. ** 0.0375 1.375	11.925**	0. **
III	Dexamethasone and Imipramine	7.0	7:45 a.m. 11:45 a.m. 10:00 p.m.	1.95 3.0 1.65	3.975 2.15 0.45	13.5	2.7

* Mean Score of 4 Weekly Ratings,
Range 72 - 34.

** Mean Score of 4 Weekly Estimations

PATIENT NO. 2

TABLE "B"

PHASE	MEDICATION	HAMILTON RATING SCALE	TIME	PLASMA		URINE	
				FULLERS	FLORISIL	FULLERS	FLORISIL
I	Placebo	68	7:45 a.m. 11:45 a.m. 10:00 p.m.	16.5 5.3 2.8	7.0 5.1 4.6	117.0	119.7
II	Dexamethasone	48.5*	7:45 a.m. 11:45 a.m. 10:00 p.m.	1.0** 0.95 1.7	0.86** 0.5 1.55	3.825**	11.7**
III	Dexamethasone and Imipramine	6	7:45 a.m. 11:45 a.m. 10:00 p.m.	1.7 1.2 2.2	2.3 1.9 2.0	3.6	0.

* Mean Score of 4 Weekly Ratings, Range 64 - 42.

** Mean Score of 4 Weekly Estimations

PATIENT NO. 3

TABLE "C"

PHASE	MEDICATION	HAMILTON RATING SCALE	TIME	PLASMA		URINE	
				FULLERS	FLORISIL	FULLERS	FLORISIL
I	Placebo	80	7:45 a.m. 11:45 a.m. 10:00 p.m.	15.8 13.9 10.5	21.4 12.7 14.0	189.0	242.1
II	Dexamethasone	43*	7:45 a.m. 11:45 a.m. 10:00 p.m.	11.9** 5.56 6.4	17.63** 10.33 10.2	72.6**	90.0**
III	Dexamethasone and Imipramine	11	7:45 a.m. 11:45 a.m. 10:00 p.m.	7.7 2.975 2.75	9.2 4.65 3.6	27.45	19.35

* Mean Score of 3 Weekly Ratings, Range 76 - 40.

** Mean Score of 3 Weekly Estimations

PATIENT NO. 4

TABLE "D"

PHASE	MEDICATION	HAMILTON RATING SCALE	TIME	PLASMA FULLERS	PLASMA FLORISIL	URINE FULLERS	URINE FLORISIL
I	Placebo	68	7:45 a.m. 11:45 a.m. 10:00 p.m.	11.0 9.8 4.1	8.0 5.3 2.3	24.3	43.2
II	Dexamethasone	51*	7:45 a.m. 11:45 a.m. 10:00 p.m.	0.7** 1.85 1.15	0.45** 0.25 0.65	7.2**	22.05**
III	Dexamethasone and Imipramine	18	7:45 a.m. 11:45 a.m. 10:00 p.m.	0.4 0.5 0.5	0.4 0.5 0.8		

* Mean Score of 2 Weekly Ratings, Range 58 - 44.

** Mean Score of 2 Weekly Estimations

AVERAGE SCORES FOR ALL PATIENTS ON THE HAMILTON RATING SCALE

PHASE I	PHASE II	PHASE III
72.5*	48.5**	8.5***

*	Mean Score of 4 Estimations:	Range 80 - 74
**	Mean Score of 13 Estimations:	Range 76 - 34
***	Mean Score of 4 Estimations:	Range 11 - 6

given at midnight for two weeks. Imipramine (or one
of the other antidepressant medications such as
amitriptyline or an MAO inhibitor, for example isocarboxazid) is then added in the second week of
treatment. The dexamethasone together with the antidepressant medication is continued during the second
week. The dexamethasone is then discontinued and
the antidepressant medication is continued for the
third and fourth weeks alone. Weekly estimations
of plasma and urinary corticosteroids and clinical
estimations with the Hamilton Rating Scale are
carried out. So far, all patients have responded
favourably. During the second week of the treatment
plan when the combined therapy starts there is a
marked improvement in the mood of the patient. All
patients have been discharged within four weeks from
the hospital clinically recovered. Maintenance
therapy with the antidepressant medication is continued after discharge. The response is much swifter
than that using imipramine alone. It would appear
that the dexamethasone is a primer for imipramine in
some way. We are at present starting our control
series of patients in which a placebo will be used
with dexamethasone in a double blind trial.

To explain these findings let us consider what might be happening to these patients. From animal and clinical studies previously mentioned in this text, we know that there is no direct evidence that the steroids affect the activity of the brain amines. However, there is a great deal of indirect evidence that they do. The presence of steroids is essential for the proper functioning of compounds considered to be the chemical transmitter substances in the brain, namely norepinephrine, epinephrine, serotonin and acetylcholine, and hence by implication they are thought to play a role in the transmission of nerve impulses by potentiating the action of these neurohormones.

The findings from animal experiments show that the adrenal cortex is the source of brain steroids and that the central concentration varies directly with the peripheral concentration of these hormones. This has suggested to some investigators that small changes in plasma cortisol levels might significantly change the steroid concentration in human brain. A considerable amount of knowledge now indicates that depressive illness per se does not cause the adrenocortical overactivity that was concluded from earlier studies and that overactivity, if it does occur, is most probably due to the presence of anxiety brought on by environmental stresses. Furthermore there are reports of increased degradation of cortisol or an alteration in normal cortisol synthesis producing a deficiency of steroid rather than an excess in depressive illnesses. If such a deficit does occur in the peripheral circulation, it will be reflected markedly in the brain and this shortage of steroid will affect the functioning of the neurohormones. We know however from an important school of thought that there is probably a lack of central amines in depression. We feel then that there is a reasonable possibility that the biochemical state of the depressed patient might be a deficiency of both central adrenocorticosteroids and catecholamines.

If our hypothesis is correct, the replacement of physiological amounts of steroid will produce a favourable milieu for the proper functioning of the brain amines. This is done by the administration of dexamethasone. The imipramine (or other anti-depressant medication) provides the necessary central catecholamine which is then able to act immediately at the brain tissue level.

The results so far indicate that in psychotic depressive illness we are obtaining a much swifter clinical response with dexamethasone and imipramine combined than one normally does with imipramine alone. The explanation for these findings appears to be in the favourable interaction between the steroids and the brain catecholamines.

REFERENCES

1. Addison, T., On the constitutional and local effects of disease of the suprarenal capsules. In: A Collection of the Published Writings of Thomas Addison, London. New Sydenham Society, 36, 1868.

2. Armstrong, C.W.J., Austen, D.C. and Cleghorn, R.A., Blood Pressure Studies in Adrenalized Dogs, Am. J. Physiol. 123, 40, 1938.

3. Ashcroft, G.W. and Sharman, D.F., Hydroxyindole: in Human Cerebrospinal Fluids, Nature, 186, 1050-1051, 1960.

4. Barnard, R.D. and Fox, H.L., The Affect of Cholase Administration in Some Dermatologic Conditions, N.Y. State J. Med., 53, 2826-2830, 1953.

5. Bergsman, A., The Urinary Excretion of Adrena-line and Noradrenaline in Some Mental Diseases, ACTA Psychiat. Neurol. Scand. Suppl., 33, 1959.

6. Bleuler, M., Psychiatry of Cerebral Diseases, Brit. Med. J., 2, 1233-1238, 1951.

7. Bliss, E.L., Sandberg, A.A., Nelson, D.H. and Eik-Nes, K., The Normal Levels of 17-hydroxy-

corticosteroids in the Peripheral Blood of Man,
J. Clin. Invest. 32, 818, 1953.

8.   Bliss, E.L., Reaction of Adrenal Cortex to
      Emotional Stress, Psychosom. Med., 18, 56-76,
      1956.

9.   Board, F., Persky, H. and Hamburg, D.A., Psy-
      chological Stress and Endocrine Function, Psy-
      chosom. Med. 18, 324-333, 1956.

10.   Board, F., Wadeson, R. and Persky, H., Depres-
      sive Affect and Endocrine Functions, AMA Neu-
      rol. Psychiat., 78, 612-620, 1957.

11.   Buchan, W., Domestic Medicine: or Treatise on
      the Prevention and Cure of Diseases, 6th Edi-
      tion, 119, London, 1779.

12.   Bunney, W.E. Jr., Mason, J.W. and Hamburg, D.A.,
      Correlations Between Behavioural Variables and
      Urinary 17-hydroxycorticosteroids in Depressed
      Patients, Psychosom. Med., 27, 299-308, 1965.

13.   Bunney, W.E. Jr., Mason, J.W., Roatch, J.F.
      and Hamburg, D.A., A Psychoendocrine Study of
      Severe Psychotic Depressive Crises, Am. J.
      Psychiat., 122, 72-80, 1965.

14.   Bunney, W.E. Jr., Hartmann, E.L. and Mason, J.
      W., Study of a Patient with 48 Hour Manic De-
      pressive Cycles, Arch. Gen. Psychiat., 12, 619-
      625, 1965.

15.   Bunney, W.E. Jr. and Davis, J.M., Norepinephrine
      in Depressive Reactions: A Review, Arch. Gen.
      Psychiat., 13, 483-494, 1965.

16.   Cleghorn, R.A. and Austen, D.C., Effect of
      Autonomic Nerve Stimulation and Pressor Drugs
      on the Cardiovascular System of Adrenalized
      Dogs, (Can. Psysiol. Soc. Proc.), Canad. Med.
      Assn. J., 39, 189, 1938.

17.   Cleghorn, R.A., Fowler, J.L.A., Greenwood, W.F.,

and Clarke, A.P.W., Pressor Responses in Health Adrenalectomized Dogs, Am. J. Physiol. 161(1), 21-28, 1950.

18. Cleghorn, R.A., Adrenal Cortical Insufficiency: Psychological and Neurological Observations, Canad. Med. Ass. J., 65, 449-454, 1951.

19. Cleghorn, R.A., Alterations in Psychological States by Therapeutic Increase in Adrenal Cortical Hormones, CIBA Found. Colloquia. Endocrinol., 3, 187-196, 1952.

20. Cleghorn, R.A., Steroid Hormones in Relation to Neuropsychiatric Disorders, Conference on Neuroendocrinology, Harriman, New York, May, 1956. Published in Book: Hormones, Brain Function and Behaviour, H. Hoagland, Ed., Academic Press Inc., New York, 1957.

21. Cochran, B. Jr. and Marbach, E.P., Some Acute and Chronic Biochemical Responses to Electroconvulsive Therapy, Recent Advan. Biol. Psychiat., 4, 154-169, 1962.

22. Cumming, J. and Kort, K., Parent reversal by cortisone of an electroconvulsive refractor state in a psychotic patient with Addison's disease, Canad. Med. Ass. J., 74, 291, 1956.

23. Dencker, S.J., Malm, U., Roos, B.E. and Werdinius, B., Acid Monoamine Metabolites of Cerebrospinal Fluid in Mental Depression and Mania., J. Neurochem., 13, 1545-1548, 1966.

24. Eik-Nes, K.B. and Brizzee, K.R., Concentration of tritium in brain tissue of dogs given $(1,2,{}^3H_2)$ cortisol intravenously, Biochem. Biophys. Acta, 97, 320-333, 1965.

25. Engel, G.L. and Margolin, S.G., Neuropsychiatric disturbances in Addison's disease and the role of impaired carbohydrate metabolism in production of abnormal cerebral function, Arch. Neurol. Psychiat., 45, 881-884, 1941.

26. Fawcett, J.A. and Bunney, W.E. Jr., Pituitary Adrenal function and depression, Arch. Gen. Psychiat, 16, 517-535, 1967.

27. Gellhorn, E. and Loofbourrow, G.N., Emotions and emotional disorders, Harper and Row, Publishers, Inc., New York, 1963.

28. Gibbons, J.L., Gibson, J.G., Maxwell, A.E. and Willcox, D.R.C., An endocrine study of depressive illnesses, J. Psychosom. Res., 5, 32-41, 1960.

29. Gibbons, J.L. and McHugh, P.R., Plasma cortisol in depressive illness, J. Psychiat. Res., 1, 162-171, 1962.

30. Gibbons, J.L., Cortisol secretion rate in depressive illness, Arch. Gen. Psychiat., 10, 572-575, 1964.

31. Halberg, F., Eosinopenic effects of tryptamines in mice., Am. J. Physiol., 179, 309, 1954.

32. Hamilton, M., A rating scale for depression, J. Neurol. Neurosurg. Psychiat., 23, 56-62, 1960.

33. Holmberg, G., Biological aspects of electroconvulsive therapy, Inter. Rev. Neurobiol., 5, 389-412, 1963.

34. Jakobson, T., Steinback, A., Strandstrom, L. and Rimon, R., The excretion of urinary 11-deoxy- and 11-oxy- 17-hydroxy-corticosteroids in depressed patients during basal conditions and during the administration of methopyrapone, J. Psychosom. Res. 9, 363-374, 1966.

35. Kaplan, A., Philosophical point of view, Psychiat. Res. Repts., 6, 199-211, 1956.

36. Karki, N.T., The urinary excretion of noradrenaline and adrenaline in different age groups, its diurnal variation and the effect of muscular

work on it, ACTA Physiol. Scand. Suppl., 132 (39), 1-96, 1956.

37. Kety, S.S., Catecholamines in neuropsychiatric states, Pharmacol. Rev., 18, 787, 1966.

38. Klerman, G.L., Schildkraut, J.J., Hasenbush, L.L., Greenblatt, M., Friend, D.G., Clinical experience with dihydroxyphenylalanine (DOPA) in depression, J. Psychiat. Res., 1, 289-297, 1963.

39. Kopin, I.J. and Gordon, E.K., Metabolism of norepinephrine-H^3 Released by tyramine and reserpine, J. Pharmacol. Exp. Ther., 140, 207-216, 1963.

40. Kopin, I.J. and Gordon, E.K., Metabolism of administered and drug released norepinephrine 7-H^3 in the rat., J. Pharmacol. Exp. Thera. 140, 207-216, 1963.

41. Kurland, G.S. and Freedberg, A.S., The potentiating effect of ACTH and of cortisone pressor response to intravenous infusion of L-norepinephrine, Proc. Soc. Exp. Biol. Med., 78, 28-31, 1951.

42. Kurland, H.D., Steroid excretion in depressive disorders, Arch. Gen. Psychiat., 10, 554, 1964A.

43. Kurland, H.D., Urinary steroids in neurotic and manic depression, Proc. Soc. Exp. Biol. Med., 115, 723-725, 1964B.

44. Kurland, H.D., Physiologic treatment of depressive reaction: a pilot study, Am. J. Psychiat., 122, 457-458, 1965.

45. Lewin, B., Psychoanalysis of elation, W.W. Norton and Company, New York, 1950.

46. Maclay, W.S., Stoakes, A.B. and Russel, D.S., Mental Disorder in Cushing's syndrome with pathological report, J. Neurol. Psychiat., 1,

110-119, 1938.

47. Mayer-Gross, W., Slater, E. and Roth, M.,
    Clinical psychiatry, Cassell and Co., Ltd.,
    London, 1960.

48. Murphy, B.E.P., Cosgrove, J.B., McIlquham, M.C.
    and Pattee, C.J., Adrenal corticoid levels in
    human cerebrospinal fluid, Canad. Med. Ass. J.,
    97, 13-17, 1967.

49. Murphy, H.B.M., Wittkower, E.D. and Chance, N.A.,
    Crossculture inquiry into the symptomatology of
    depression, Transcultural Psychiat. Res., 1,
    5-18, 1964.

50. McClure, D.J., The diurnal variation of plasma
    cortisol levels in depression, J. Psychosom.
    Res., 10, 189-195, 1966A.

51. McClure, D.J., The effects of antidepressant
    medication on the diurnal plasma cortisol
    levels in depressed patients, J. Psychosom.
    Res., 10, 197-202, 1966B.

52. Nichols, T., Nugent, C.A. and Tyler, F.H.,
    Diurnal variation in suppression of adrenal
    function by glucocorticoids, J. Clin. Endocrin.
    Metab., 25, 343-349, 1965.

53. Persky, H., Adrenocortical function in anxious
    human subjects: the disappearance of hydro-
    cortisone from plasma and its metabolic fate,
    J. Clin. Endoc., 17, 760-765, 1957.

54. Persky, H., Introduction and removal of hydro-
    cortisone from plasma, Arch. Gen. Psychiat., 7,
    93-97, 1962.

55. Peterson, N.A. and Chaikoff, I.L., Uptake of
    intravenously injected (4-^{14}C) cortisol by
    adult rat brain, J. Neurochem., 10, 17-23, 1963.

56. Pincus, G., Diurnal rhythm in the excretion of
    urinary ketosteroids by young men., J. Clin.

Endocrin. Metab., 3, 195, 1943.

57.   Porter, R.W., The central nervous system and stress-induced eosinopenia, Recent. Prog. Hormone Res., 10, 1-27, 1954.

58.   Prange, A.J. Jr., McCurdy, R.L. and Cochrane, C.M., The systolic blood pressure response of depressed patients to infused norepinephrine, J. Psychiat. Res., 5, 1-13, 1967.

59.   Ramey, E.R. and Goldstein, M.S., The adrenal cortex and the sympathetic nervous system, Physiol. Rev., 37, 155-195, 1957.

60.   Reiss, M., Psychoendocrinology, J. Ment. Sci., 101, 686-695, 1955.

61.   Remington, J.W., Collings, W.D., Hays, H.W., Parkins, W.M. and Swingle, W.W., The response of the adrenalectomized dog to renin and other pressor agents, Am. J. Physiol., 132, 622-628, 1941.

62.   Rizzo, N.D., Fox, H.M., Laidlaw, J.C. and Thorn, G.W., Concurrent observations of behaviour changes and of adrenocortical variations in a cyclothymic patient during a period of twelve months, Ann. Intern. Med., 41, 798-815, 1954.

63.   Rome, H.P. and Braceland, F.J., Psychological response to ACTH, cortisone, hydrocortisone and related steroid substances, Am. J. Psychiat., 108, 641-650, 1952.

64.   Rosenblatt, S., Chanley, J.D., Sobotka, H. and Kaufman, M.R., Interrelationships between electroshock, the blood brain barrier and catecholamines, J. Neurochem., 5, 172-176, 1960.

65.   Sachar, E.J., Corticosteroids in depressive illness, I. A revaluation of control issues and the literature, Arch. Gen. Psychiat., 17, 544-553, 1967A.

66. Sachar, E.J., Corticosteroids in depressive
     illness, II. A longitudinal psychoendocrine
     study, Arch. Gen. Psychiat., 17, 554-567, 1967B.

67. Schildkraut, J.J., Klerman, G.L., Hammond, R.
     and Friend, D.G., Excretion of 3-methoxy-4-hy-
     droxymandelic acid (VMA) in depressed patients
     treated with antidepressant drugs, J. Psychiat.
     Res., 2, 257-266, 1964.

68. Schildkraut, J.J., The catecholamine hypothesis
     of affective disorders: a review of supporting
     evidence, Am. J. Psychiat., 122, 509-522, 1965.

69. Schildkraut, J.J., Gordon, E.K. and Durell, J.,
     Catecholamine metabolism in affective disorders:
     I. Normetanephrine and VMA excretion in de-
     pressed patients treated with imipramine., J.
     Psychiat. Res., 3, 213-228, 1965.

70. Schildkraut, J.J., Green, R., Gordon, E.K., Du-
     rell, J., Normetanephrine excretion and affect-
     ive state in depressed patients treated with imi
     pramine., Am. J. Psychiat., 123, 690-700, 1966.

71. Schildkraut, J.J. and Kety, S.S., Biogenic
     amines and emotions, Science, 156, 21-30, 1967.

72. Schwartz, M., Mandell, A.J., Green, R. and
     Ferman, R., Mood, motility and 17-hydroxycor-
     ticosteroid excretion: a polyvariable case
     study, Brit. J. Psychiat., 112, 149-156, 1966.

73. Selye, H., The story of the adaptation syndrome,
     ACTA Inc., Medical Publishers, Montreal, Canada,
     1952.

74. Sourkes, T.L., (Editorial), Canad. Med. Ass. J.,
     96, 550-551, 1967.

75. Stoll, W.A., Die psychiatrie des morbus Addison,
     Thieme, Stuttgart, 1953.

76. Strom-Olsen, R. and Weil-Malherbe, H., Humoral
     changes in manic depressive psychosis with

particular reference to the excretion of catecholamines in urine, J. Ment. Sci., 104, 696-
704, 1958.

77. Touchstone, J.C., Kasparow, M., Hughes, P.A.
and Horwitz, M.R., Corticosteroids in human
brain, Steroids, 7, 205-211, 1966.

78. Weil-Malherbe, H., The effect of convulsive
therapy on plasma adrenaline and noradrenaline,
J. Ment. Sci., 101, 156-162, 1955.

79. Wittkower, E.D. and Cleghorn, R.A. Editors,
Recent developments in psychosomatic medicine,
J.B. Lippincott Co., Philadelphia, 1954.

80. Woodbury, D.M., Relation between the adrenal
cortex and the central nervous system, Pharmacol
Rev., 10, 275-357, 1958.

81. Wurtman, R.J. and Axelrod, J., Adrenal synthesis
Control by the pituitary gland and adrenal glucocorticoids, Science, 150, 1464-1465, 1965.

D. THE BIOCHEMISTRY OF PSYCHOTIC STATES

A PSYCHOPHARMACOLOGICAL APPROACH TO BIOCHEMICAL RESEARCH IN PSYCHOSES

Carlo L. Cazzullo

A good deal of metabolic research has been carried out in different psychoses, particularly depression and schizophrenia, and hypotheses have been put forward as to changes of enzymatic systems, energy metabolism, electrolyte balance, hormones, etc., in these illnesses.

From a general view-point, two main experimental pathways have been followed: one on the metabolic differences between psychotic patients and normal subjects, the other on metabolic changes in the same patient at different stages of the illness. The latter approach allows a follow-up of metabolic changes in the same subject with a reduction of the scatter due to a large population and gives greater information on metabolic changes during the development of the clinical picture. A similar study can be carried out in patients before and after treatment with psychotropic drugs or other types of therapy. Here one should discriminate between the metabolic changes due to clinical modifications from the changes which are only secondary to treatment. At this regard a greater meaning should be given to metabolic changes observed only in patients who have clinically responded to therapy. At any rate, in spite of the uncertainties, it appears that biological therapies are contributing a great deal to the knowledge of basic biochemistry of mental illness probably because, if a metabolic derangement occurs in the major psychoses, this may be so small that it must be stimulated with proper experimental devices before evidencing it.

One area of research which seems to have produced interesting results concerns the changes of electrolyte balance after treatment. It should be said that no definitive report exists on the biological processes governing ion concentration in cell compartments and their distribution in extracellular fluids. However interesting studies have been carried out on sodium balance and the volume of extracellular and intracellular fluids in psychoses.

As early as in 1949, Altschule and his co-workers showed that the rapid increase of body weight in depressed patients during electroconvulsive therapy was the result of the increased volume of extracellular fluids (1).

With the introduction of isotopes in clinical research other experiments were carried out in depressed patients before and after electroconvulsive treatment. It was reported (2) that clinical depression was coupled to sodium retention, whereas, with the use of the long half-life isotope Na22, it was indicated by others (3) that a redistribution of sodium among different cell compartments is likely to take place in depressed patients.

Other studies have shown that whereas no substantial differences have been found as to electrolyte balance between depressed and manic patients, an increase of extracellular fluids volume has been observed in the manic phase of manic-depressive psychoses as compared with the depressive phase (4). And depressed patients who have responded to electroshock or imipramine treatment show an increase of extracellular fluids volume after recovery (5, 6). This is a further indication of the experimental advantage given by studies on the

same patients examined at various times of the illness.

Other biochemical changes have been observed
in our laboratory in depressed patients treated
with some antidepressant drugs (7, 8), indicating
a marked increase in the conversion of tryptophan
into xanthurenic acid as a result of drug treatment similarly to that observed in manic patients
before therapy. Thus, by means of therapeutic
agents, it has been possible to modify electrolyte
balance on one hand and tryptophan metabolism on
the other in depressed patients towards the values
observed in mania. Further interesting implications are in the observation of a decreased excretion of xanthurenic acid during chlorpromazine
treatment (9) as compared to the increase of xanthurenic acid induced by antidepressant drugs.

Data have also been obtained as to the biological effects of neuroleptic drugs in schizophrenic
patients showing an increase of the volume of extracellular fluids during treatment with either
chlorpromazine or haloperidol (9), with a greater
increase of extracellular fluid volume during haloperidol treatment and a more pronounced clinical
improvement with this drug.

In normal subjects the volume and composition
of body fluids are maintained within physiological
values in spite of the daily variations in their
intake. With regard to the regulatory mechanism,
more than one hormone is involved, i.e. the anti-
diuretic hormone of hypophysis, aldosterone and
adrenal glycocorticoids. Tryptophan metabolism
through the nicotinic acid pathway is also influenced by adrenal glands at level of tryptophan-
pyrrolase and animal experiments indicate that both
hormones and psychotropic drugs may interfere with
the activity of this enzyme (10,11,12).

Endocrine changes occurring in psychotic illness, in basal conditions as well as after treatment, may prove to be a fruitful area of research
on electrolyte and amino acid metabolism. In this

connection, a marked decrease of hypothalamus-hypophysis activity has been observed in schizophrenic patients (13), whereas haloperidol treatment has been shown to raise hypophysis activity towards normal values (14); that is yet another case where treatment has modified a biological parameter to values of normal subjects.

Certainly these are only a few examples taken from literature and from data of our laboratory, which seem however to be indicative of a fruitful approach to psychiatric research. Indeed, psychotropic drugs may reveal some latent metabolic picture not caught up by current laboratory techniques unless evidenced by proper experimental devices such as drug or other types of treatment. If such a view-point is correct, starting from the biochemical changes induced by treatment, we might witness in the near future important discoveries into the metabolic background of psychoses.

REFERENCES

1. Altschule, M. D., Ascoli, I. and Tillotson, K. J., Extracellular fluid and plasma volumes in depressed patients given electric shock therapy, Arch. Neurol. Psychiat., 62, 618, 1949.
2. Gibbons, J. L., Total body sodium and potassium in depressive illness, Clin. Science, 19, 133, 1960.
3. Coppen, A., Shaw, D. M. and Mangoni, A., Total exchangeable sodium in depressive illness, Brit. Med. J., ii, 295, 1962.
4. Hullin, R. P., Bailey, A. D., McDonald, R., Dransfield, G. D. and Milne, H. B., Body water variations in manic-depressive psychosis, Brit. J. Psychiat., 113, 584, 1967.
5. Brown, D. G., Hullin, R. P.,and Roberts, J. M., Fluid distribution and the response of depression to E.C.T. and imipramine, Brit. J. Psychiat., 109, 395, 1963.
6. Cazzullo, C. L, Lanzara, D. and Mangoni, A., Variazioni biologiche nella depressione e nella schizofrenia nel corso di trattamento con imipramina, 1° Natl. Congr. Ital. Soc. Neuropsychopharm., Naples, 1967.

7. Cazzullo, C. L., Mangoni, A. and Mascherpa, G., Tryptophan metabolism in affective psychoses, Brit. J. Psychiat., 112, 157, 1966.

8. Cazzullo, C. L., Penati, G. Bozzi, A. and Mangoni, A., Sleep patterns in depressed patients treated with MAO inhibitors, Correlation between EEG and metabolites of tryptophan, Proc. VI Intern. Congr. C.I.N.P., Tarragona, 1968. In press.

9. Mangoni, A., Cazzullo, C. L., Bernetti, M. G. and Lanzara, D., Longitudinal clinical and biochemical studies in schizophrenic patients treated with chlorpromazine and haloperidol, Proc. IV World Congr. Psychiat., Madrid, 1966.

10. Knox, W. E. and Auernach, V. H., The hormonal control of tryptophan-peroxidase in the rat, J. Biol. Chem., 214, 307, 1955.

11. Mangoni, A. and Paracchi, G., L'attività triptofano-pirrolasica in ratti trattati con farmaci antidepressivi, Boll. Soc. Ital. Biol. Sper., 41, 1552, 1965.

12. Paracchi, G., Lanzara, D., Bernetti, M. G. and Mangoni, A., Effects of antidepressant drugs on tryptophan metabolism, Proc. IV World Congr. Psychiat., Madrid, 1966.

13. Brambilla, F., Cazzullo, C. L. and Riggi, F., Endocrinology in chronic schizophrenia, Dis. Nerv. Syst., 28, 745, 1967.

14. Brambilla, F. and Riggi, F., Attività dell'haloperidolo sulla secrezione ipofisarica, surrenalica e gonadica di pazienti schizofrenici cronici, Ann. Neurol. Psichiat., 59, 1965.

ON THE PREDOMINANT AFFECT AND THE PHASE OF
PSYCHOTIC ILLNESS

DR. HIMWICH: I noticed that in Dr. McClure's
presentation that there was a great variability
not only between the various catecholamines but
also even more so among the corticosteriods, in
regard to their correlation with psychotic behavior. Could one have achieved better correlations
with anxiety than with overt psychotic behavior as
your criterion of comparison?

DR. SEMRAD: I would like to say a word in
relation to Dr. Himwich's question about anxiety.

In the schizophrenic patients that we have
seen over the years in psychotherapy, we regularly
observed, not anxiety, but the affect of sadness,
intolerable sadness of failure in life, i.e. of
not being able to maintain his optimum integration.
In the acute situation as the precipitating factor
is identified and worked with, the patient's ego
compensates and the schizophrenic patient gives
up his symptoms of catatonia or paranoia, or acute
turmoil deluded hallucinated states; he becomes a
very severe psychotic depression.

The therapist is no longer dealing with a
schizophrenic patient, but a severe depression.
Only then symptoms of anxiety and agitation appear, especially if the patient has a strong development in his personality of the compulsive
obsessive patterns of reaction; the anxiety is
more manifest in them than it is in those who have
a strong development in the neurasthenic or hypochondriacal patterns of adapting.

When one gets to this phase, then one deals
essentially with a character disorder who functions
with all the characteristics and limitations of
his basic personality structure.

Chronicity of ego decompensation so often
relates to intolerable sadness and the patient's

persistent avoidance behavior designed not to
experience it i.e. behavior patterns of denial,
projection and distortion; the core action behavior of schizophrenics.

ON THE NEED FOR SPECIFICITY IN THE DESCRIPTION OF CLINICAL FUNCTION IN SOMATIC STUDIES OF PSYCHOTIC PATIENTS

DR. SEMRAD: This progress in the basic brain
sciences is not only gratifying but has all the
promise of being rewarding. Many of the people
working in it I have met before. I recognized
several names in Dr. McClure's review of the work
as some of my former associates.

The integrating functions of the brain are a
primary interest and responsibility of the clinician; to aid and abet his patients integration at
optimum level by whatever means he has learned
and/or hopes to learn. The task of identifying
and understanding what does not work or why it
doesn't work, and how it came about that it doesn't
work; in order to enable optimal person functioning;
these factors have to be brought together. So far,
it seems that the workers in the social, psychological, and somatic areas have not found a common way
of keeping observations adequately correlated.
Extrapolations beyond the data provided by our
investigative instruments are a major pitfall and
often the source and generation of irrelevant
hypotheses.

We first became interested in correlative
studies in the early '50's and did such elementary
things as using the eosinophile counts and corticosteroids as indices; you know how much trouble
one can get into trying it this way. Not only
that, but what was more fundamental was that there
was very little satisfaction in talking about
schizophrenia, or manic depressive illness, etc.,
because this gave an unclear picture of how precisely a given person was functioning in his ad-

aptation at the moment that the physical indices
were being measured.

It is precisely this point that I want to
say a few words about.

Our interest in the development of an Ego
Profile Scale (10) was stimulated by our dissat-
isfaction with correlative psychophysiological
studies. To correlate by diagnosis proved a very
difficult problem. To correlate by symptom change
was not satisfactory....In our early work (1,2),
it became clear that we needed an instrument for
measuring psychological function which would enable
us to correlate a given state of functioning of a
person (ego functioning) with somatic function indices. It was our hope that this would enable us
to have a better understanding of clinical states,
especially schizophrenia as observed during states
of regression and return to optimum functioning.
Some of our associates have done further work in
this field (3,4,5). The somatic function indices
are in a more advanced state of development than
are the psychological function indices. The development of an Ego Profile Scale in our pilot
efforts has provided considerable advantage and
promises results which may make correlation studies
more precise and even enable us to develop a functional classification for purposes of making diagnoses, making judgements on the progress of therapy, providing guidance to a given therapist and
enabling better statements of prognosis. There
are many schizophrenias: there are those who deny,
there are those who project, there are those who
do not think straight, that is, who distort. There
is a whole group of very chronic patients who have
stayed in a state of denial for years and are
sometimes thought of as basically organic problems.
What is cause and what is effect is still an un-
settled problem (6). In terms of the way they
function in their personal relationships, in their
personal efficiency as work units, and in their
maintenance of a subjective state of comfort, at
various levels of ineptitude and inability, it
becomes crucial to identify and know the level at

which the patient is functioning for fruitful
correlative measurement. It is important to know
whether one is talking about a person who is in
an acute state of denial precipitated by an over-
whelming loss in his life, or whether one is talking about a person in an acute state of denial who
gradually has regressed because of frustration of
object need to this position of catatonia (7).
These are two different groups of problems; certainly therapeutically they are two different
groups of problems.

We (7,8) have elucidated the recovery process
in schizophrenia by ordering the defenses along a
pathological continuum. During convalescence the
patient employs progressively less pathological
defenses to recovery of ability to experience
relatively unmodified sadness and anxiety. The
defense mechanisms that any individual chooses
for coping with or defending against such stress
may be correlated with the severity of his overall psychopathology. We stress that some defenses
are more mature than others and that immature
defenses evolve into more mature defenses.

Fig. 1 The SEMRAD-GRINSPOON EGO PROFILE SCALE
records the changes of a thirteen weeks course
of a schizophrenic girl on Thioridazine. The
scale consists of three triads: Narcissistic,
Affective, and Neurotic. The Narcissistic triad
is made up of the following categories: denial,
projection, and distortion. The Affective triad
includes obsessive compulsiveness, hypochondria-
sis, and neurasthenia. The Neurotic triad consists of dissociation, somatization, and anxiety
alerts. Each point on the graph represents the
sum of the 5 items which combine to form one of
the nine categories. Since each item is rated on
a scale from 0-6, the total score may range from
0-30. This patient was rated weekly during her
13 week stay in the hospital.

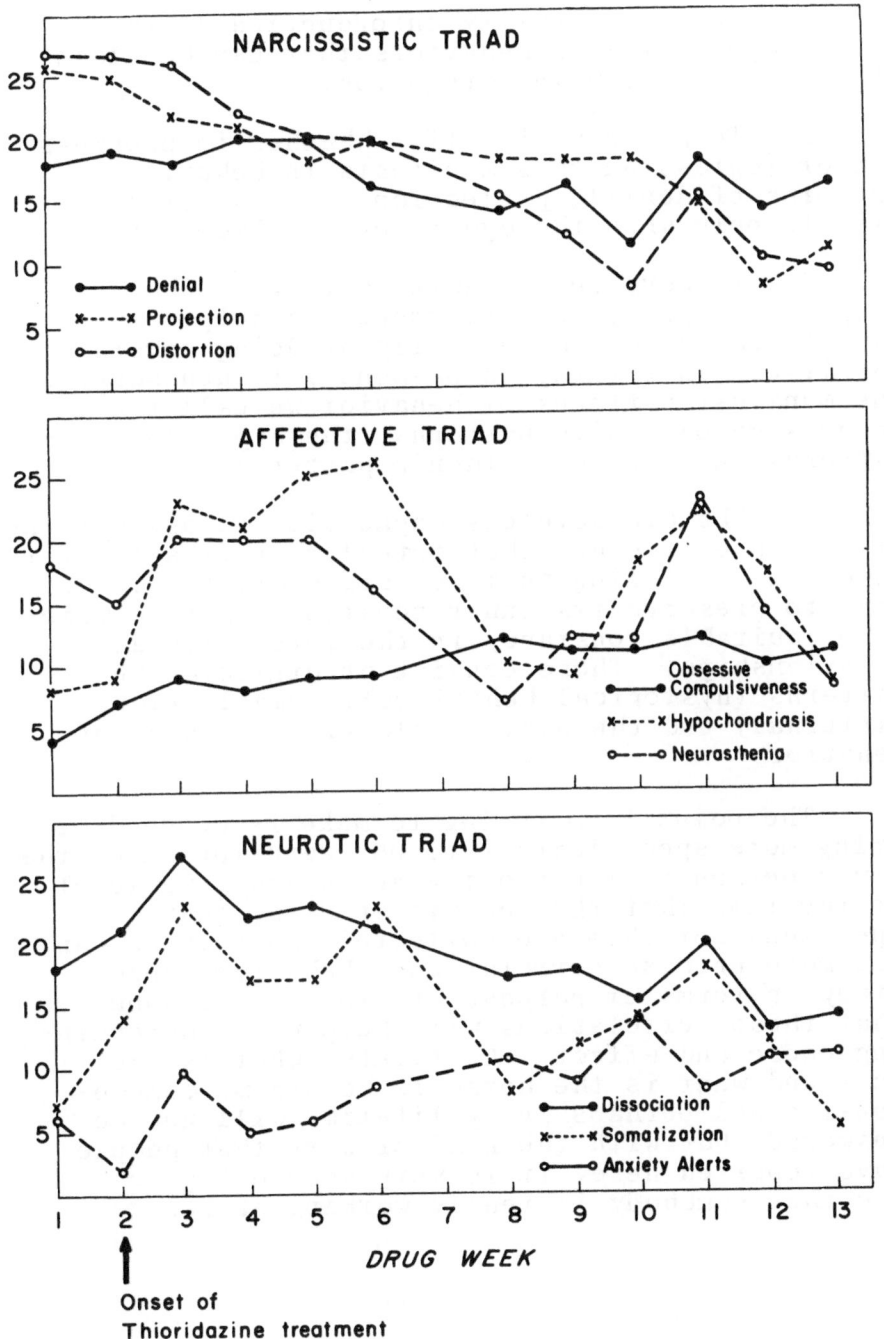

In order of maturity ontogenetically the
extant ego manifests behaviors that can be classified into three main categories.

1. To preserve itself. Whether in progression or regression this manifests in behavior
patterns of denial, projection and distortion
visible only when the ego is overwhelmed.

2. It develops capacities to get other
people's support, encouragement and help. In a
way it learns this essentially by doing something
for itself in return, in payment for such services.
The manifest patterns of behavior we call the
compulsive obsessive patterns, the hypochondriacal
patterns and the neurasthenic patterns.

3. The ego develops capacities to ask nothing
of its objects other that mutual gratification for
which it is willing to make self sacrifices, not
only to preserve its inner constancy, but to maintain equitable constancy in the interpersonal
relationships. These consist of dissociation
patterns (hysterical behaviors), somatizations
(internal) and the anxiety alerting patterns of
behavior.

The point I am trying to make is we need to
bring more specificity into our description of the
way a person is functioning at an adaptive level
at the time that the somatic studies, that is,
specimens for them are collected, so that we can
get more precise correlation. Behavior needs
study in terms of purpose; to insure the hope
that these correlations will help us in determining cause and effect. Certainly, what is the
cart and what is the horse is beyond my comprehension and perhaps in my lifetime will not be
answered, but with the kind of work that people
have reported here, it is very heartening to
see that somebody is really working at it.

REFERENCES

1. Menzer, D., Shands, H.D. & Swartz, J., On Psychosomatic Studies and Psychosis, Abstract in Jrnl. Nerv. & Ment. Dis., 117, 2, 165, Feb., 1953.

2. Shands, H. & Menzer, D., Eosinophile Variation in its Course of Insulin-Coma Therapy, Amer. Jrnl. Psy., 109, 1953; 757, 1966.

3. Sachar, E.J., Mason, J.W., Holmes, H.B. Jr. & Artiss, K.L., Psychoendocrindo Aspects of Acute Schizophrenic Reaction, Psychosomatic Med., 25, 510-537, 1963.

4. Sachar, E.J. et al, Corticosteroid Responses to Psychotherapy of Depressions, Arch. Gen. Psychiat., 16, 461-470, Apr. 1967.

5. Sachar, E.J., Corticosteroids in Depressive Illness, Arch. Gen. Psychiat., 17, 554-567, Nov. 1967.

6. Semrad, E.V. & Finley, K., Note on the Pneumonoencephalogram and Electroencephalogram Findings in Chronic Mental Patients. Psych. Qrtly., 17, 76-80.

7. Semrad, E.V., Long Term Therapy of Schizophrenia, Formulation of Clinical Approach, 155-173: Psychoneurosis and Schizophrenia, Usidin, G.E. (ed.), Lippincott, Phila., 1966.

8. Semrad, E.V., The Treatment Process, Amer. Jrnl Psych., 3, 426-428, Dec. 1954.

9. Semrad, E.V., Hierarchy of Psychotic Defenses, to be published.

10. Semrad, E.V. & Grinspoon, L., An Ego Profile Scale, to be published.

SCHIZOPHRENIA AS A GROUP OF DISEASES

DR. SIMPSON: I would also like to address myself to the people who have discussed or who gave, papers relating to schizophrenia. I think that anyone who works in this field of biological psychiatry has to be complimented, and if I were king, then my royal coffers would open wide, and they would have bounteous largesse.

I would like to introduce a thought (in relation to clinical aspects of this subject) that might seem trite and old-fashioned. Schizophrenia to me is indeed a group of diseases, and if it was discovered that it included ten illnesses or subgroups, it would not particularly surprise me. Therefore, diagnosis, however inaccurate, has to be discussed. I think that was why the Liverpool study that Dr. Friedhoff mentioned was so impressive, because they used the carefully defined Schneiderian classification for the diagnosis of schizophrenia, and they also separated paranoid schizophrenia as a group. These latter subjects were clearly defined, not merely people who are sensitive, who have ideas of reference or who may have some minor psychopathology. The positive results, particularly the marked differences between the paranoid and the non-paranoid group, is therefore very impressive in this study.

Perhaps related to the problem of diagnosis is the fact that most of biological psychiatry until recently has been unable to show consistent significant differences between normal and schizophrenic subjects. The most frequent finding has been that the variance in the schizophrenic subjects was much greater. This perhaps illustrates once more that we are probably dealing with more than one illness. There is a further complicating feature, and that is the actual stage of the illness.

A few years ago we studied histamine sensitivity in schizophrenia and showed quite clearly that there were no differences between acute schizophrenics and normal subjects in this test, but that chronic schizophrenics differed from both normals and acute schizophrenics in their

reaction to intra-dermal histamine. Our acute
subjects were newly-admitted patients with acute
symptomatology. In a further (unpublished) study,
we used only subjects with a first schizophrenic
episode, and we found that not only did chronic
schizophrenics differ from normals and acute
subjects, with intermediate chronic groups showing
an intermediate response to histamine, but the
acute schizophrenic subjects differed from the
normals. The acute schizophrenics were hyper-
sensitive to histamine, and the chronic schizophrenics were hypo-sensitive. So that one could
get different results, depending on the stage of
the illness of the group studied. In the first
study, the hyper-sensitive first episode patients
were perhaps mixed with some recurrent episodes
in subjects suffering from a more chronic form
of the illness and this neutralized the effect.

This therefore raises a note of caution,
and if I might address myself to Dr. Heath, I
would like to express my admiration for the
courageous work that he has been doing for such
a long time. However, I am a trifle worried
about the cleanness of his data. From what we
have said previously, an across-the-board result
would surprise me in any measure in "schizophrenia".
It may well be that only chronic schizophrenic
subjects were used, and certainly there is a
tendency (and we are all guilty of this) to use
such subjects, and I have no easy answers to
prevent it, except that obviously controls are
required. A similar response in all subjects
would certainly suggest some non-specific factor,
a factor which might or might not be related to
schizophrenia. Perhaps we should focus more on
the patients who differ rather than on those who
are similar.

Stimulated by Dr. Heath's work some ten
years ago, we carried out a cross transfusing
experiment using a normal control and a chronic
schizophrenic subject with marked psychotic
symptoms. The hypothesis was that the patient
would improve and the control become worse. In
effect, what happened was that the patient got

worse and the control got married. Perhaps if
we had used someone more acutely ill the results
might have been different.

ON REPRODUCTIVE FUNCTIONS IN SCHIZOPHRENIC PATIENTS

DR. SIMPSON: I would like to comment on
Dr. West's somewhat pessimistic observation concerning the dangers of discharged schizophrenics
spreading their genes in the community.

Many schizophrenic patients, as part of their
illness, are seclusive, withdrawn and show lack
of interest in other people. More recent studies
have shown that many schizophrenic patients who
were discharged and who lived on their own did
better than patients forced to socialize. Thus,
they are not the most sexually promiscuous subjects.

There are many patients who are discharged
from mental hospitals, whether receiving psycho-
pharmaceuticals or not, who have no sexual interest
at all. There are many males who, despite sexual
interest, are incapable of sexual performance.
There are many male patients receiving psychophar-
maceuticals who have sexual interests, are sexually
potent, but who, because of their medication, are
unable to ejaculate. They are, therefore, not the
most likely subjects for begetting large families.

One of the more interesting innovations in
treatment of psychiatric patients has been the
introduction of long-acting phenothiazines, which
have provided a range of control not previously
available. An injection every two weeks of a
drug like fluphenazine enanthate is often sufficient treatment for schizophrenic subjects. It is
obvious that we are about to see a new range of
long-acting drugs, not only in psychiatry, but
probably also as anti-fertility agents.

So while this is a problem we should be
concerned about, I think that if we are able to
discharge schizophrenic patients to the community,
the control of symptoms is the major difficulty,
the control of their fertility a lesser one.

INCREASING VALIDITY OF BIOCHEMICAL STUDIES IN
MENTAL DISORDERS

DR. CAMPBELL: I am in the fortunate position
of having most of the points that I might have
raised concerning the relationship of Clinical Psychiatry and the Brain Sciences already discussed
and I need not, therefore, belabor the points already made. I would certainly echo Dr. Simpson's
reminder that the schizophrenias are most likely
a group of diverse diseases and that more attention
should be paid to that consideration in the evaluation of neurophysiologic and biochemical studies
of the psychoses.

Debate continues as to what factor or factors
might be primarily responsible for the development
and manifestation of schizophrenic disorders, but
it is generally agreed that the evidence for family
factors, alone, producing what is called schizophrenia is unconvincing. Incidence studies to date
suggest that while the expectancy rate in the general population may fall somewhere between one and
two percent, the observed incidence in most populations has usually been found to be less than one
percent. Admittedly, it is difficult to compare
incidence surveys in different cultures and loci
that were done by different workers, but, as it
turns out, there is marked consistency among studies that further attempt to report variations in
incidence in different sub-populations. The one
factor directly related to the observed variations
in the schizophrenia rate is the degree of con-
sanguinity to the known schizophrenic and his family. One of the main findings of reports from
the United States and Europe over the last fifty
years has been that age corrected risk figures for
properly grouped relatives of schizophrenics have
invariably been many times higher than for unrelated contemporaries in the general population.
Because of these findings, many believe that at
least one of the essential requirements for the development of clinical manifestations of the schizophrenias is a genetically determined disorder of
metabolism or, at least, some kind of inborn physiologic defect. Attempts to isolate and identify
the nature of that defect have constituted a re-

current theme of American and European psychiatry throughout the century.

Although different organs and body fluids have been examined and assessed by various methods, most claims proceeding from such examinations have been discredited. The problems in most of the investigations are many, and much of the lack of agreement on the findings may reflect a tendency or workers to treat schizophrenia as a homogenous group. This may be the reason that the major finding of some investigations has been the marked variability of schizophrenics' responses, rather than consistent and uniform deviations from the normal. By failing to differentiate between sub-groups, such studies may well have obliterated the real differences between some of those sub-groups and the normal. It is suggested that future investigations try as much as possible to separate out discrete sub-enties on the basis of more objectively distinguishable differences; viz., paranoids from non-paranoids, catatonics from non-catatonics, acute from chronic forms, process from reactive types, DMPEA excretors from non-excretors, and high cerebrospinal fluid glycoprotein neuraminic acid levels from low levels (Campbell et al., Amer. J. Psychiat., 123, 952, 1967 Furthermore, the sub-groups so identified should be matched on such factors as age and sex of patient, duration of illness, number of previous episodes, degree of family loading, and types and effects of previous and current treatments--including something that is often overlooked, self-administered drugs such as psychostimulants, tranquilizers, and hallucinogens. That we are meeting here at all is an indication that most of us feel that recent findings of disordered metabolism, such as abnormal transmethylation processes, hold more validity than some of the earlier studies. It is suggested that if more careful attention is paid to the heterogeneity of schizophrenics, the validity of these investigations for at least some of the identified sub-groups might be even greater.

Finally, I would like to make a comment on Doctor West's observation of the likelihood that as we know more about the schizophrenias and begin more adequately to treat more schizophrenic patients, we are likely to end up with more schizophrenics in our society. Doctor Simpson has already detailed several of the factors that tend to minimize or reduce the likelihood of such an occurrence. While it may be that there are many factors interfering with reproduction rates in schizophrenics, it must nonetheless be recognized that studies already performed have indicated that these factors alone will not hold the disorder in check. Kallman and his associates, for instance, found many changes in the 1954-56 sample of hospitalized patients as compared with the 1934-36 sample. More patients were married at the time of admission (a greater increase than in the general population during that period), and this increase in marriage rate was associated with an increase in productivity that was almost double the increase in the general population, and with a disproportionate rise in the number of marriages that ended in separation or divorce. There was a similar upward trend in dual matings. Finally, the study suggested that the increasing number of children born to schizophrenics will come from a preponderance of broken homes and from those steadily growing contingents of battered children who require protective custody because of neglect or maltreatment. Nearly half of dual marriage children born after the mother's first hospitalization required the intervention of child care agencies. Such findings suggest that the proportion of schizophrenics in future generations can no longer be expected to decrease gradually because of limited reproductive capacities. With the reproduction of schizophrenics in our present population approaching that of their non-schizophrenic contemporaries, any selective disadvantage that schizophrenic illness may previously have had in terms of reproduction is now disappearing. And obviously, that conclusion is of enormous significance in family counseling.

MANFRED SAKEL, KARL DUSSIK, MAX RINKEL AND THE DEVELOPMENT OF BIOLOGICAL PSYCHIATRY

H. Peter Laqueur

From this rostrum, 22 years ago, my father delivered the Harvey Lecture describing progress in Scientific Endocrinology during the first 40 years of this century. From him I learned the importance of quantitative multi-factor analysis in biology and today I feel privileged to honor his teachings in their application to the work of three excellent men who, like my father, have passed into the colorful realm of memories.

I was a budding clinical endocrinologist when, in the early part of 1936, I saw my first Insulin Ward in the provincial mental hospital in Vught in Noord Brabant, the Netherlands. A very active, carefully supervised group of physicians and nurses, under the direction of A. C. A. J. Wubbe, assistant director of the institution, treated about 20 patients according to what is now called the "classical" technique of Manfred Sakel. Only three years before, Manfred Sakel had done the basic work on this treatment in Poetzl's Clinic in Vienna, changing thereby the fate of schizophrenic patients all over the world.

Manfred Sakel and Karl T. Dussik published their results with this new hypoglycemic coma treatment in 1938 and we may say today that these two men have done for schizophrenia what Wagner-Jauregg and Dattner did for syphilitic brain disease first with malaria-fever therapy and later with penicillin.

Max Rinkel, former chairman of the Scientific and Medical Board of the Manfred Sakel Foundation, was interested in biological aspects of psychiatry since his early years as a medical practitioner. His book on the Chemistry of Psychoses, and his masterful editing of the Proceedings of the First and the Second International Conference of the Manfred Sakel Institute helped to move Biological Psychiatry into the foreground of professional attention.

It is fitting and proper that Manfred Sakel, Karl Dussik and Max Rinkel should be honored at this Third International Conference on the Future of the Brain Sciences and I ask you to rise in honor of these three men.

Insulin therapy, in its early days, had its risks and misfortunes that accompanied the stupendous jump forward in hope and better prognosis for schizophrenia patients. It has been my privilege to help make this method safer so that in the last 18 years not a single fatality occurred in the about 80,000 insulin comas given in our institution.

It is, however, a much larger phase-jump to teach man to fly as Lilienthal and the Wright Brothers did, than to develop aircraft from the two-decker plane to today's high altitude jet planes.

Similarly, in the early thirties, it was a fantastic jump from the "hopeless outlook" for dementia praecox patients, as they were called, to the 70 to 80% chance for recovery for schizophrenia patients after Manfred Sakel had introduced his insulin treatment.

Yet, progress in psychiatry, as it has been made through the efforts of such men as Manfred Sakel, Karl Dussik and Max Rinkel, could not have taken place without simultaneous interdisciplinary work in the allied fields of medicine, biology, biochemistry, biophysics, psychology and sociology.

If Bleuler had not changed the definition of the condition of patients from dementia praecox to schizophrenia, if Freud had not shown the correlation between disturbances of mental function and conflicts repressed into the hitherto unknown field of the unconscious, if Kurt Lewin had not applied field theory to social interaction, if Kurt Goldstein had not emphasized that the whole organism is more important than the sum of its parts, and, last but not least, if Banting and Best, Houssay, and even earlier Mehring and Minkowski had not done their crucially important work on the chemical factor Insulin, which interferes so powerfully with all enzymatic reactions regulating carbohydrate metabolism, no mental patient could have genuinely hoped to find a better fate and prognosis when Manfred Sakel talked to Dr. Poetzl in Vienna about trying insulin on mental patients after he had, at first, erroneously speculated that it might be useful in the treatment of drug addiction.

If Sakel had not come to the United States, where grateful friends and patients enabled him and later Dussik, Rinkel and myself to carry on research in the field of insulin therapy, the ugly wave of National-Socialism in Central Europe might have killed the chance of further progress in this field, although Max Mueller in Switzerland, and Sargent and Slater in England had already begun to develop an interest in the biochemical therapies which so far had consisted only of Klaesi's sleep therapy and Meduna's pentamethylenetetrazol shock treatment.

Bond and Shurley, Kalinowsky, Hoch, and lastly the Manfred Sakel Foundation with Max Rinkel did their best to help biological treatments to take root in the United States. These treatments were studied and promoted by Leo Alexander, Jonathan Cole, Leonard Cammer, Joseph Wortis, Linus Pauling, Ralph Gerard, Harold Himwich, and others.

During the last 20 years, the American Society of Medical Psychiatry and the Society of Biological Psychiatry have contributed to such fruitful co-operation through many publications and symposia.

Hippius and Selbach in Berlin, Mueller in Lausanne, Jean Delay in Paris, Hoff and Arnold in Vienna, Nathan Kline and Henry Brill and many others in the U.S.A., many Russian and Czechoslovak authors, and many more, too numerous to name, have carried the tradition of biological psychiatry into the scientific world where it has now a firmly rooted position.

However, I would not honor Sakel, Dussik and Rinkel properly, if I ignored the fact that biological psychiatry is only one facet of the "whole system" called Psychiatry.

It is a mistake to talk about psychological psychiatry as if it were an independent entity, and we also have come away from looking at social psychiatry as a thing by itself. We should not now make the same mistake of looking at biological psychiatry as a science independent from the other psychological and social parameters.

The World Health Organization has declared health "a state of complete physical, mental and social well-being", without defects. Psychiatry deals with illness, nervous, mental and emotional, and thus with any deviation from such a complete state of biological, psychological, and social health. Psychiatry takes the whole brain as part of the whole organism, which in turn is part of a whole environment, as its proper object of study and tries to determine how unhealthy medical, mental and emotional responses can be reversed to healthy ones.

While during a half century in which psychological and social psychiatry held a dominant position on psychiatry as a whole, those interested in the biological aspects of illness understandably felt pushed somewhat into the defensive, today there must be no more contrast between psychological and biological psychiatry. Freud's friend Glover once compared the contrast between Freud and Jung with the diametrical opposition of Ptolemy's and Kopernicus' viewpoints. This is absurd because these allegedly diametrically opposed and unresolvable viewpoints have been resolved by Isaac Newton's eclectic higher level of understanding of common points of gravity around which heavenly bodies revolve.

In the same way, psychological and biological psychiatry are special aspects of the whole field of psychiatry in which health must be defined as a continuously changing condition of stability of shape and function of the organism within a changing world. Ross Ashby once defined this as "stability being the ability to change with change".

Results in psychiatry, by which we mean the restoration of health so defined, depends on the psychiatrist's ability to take a general holistic view before analyzing special systems. We might express this in a formula:

$$R = F \frac{(x_1, x_2, \ldots x_n) \; (y_1, y_2, \ldots y_n) \; (z_1, z_2, \ldots z_n)}{T \quad S}$$

in which x represents biological, y represents psychological, z represents social and sociological factors, while T stands for time and S for situation in their variable aspects in treatment.

Some feel that the biological parameter (x) encompasses the other factors (y and z), but this same claim could also be made by those interested in the y's and the z's.

Karl Dussik was in good company with Kurt Goldstein with his Ganzheits Theory, and Ludwig von Bertalanffy with his General Systems Theory, when he demanded that both our understanding and our care and treatment of psychiatric patients should be holistic and that any attempt to cure one aspect while ignoring the others can only lead to misunderstanding and possible disaster.

Karl Theo Dussik was not only Manfred Sakel's first and most loyal coworker but an excellent neurologist and psychiatrist in his own right. He is credited with having introduced ultrasonic medical investigation in this country and he can certainly rightfully be called the father of the holo-therapeutic approach to mental patients. The patients' recovery and well-being were much more important to Karl Dussik than any "special research parameters". He saw no justification for withholding one tool that might help while experimenting with another. He had no use for the "phase-in", "phase-out" and "controlled experimental" approach to medical care and believed that good clinical observation while we do all we can for our patients would in the long run teach us just as much as double-blind experiments with placebos instead of presumably active pharmacological agents.

I could not fully agree with him on this point in spite of my warm friendship with him, but I always could respect his motives and the gratitude of hundreds of his patients and their families in his patients and their families in his Association for the Rehabilitative Treatment of State Hospital Patients must have more than compensated him for any professional differences with his colleagues.

The methods of Sakel, Dussik and Rinkel dominated the first two International Conferences of the Manfred Sakel Institute while this Third Conference, showing the way into the future, is a fulfillment of the wishes of these three men whom we are honoring here today. Today's conference is the beginning of a new psychiatric outlook which, though necessarily founded in General Systems Theory, may lead us to many new special systems approaches but must never lose sight of the three parameters of physical, mental and social health.

The focus of the First Conference was on Insulin Treatment, the focus of the Second Conference

on Biological Treatment, and the focus of this
Third Conference is on the Future of the Brain
Sciences, a logical development from the more specific to the more general aspects of psychiatry
for which these three men, Sakel, Dussik and Rinkel,
helped to build the bridge.